Learning Through Teaching Mathematics

MATHEMATICS TEACHER EDUCATION

VOLUME 5

SCOPE

The *Mathematics Teacher Education* book series presents relevant research and innovative international developments with respect to the preparation and professional development of mathematics teachers. A better understanding of teachers' cognitions as well as knowledge about effective models for preservice and inservice teacher education is fundamental for mathematics education at the primary, secondary and tertiary level in the various contexts and cultures across the world. Therefore, considerable research is needed to understand what facilitates and impedes mathematics teachers' professional learning. The series aims to provide a significant resource for teachers, teacher educators and graduate students by introducing and critically reflecting new ideas, concepts and findings of research in teacher education.

For other titles published in this series, go to
http://www.springer.com /series/6327

Roza Leikin · Rina Zazkis
Editors

Learning Through Teaching Mathematics

Development of Teachers' Knowledge and Expertise in Practice

Springer

Editors
Roza Leikin
University of Haifa
Faculty of Education
31905 Haifa
Mount Carmel
Israel
rozal@construct.haifa.ac.il

Rina Zazkis
Simon Fraser University
Faculty of Education
8888 University Drive
Burnaby BC V5A 1S6
Canada
zazkis@sfu.ca

ISBN 978-90-481-3989-7 e-ISBN 978-90-481-3990-3
DOI 10.1007/978-90-481-3990-3
Springer Dordrecht Heidelberg London New York

Library of Congress Control Number: 2010922286

Printed on acid-free paper

Springer is part of Springer Science+Business Media (www.springer.com)

Contents

Contributors

Lara Alcock Loughborough University, Loughborough, Leicestershire, UK, L.J.Alcock@lboro.ac.uk

Marcelo C. Borba GPIMEM and UNESP, Rio Claro-SP, Brazil, mborba@ms.rc.unesp.br

Daniel Chazan University of Maryland, College Park, MD, USA, dchazan@umd.edu

Helen Doerr Syracuse University, Syracuse, NY, USA, hmdoerr@syr.edu

Shulamit Elikan University of Haifa, Haifa, Israel

José Guzmán Instituto Politécnico Nacional, Mexico City, México, jguzman@cinvestav.mx

Dave Hewitt University of Birmingham, Birmingham, UK, d.p.hewitt@bham.ac.uk

Nicholas Jackiw KCP Technologies, Emeryville, CA, USA, njackiw@keypress.com

Carolyn Kieran Université du Québec à Montréal, Québec, Canada, kieran.carolyn@uqam.ca

Roza Leikin University of Haifa, Haifa, Israel, rozal@construct.haifa.ac.il

Stephen Lerman South Bank University, UK, lermans@lsbu.ac.uk

Peter Liljedahl Simon Fraser University, Burnaby, BC, Canada, liljedahl@sfu.ca

Robin Marcus University of Maryland, College Park, MD, USA, marcus_robin@hotmail.com

John Mason The Open University (emeritus), Milton Keynes, Buckinghamshire; University of Oxford, Oxford, UK, j.h.mason@open.ac.uk

Nathalie Sinclair Simon Fraser University, Burnaby, BC, Canada, nathsinc@sfu.ca

Ron Tzur University of Colorado Denver, Denver, CO, USA, ron.tzur@ucdenver.edu

Michal Yerushalmy University of Haifa, Haifa, Israel, michalyr@construct.haifa.ac.il

Rina Zazkis Simon Fraser University, Burnaby, BC, Canada, zazkis@sfu.ca

Rúbia B.A. Zulatto GPIMEM, FIEL and FAAL, Limeira-SP, Brazil, rubiaba@uol.com.br

Introduction

The idea of teachers Learning through Teaching (LTT) – when presented to a naïve bystander – appears as an oxymoron. Are we not supposed to learn before we teach? After all, under the usual circumstances, learning is the task for those who are being taught, not of those who teach. However, this book is about the learning of teachers, not the learning of students.

It is an ancient wisdom that the best way to "truly learn" something is to teach it to others. Nevertheless, once a teacher has taught a particular topic or concept and, consequently, "truly learned" it, what is left for this teacher to learn? As evident in this book, the experience of teaching presents teachers with an exciting opportunity for learning throughout their entire career. This means acquiring a "better" understanding of what is being taught, and, moreover, learning a variety of new things. What these new things may be and how they are learned is addressed in the collection of chapters in this volume.

LTT is acknowledged by multiple researchers and mathematics educators. In the first chapter, Leikin and Zazkis review literature that recognizes this phenomenon and stress that only a small number of studies attend systematically to LTT processes. The authors in this volume purposefully analyze the teaching of mathematics as a source for teachers' own learning.

Research literature sometimes interprets LTT as learning by observing the teaching of others, for example, examining videotaped lessons of expert teachers (e.g., Brophy, 2003; Lampert & Ball, 1998) or having teachers complete retrospective analyses of their own teaching (e.g., Lampert, 2001). However, this book specifically addresses what teachers learn *while* they are teaching.

The chapters in this volume are written by authors from different countries: Brazil, Canada, Israel, Mexico, UK, and USA. They address teaching diverse contents: numerical literacy (Doerr & Lerman, Liljedahl), geometry (Borba & Zulatto, Leikin, and Jackiw & Sinclair), algebra (Yerushalmy & Elican, Marcus & Chazan, and Kieran & Guzmán), and Real Analysis (Alcock). The focus of analyses involves teaching which occurs at a variety of levels: elementary school (Liljedahl, Doerr, & Lerman), secondary school (Yerushalmy & Elican and Marcus & Chazan), university undergraduate mathematics courses (Alcock), and teacher education courses (Zazkis, Borba & Zulatto, and Hewitt). These authors employ different methodological tools and different theoretical perspectives as they consider teaching in different learning environments: lecturing (Alcock), small group work on problems and tasks (Hewitt, Liljeahl), mathematical explorations with the support of technological software (Jackiw & Sinclair, Kieran & Guzmán), or e-learning (Borba & Zulatto). However, despite these differences, each author discusses issues that support or impede teachers' learning and exemplify teachers' learning that occurred during their professional practice.

Research on teacher education, teacher knowledge, and teacher practice is an explicit focus of contemporary research in mathematics education. Numerous studies (too numerous to give justice by citing a few) describe the complexity of teachers' work, the fragility of teachers' knowledge, and the deficiencies and strengths within teacher education practices. The important and original contribution of this book is that it ties these notions together—presenting them through the lens of a relatively unexplored phenomenon: Learning through Teaching.

When conceiving this book, our initial idea was to focus on teachers' learning of mathematics. That is, we were interested in whether and how teachers' mathematical knowledge is enhanced as a result of their teaching practice. In Leikin and Zazkis' chapter we characterize the mathematics that teachers learn when teaching. However, whenever a case of LTT was considered, we asked ourselves: "Is it mathematics or is it pedagogy?" We quickly realized that answering this question was very complex. Teachers' learning of mathematics through teaching is usually either embedded in their pedagogical choices or results in pedagogical considerations. We found that even when the authors do not discuss pedagogical issues explicitly, the mathematics that is learned by the teachers is often mathematics *for* teaching. That is, chapters that focus on teachers' learning of mathematics are not devoid of pedagogical considerations.

In addressing this complexity we found the distinction between *mathematical pedagogy* and *pedagogical mathematics* as introduced by Mason (2007) to be very helpful. In his view, mathematical pedagogy involves strategies and useful constructs for teaching mathematics. In contrast, pedagogical mathematics involves mathematical explorations useful for, and arising from, pedagogical considerations. We use this distinction as a lens for organizing the chapters in this volume: Part II includes chapters that address mainly pedagogical mathematics, while the focus of chapters in Part III is mainly on mathematical pedagogy. The chapters in Part II and Part III are introduced by the editors in the beginning of each section (see Interlude 1 and Interlude 2). Here we focus on Part I.

Part I of this volume addresses issues related to the theory and the methodology of research on LTT. In particular, the first chapter, co-authored by Leikin and Zazkis, exemplifies and examines factors that contribute to or impede LTT – issues that are then echoed in subsequent chapters. However, in order to explore these factors, it is essential to understand what is meant by teachers' learning or by learning, in general. Mason, in his chapter, offers a concise definition: Learning is a transformation of attention. He elaborates further, stating that this transformation involves both "shifts in the form as well as in the focus of attention." Mason substantiates and instantiates this view using a series of phenomena in which learning occurred while teaching or doing mathematics. A common theme in examining these presented episodes is that of reflection, which is considered an essential condition for learning from experience.

Mason's theoretical discussion and the analysis of the phenomena are followed by some practical considerations which address the question of "What can teachers do?" Some suggestions are provided regarding what teachers can do for their students, for themselves, and for each other in order to enhance their learning. The theme of reflection is also at the heart of Tzur's chapter, who presents an explicit theoretical model entitled, "reflection on activity-effect relationship." This model is used to elaborate upon *what* and *how* teachers can learn from their teaching.

Mason suggests that "Learning about teaching from teaching is a lifetime process of refining sensitivities to students and to the conditions in which learning is fostered and sustained." Tzur elaborates further on this idea, claiming that the experience of student-teaching, or "practicum," is usually stressful and insufficient to develop efficient ways of engaging with students and to notice an impact on students' learning. As such, on-going LTT is the only possible solution for developing effective teaching practices. However, while Tzur believes that every teaching activity is a potential source for learning, he refers to LTT as "unrealized potential" and elaborates in great detail on the reasons for that. Based on the work of Simon (2006, 2007), both Tzur and Leikin and Zazkis identify "perception based perspective" – in contrast with "conception based perspective" – as one of the major factors that impede learning in general, and LTT, in particular. Tzur concludes with a detailed list of ideas and questions for further research on LTT. These ideas include explicating what "counts" as evidence of LTT and examining how it can be measured or occasioned.

The study by Leikin – that concludes Part I – provides a partial answer to the question of occasioning LTT. She describes a methodology of systematic exploration of LTT through employing multiple solution tasks in teaching experiments and longitudinal teacher-development experiments. Multiple solution tasks are used both as a didactical tool to engage teachers' learning, as well as a research tool used to intensify unforeseen (for teachers) situations and to analyze teachers' LTT. Leikin's conclusions point to an interrelationship between teachers' mathematical and pedagogical knowledge – the issue that is explored in further detail in all the chapters of this volume.

References

Brophy, J. (Ed.). (2003). *Using video in teacher education*. Elmont, NY: Elsevier.

Lampert, M., & Ball, D. (1998). *Teaching, multimedia, and mathematics: Investigations of real practice. The Practitioner Inquiry series*. New York: Teachers College Press.

Lampert, M. (2001). *Teaching problems and the problems of teaching*. New Haven: Yale University Press.

Mason, J. (2007, June). Mathematical pedagogy and pedagogical mathematics. Two talks at the pacific institute for mathematical studies 10th anniversary. Retrieved in March 2009 from http://mcs.open.ac.uk/jhm3/Presentations/%20Presentations%20Page.htm

Simon, M. A. (2006). Pedagogical concepts as goals for teacher education: Towards an agenda for research in teacher development. In S. Alatorre, J. L. Cortina, S. Mariana, & A. Méndez (Eds.), *Proceedings of the twenty eighth annual meeting of the North American chapter of the international group for the psychology of mathematics education* (Vol. 2, pp. 730–735). Merida, Yucatán, Mexico: Universidad Pedagógica Nacional.

Simon, M. A. (2007). Constrains on what teachers can learn from their practice: Teachers' assimilatory schemes. In J-H Woo, H-C Lew, K-S Park & D-Y Seo (Eds.), *Proceedings of the 31st international conference for the psychology of mathematics education* (Vol. 1, pp. 137–142). Korea: The Korea Society of Educational Studies in Mathematics.

Part I
Theoretical and Methodological Perspectives on Teachers' Learning Through Teaching

Teachers' Opportunities to Learn Mathematics Through Teaching

Roza Leikin and Rina Zazkis

Every teacher's greatest opportunity for further learning in mathematics education is her classroom teaching. (Simon, 2006, p. 137).

Introduction

Numerous studies on mathematics teacher development have demonstrated that mathematics teachers learn through their teaching experiences (e.g., Cobb & McClain, 2001; Kennedy, 2002; Lampert & Ball, 1999; Lesh & Kelly, 1994; Mason, 1998, 2002; Ma, 1999; Shulman, 1986; Wilson, Shulman, & Richert, 1987). Several articles explored teachers' learning in the course of their teaching careers, among other elements of their professional growth (e.g., Franke, Carpenter, Levi, & Fennema, 2001; Chazan, 2000; Kennedy, 2002; Lampert, 2001; Ma, 1999; Mason, 1998; Schifter, 1998). Other studies investigated different approaches to professional development and stumbled upon learning through teaching (LTT) while focusing on other issues (e.g., Cobb & McClain, 2001; Lesh & Kelly, 1994). Only a few studies have been devoted to the systematic investigation of teachers' learning in their own classrooms (Leikin, 2006; Leikin, 2005; Leikin & Rota, 2006; Nathan & Knuth, 2003).

The main body of research that addresses LTT focuses mainly on teachers' continual inquiry into the students' thinking and learning, but little is known about the teachers' learning of mathematics in their own classrooms. Perrin-Glorian, DeBlois, and Robert (2008) stressed that research concerning LTT (they used the term "indirect learning") is scant and that the educational community has a relatively limited understanding of what it is that changes in teachers' knowledge, or how these changes come about in an authentic classroom situation, especially in the domain of mathematical knowledge.

R. Leikin (✉)
University of Haifa, Haifa, Israel

R. Leikin, R. Zazkis (eds.), *Learning Through Teaching Mathematics*, Mathematics Teacher Education 5, DOI 10.1007/978-90-481-3990-3_1,

Evidence for Learning Through Teaching

Theories of Teacher Knowledge and Teaching

In their seminal work that analyzes the epistemological structure of teachers' knowledge, Wilson et al. (1987) described the complexity of teachers' knowledge and of its sources. The researchers argued that teachers' reasoning begins with comprehension of the set of core ideas related to the topic to be taught, and that teachers' knowledge related to the topic is transformed while teaching. This process of transformation is associated with the planning and design of instructional activities, evaluation, and reflection. As a result, after teaching, teachers attain new comprehension, enriched by fresh understanding and increased awareness of the purposes of instruction, its subject matter, and the participants.

The teaching/learning process can be modeled as a sequence of situations that result in new knowledge construction by students (Brousseau, 1997). So-called *a-didactic situations,* in which the teacher passes some of the responsibilities for the learning process onto the students, are considered to be most effective for students' learning. We consider such situations to be most effective for teachers' learning as well. The teacher is responsible for the *devolution* of a meaningful task that supports the design of an a-didactic situation. When an a-didactic situation is created, students are responsible for realizing the learning purposes by approaching the task, and the teacher's role is to facilitate this realization. In a situation of this type, the teachers adjust their plans to the students' ideas and learn together with the students.

Based on the theory of didactic situations, Simon (1997) suggested a Mathematics Teaching Cycle model for the teaching process. According to the model, teachers design a hypothetical learning trajectory based on the various types of knowledge they possess. The trajectory includes three interrelated elements: learning goals, the teacher's plan for learning activities (tasks), and the teacher's hypothesis of the learning process. When implementing this hypothetical trajectory in the classroom, teachers need to adjust it based on their interactions with students. These adjustments lead teachers to new understandings that precede the next hypothetical learning trajectory they design.

Following the analysis of elements of epistemological knowledge of mathematics teachers, Steinbring (1998) devised a two-ring model of teaching and learning mathematics as autonomous systems. According to this model, teachers use their knowledge of content and of students to design and devolve learning tasks for the students, and the students cope with these tasks by activating their own knowledge and devising their own interpretations of the tasks. Students approach the tasks, reflect on the process, and, as a result, construct their knowledge. Simultaneously, the teacher observes and supports the learning process, reflects on the learning situation, adjusts the task to the situation, and transforms the teacher's own knowledge. New learning opportunities are based on knowledge enriched by the teaching experiences. The cyclic view of teaching (e.g., Artzt & Armour-Thomas, 2002; Steinbring, 1998; Simon, 1997; Wilson et al., 1987) does not claim that teachers learn through teaching, but demonstrates that teaching has a great *potential* for teachers' learning.

The research literature suggests that the key feature of teachers' learning from teaching practice is their reflective ability (e.g., Dewey, 1933; Schön, 1983; Jaworski, 1998; Berliner, 1987, 1994). Jaworski (1998) and Mason (2002) distinguished between reflection-on-action as thinking back after the fact and reflection-in-action as being aware of inner thoughts while engaging in an activity. Mason added another type of reflection – reflection-through-action – that concerns the teachers' awareness of their own practice through the act of engaging in that practice. But Mason argues that these distinctions are rather vague, like the reflection itself. Reflection may be of varied scope and may have different purposes. Correspondingly, effectiveness of reflection in the LTT process may differ, depending on its goal, on the scope and focus of the reflection, and on the attention and awareness of the teacher.

In addition to reflection, cyclic models of teaching include the components of lesson planning, the choosing of instructional tasks, and the teachers' interactions with the students. In this view of teaching (e.g., Artzt & Armour-Thomas, 2002; Steinbring, 1998; Simon, 1997), it is reasonable to expect development in teachers' knowledge when planning the lesson and working on the tasks chosen for students, and during interactions with the students. Further support for the importance of classroom interactions for LTT can be found in the argument that norms and practices are being formulated and are often implicit in everyday classroom mathematical activity (McNeal & Simon, 2000). Meaning is constructed by teachers and students alike as a result of the participants' interpretation of classroom interactions. Although students are supposed to learn *deliberately*, and the teachers' main purpose is to support student learning, the teachers themselves learn *unintentionally* through teaching (or *indirectly,* to use the terms of Stigler & Hiebert, 1999).

Teachers' Experience and Expertise

Additional evidence for LTT can be found in the research literature on teachers' expertise and the role of experience in the development of expertise. Berliner (1987, 1994) connected experiences and expertise, noting that not only is the relation complex and diverse, but also that experienced teachers differ significantly from novices in the effectiveness of their teaching. In his works, Berliner stressed that expertise grows through personal experience, even if different experiences lead to different levels of expertise. He distinguished five stages of expertise: novice, advanced beginner, competent performer, proficient teacher, and expert. The stages differ in the teachers' "interpretive abilities, their use of routines and their emotional investments that they make in their work" (p. 113).

Further connections between expertise and experience are identified in cognitive psychology studies. Pressley and McCormick (1995) defined expert teachers by *self-regulated learning ability* that includes strategies for knowledge acquisition, procedures for problem solving, and transfer of prior experience to new tasks. Additional characteristics of expert teachers suggested by Pressley and McCormick (1995) include being well-organized, *alert to classroom events,* showing concern for individuals and groups, and having command of subject matter delivery. Sternberg and Horvath (1995) identified three ways in which experts differ from novices: knowledge, efficiency, and insight. They claimed that experts construct connections between different elements of knowledge more effectively, solve problems more efficiently, accomplish more in less time, and are more likely than novices to arrive at novel and appropriate solutions to problems.

Berliner (1987) emphasized that teachers with similar experiences could have different levels of expertise and claimed that progress along the stages of expertise depends on the teachers' reflective abilities. Sternberg and Horvath (1995) argued that distinctions between expert and experienced teachers should inform teacher education practices. Teachers have different opportunities for LTT, and different abilities to use these opportunities. Berliner (1994) noted the shortage of scientific knowledge of how novices *become* experts. In an attempt to address this deficiency, in this chapter we discuss teaching experiences in which teachers used opportunities to learn mathematics, and we analyze the factors and mechanisms that supported their learning.

Teaching Experiments and Changing Approaches to Teaching

We find multiple examples of teachers' LTT in research literature dealing with *teaching experiments* (Cobb, 2000, Steffe & Thompson, 2000). A teaching experiment (TE) is a "transformational research that has as its goal the development and investigation of theoretically grounded innovations in instructional settings" (Cobb, 2000, p. 308). Teachers and researchers involved in the experiments share the responsibility for the quality of the students' mathematics education. Teaching within an experimental setting usually provides many discoveries, innovations, unpredicted situations, and the need to adjust initially planned procedures.

> The teacher-researcher also might interpret the anticipated language and actions of the students in ways that were unexpected prior to teaching. Impromptu interpretations occur to the teacher-researcher as an insight that would be unlikely to happen in the absence of experiencing the students in the context of teaching interactions (Steffe & Thompson, 2000, p. 276).

TEs provide rich opportunities for teachers to learn by building models and communicating with researchers involved in the experiment. In his extensive analysis of TEs as a research methodology, Thompson (1979) identified their following distinctive components: In a TE, specific attention is paid to orientations that uncover processes by which students learn school subject matter; investigation is longitudinal; researchers intervene in the students' learning processes; investigation and planning of future activities are based on gathered observations. Furthermore, TEs intensify learning opportunities for teachers by focusing their attention on the students' thinking. For example, Cobb and McClain (2001) acknowledged that while analyzing a TE designed to improve student learning, they realized how much the teachers themselves had been learning in the course of the experiment.

A TE methodology is at work when researchers invent, support, or explore new curricular approaches in school mathematics (Chazan, Yerushalmy, & Leikin, 2008). For example, when implementing a new technology-based environment in the classroom, teachers intentionally learn technological details and ways of managing student learning, and unintentionally develop sensitivity to their students and knowledge about them (e.g., Lloyd & Wilson, 1998). Chazan et al. (2008) demonstrated that changing the curricular approach to teaching algebra – from equation-based to function-based – required a transformation of the teachers' knowledge of mathematics related to definitions and instructional examples. Leikin and Rota (2006) performed a joint retrospective analysis of a teacher and a researcher considering changes in the teachers' knowledge and skills. They studied the development of one beginning elementary school teacher's proficiency in implementing inquiry-based teaching for the first time and in managing a whole-class mathematical discussion in an inquiry-based learning environment. They observed that the teacher became much more flexible and attentive to the students, without any formal professional development intervention.

Additional evidence for LTT can be found in the self-reports of researchers who are expert teachers analyzing their own pedagogical growth (Tzur, 2001; Chazan, 2000; Lampert, 2001). For example, Tzur (2001) conducted a TE in a third-grade classroom as a researcher-teacher and observed his own improvement in instructional practice. But the main body of research on LTT focuses on the teachers' continual inquiry into the students' thinking and learning (e.g. Franke et al., 2001; Schifter, 1998, 1996; Kennedy, 2002).

Teachers' Knowledge

Epistemological analysis of teachers' knowledge reveals significant complexities in its structure (e.g., Scheffler, 1965; Shulman, 1986; Wilson et al., 1987; Ball, Hill, & Bass, 2005; Kennedy, 2002). Addressing these complexities and combining different approaches to the classification of knowledge, Leikin (2006) identified the following three dimensions of teachers' knowledge:

Dimension 1, *kinds of teachers' knowledge*, is based on Shulman's (1986) classification. *Subject matter knowledge* comprises teachers' knowledge of mathematics, including knowledge of mathematical concepts, their definitions and properties, as well as different types of mathematical connections and their implementation for solving mathematical problems. *Pedagogical content knowledge* includes knowledge of how students approach mathematical tasks, the aptitude to fit learning tasks to the students' learning abilities and styles, as well as knowledge of learning setting. *Curricular content knowledge* includes knowledge of types of curricula and of various approaches to teaching mathematics and the ability to connect a mathematical task to different mathematical topics and concepts within a curriculum. These categories are not disparate but affect each other and are complementary.

Dimension 2, *sources of teachers' knowledge*, is based on Kennedy's (2002) distinctions. *Systematic knowledge* is acquired through studies of mathematics and pedagogy in colleges and universities, in-service and pre-service programs for teachers, as well as reading research articles and professional books. *Craft knowledge* is developed through classroom experience and is based mainly on the teachers' interactions with their students and their reflections on these interactions. *Prescriptive knowledge*, acquired through institutional policies is manifest in tests, accountability systems, and texts of a diverse nature; it is motivated mainly by the teachers' sense of responsibility toward students and the community.

Dimension 3, *forms of knowledge*, follows Atkinson and Claxton (2000) and Fischbein (1984). This dimension distinguishes between *intuitive knowledge* that determines the teacher's actions that cannot be planned in advance, and *formal knowledge,* which is connected mostly to planned teacher's actions. Scheffler (1965) distinguishes between *knowledge* and *beliefs*. Knowing has "propositional and procedural nature" whereas believing is "construable as solely propositional" (ibid., p. 15). According to Scheffler, believing is one of the conditions of knowing.

Kennedy (2002) suggests a correspondence between various sources of knowledge and different forms of knowledge (e.g., craft knowledge is mainly intuitive, systematic knowledge is mainly formal). Moreover, we consider LTT to be learning in a craft mode (within Dimension 2) and argue that the transformation of forms of knowledge (Dimension 3) is one of the indicators of LTT. These transformations can occur in teachers' knowledge of various types (Dimension 1). Our main focus, however, is on changes in the teachers' subject matter knowledge.

A View of Teaching and Learning

To address teachers' LTT, it is essential to clarify our view of both teaching and learning. We consider both notions rather broadly. In our view, teaching includes, in addition to in-class delivery of material and interaction with students, also lesson planning, the preparation of instructional materials, out-of-class interaction with students and colleagues, preparation of assessment items, checking/correcting students' work, and reflection on instructional practice.

We consider teachers' learning to include not only gaining new components of knowledge but also refining familiar ideas (Leikin, 2005, 2006), making connections between different components of previous knowledge, achieving deeper awareness of what concepts entail, and enriching their personal repertoire of problems and solutions. Furthermore, we also consider as teachers' learning the expansion of knowledge into its additional forms, such as when intuitive knowledge is developed into formal knowledge or when subject matter knowledge emerges as pedagogical content knowledge. This list is illustrative rather than exhaustive, and we aim at extending and refining it as a result of more extensive research in LTT. Although our focus is on learning mathematics, at times there is no clear distinction between mathematics and pedagogy, where pedagogy includes awareness of the students' mathematical conceptions.

In what follows, we consider four examples of LTT. The first example is presented in detail and analyzed using the model of interaction presented below. The three other examples provide brief descriptions of teaching episodes and discuss what the teachers have learned. In all cases, we attempt to identify the sources and supporting factors of LTT.

Mechanisms of LTT

Although it is clear that people learn from their practice in general, and that teachers learn from teaching, what exactly is being learned is often not clear. Leikin (2006, 2005) explored the changes that occur in teachers' knowledge through teaching, the mutual relations between the development of teachers' knowledge of mathematics and of pedagogy in the field of mathematics, and the mechanisms supporting these changes.

Prior research has shown that the main component of the LTT process consists of instructional interactions. Leikin (2005) suggested a model for a detailed analysis of teachers' interactions in a system of six themes: (1) the *purpose* for which a teacher may interact with students; (2) *initiation* by the teacher or by the students; (3) *motives,* which may be external if they are prescribed by the given educational system, or internal, including cognitive conflict, uncertainty, disagreement, or curiosity; (4) *reflection* on the teachers' or students' previous experiences; (5) *actions* that support the interactive process, e.g., advice, presentation, questioning, and discussion; (6) *focus* of the interaction, which may be mathematics or pedagogy. The case of Einat presented below (borrowed from Leikin, 2005) is analyzed according to this model.

Example 1: Learning Mathematics with Students: The Case of Einat

Einat taught mathematics for 14 years in a secondary school, where she was regarded as an expert teacher. She was asked to teach the following problem:

Problem 1 *Find the shortest way from the bottom left corner to the top right corner of the large rectangle without passing through the small (gray) rectangle* (Fig. 1a).

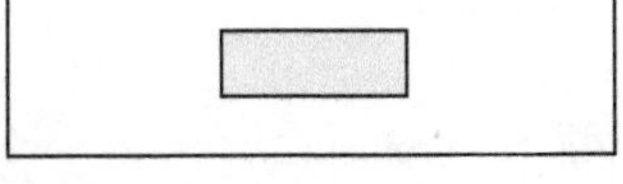

Fig. 1a Problem 1

Planning Stage

When *solving the problem* herself, Einat first considered it as a maxima–minima problem. But using calculus tools was complicated because the corresponding function had at least two variables. Therefore, she drew several possible paths and saw (by symmetry) that two of them were of minimal length (Fig. 1b). She proved this using the triangle inequality.

Einat *planned teaching the problem* in her basic level 11th grade class. She decided to begin by presenting the problem to the students in a story-like fashion, and, subsequently, applying the Pythagorean theorem.

Problem 1a *There is a rectangular pool with crocodiles in the middle of a surrounding rectangular park. Help Tom get from his house at the bottom left corner of the park to the bus station at the top right corner of the park as quickly as possible.* (Fig. 1c).

To train the students in the use of the theorem, she moved the small rectangle to the left several times and added numbers and letters to the drawing (Fig. 1d). It was clear to her that moving the rectangle broke the equality between the paths.

Problem 2 *Find the shortest way from corner A to corner B of the large rectangle without passing through the small (gray) rectangle* EHFG (Fig. 1d).

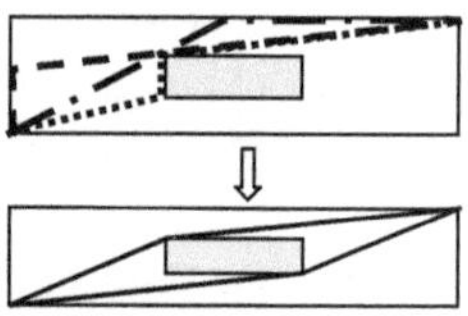

Fig. 1b Solution 1

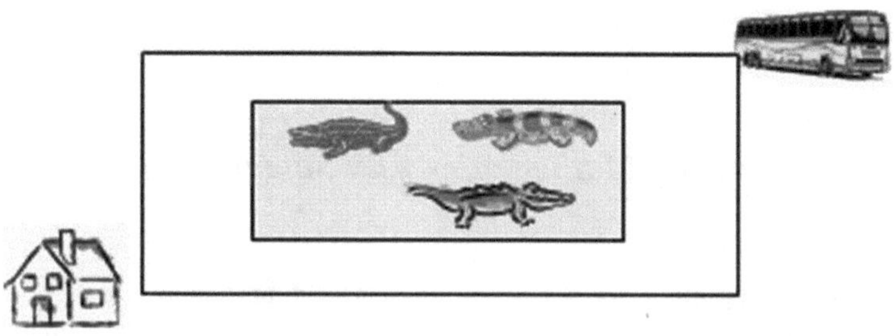

Fig. 1c Problem 1a

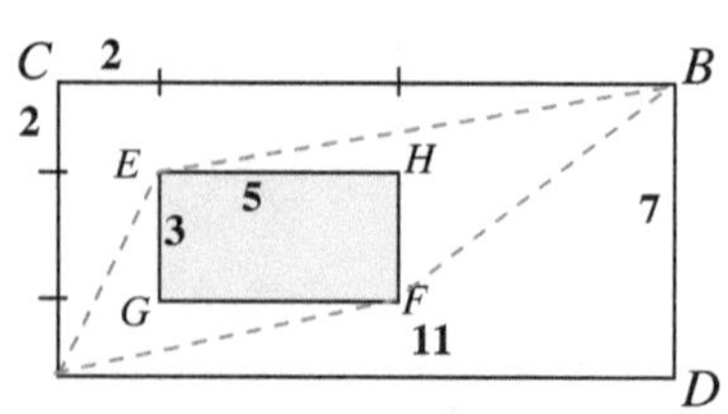

Fig. 1d Problem 2

Interactive Stage

The first phase of the lesson (solving the pool-in-the-park problem) was developed according to Einat's initial plan. When working with Problem 2, the students were surprised that the two paths (*A–E–B* and *A–F–B*) were of different lengths. One of the students (Ron) asked: "Is it possible to compare the lengths of the two paths without calculation?" This question was *unforeseen*, and both Einat and the students were *intrigued* by it. She *changed her plan*: the classroom discussion now focused on Ron's question. One of the students (Opher) raised an *unexpected conjecture*: "Obviously, if the vertex [point *F*] is closer to the diagonal AB, the path is shorter." The lesson ended at this point.

Reflection: Unexpected Questions and Answers

Einat was satisfied after the lesson because all the students had worked actively on the problem. No less important, she felt that she had learned mathematics with her students: both Ron's question and Opher's conjecture were new to her. But she was unhappy that they did not prove the conjecture during the lesson. She decided to continue the discussion and prove the conjecture in the next lesson.

As soon as she started planning the new lesson, she realized that Opher's conjecture was wrong: she knew that "of all triangles with a given side *s* and given area *A*, the isosceles triangle has the minimal perimeter." Thus, "vertices" of the two paths of different lengths may be at the same distance from *AB* (refutation 1). Furthermore, she solved the problem using a property of the ellipse: two paths, *A–F–B* and *A–E–B* (Fig. 1e), are of the same length if and only if points *E* and *F* are on an ellipse with foci at *A* and *B*, so that the vertices of two paths of equal length may be at different distances from *AB* (refutation 2).

From the didactic point of view, however, Ron's question remained open because the students were not familiar with the tools Einat used in refutations 1 and 2. She therefore designed an activity using guided inquiry with the aid of the Geometry Investigator (Dynamic Geometry Software, Schwartz, Yerushalmy, & Shternberg, 2000), which involved the construction of lines parallel to *AB* and dragging a point, *C* or *D* (Fig. 1f), along these lines, while keeping the rectangles invariant.

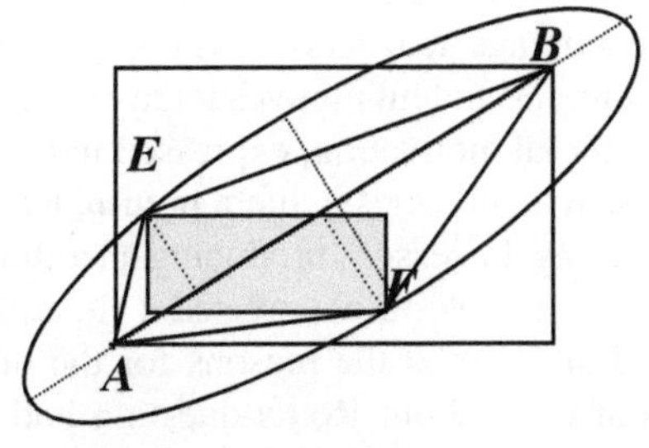

Fig. 1e General solution

Fig. 1f Exploring with the geometry investigator

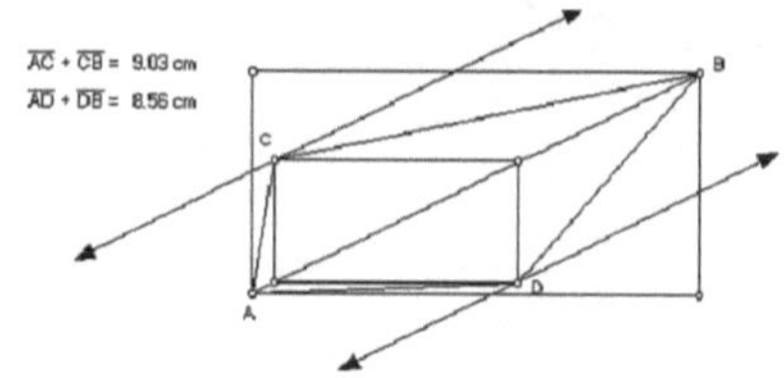

Interactions as a Source of Einat's Learning

We can show how the model of instructional interactions explains the mechanisms of teachers' LTT in Einat's case.

Purpose: The potential for teachers' learning from planning the instruction is based on the teacher having to solve unfamiliar problems before the lessons and to reason about the students' potential questions. A teacher's attempt to recognize the unforeseen elements of a problem can lead to learning. Einat missed some deep questions related to the problem when she planned her first lesson, and therefore experienced no learning during the initial planning stage. But she identified the problem's potential for student learning and transformed it in a way that allowed her to learn mathematics during the interactive stage.

When teachers analyze a past lesson for the purpose of planning the next one, their learning is based on a combination of reflective ability, critical thinking, and the need to continue teaching the topic. Einat's case clearly demonstrates this practice: When planning a follow-up lesson, Einat understood the mistake in her student's conjecture, and refuted it in several ways.

Initiation: We argue that interactions based on student initiatives are the main source of the teachers' learning. The learning of teachers (and students) depends on the teachers' noticing (Mason, 2002) the students' unexpected answers and evaluating these answers as being "worth addressing."

Einat connected the class discussion with the students' initiative, that is, thinking about an unexpected question related to the problem. Her ability to change the planned learning trajectory was an indication of her competence. As a result, she and her students "entered new [for her] mathematical territory," in Lampert's (2001) words.

Motives: Leikin (2005) differentiated between *external* and *internal* motives that govern teachers' interactions with students and claimed that internal motives encourage teachers' learning. At the same time, the teachers' ability to promote internal student motives intensifies learning by both teachers and students. Students' internal motivation, expressed in curiosity and persistence, leads them to ask unexpected questions, which, in turn, leads to the development of unpredicted situations during the lesson (producing an a-didactic situation, to use Brousseau's, 1997, term). For example, Einat created a situation in which the students were highly motivated to understand the reasons for the difference in the lengths of the paths. Her own curiosity about Ron's question and her dissatisfaction with the fact that Opher's conjecture had not been proven, served as internal motives for the development of her own mathematical knowledge.

Reflection: Every interaction is somehow related to one's previous experiences (e.g., Voigt, 1995; Schön, 1983). Teachers in the classroom may react to the students' experiences (reflection in action), or may reflect on their own experiences and plans. Teachers' ability to reflect on their learners' experiences characterizes them as showing great potential for their own learning. Note that reflection *on action* is much more effective for LTT when it is performed *with a view toward further actions*. Analyzing what has happened for the sake of continuing this happening is usually accompanied by insight derived from a motivation for reaching a better understanding of the events being considered. Einat's case demonstrates this claim.

Actions: Actions that assign an active role to students and a reflective role to the teacher in the interactive process have a greater potential for teacher's own learning. Einat designed learning actions that encouraged students to participate actively in the lesson. She listened carefully to her students' voices, and, in doing so, she enhanced her own opportunities for learning.

Focus: Einat's interaction with the students focused on mathematics, and therefore provided an opportunity for her to learn mathematics through teaching.

In this section, we attempted to exemplify the power of instructional interactions as a main source of teachers' LTT. In the next section, we address the question of what teachers learn through teaching. We start with a short summary of Einat's example and then provide several additional examples of LTT.

What Changes in Teachers' Knowledge Occur Through Teaching?

What Changed in Einat's Knowledge?

The unpredicted turn that the lesson took in relation to the solution of Problem 2 nurtured Einat's learning of mathematics. According to her plan, Problem 2 was aimed at performing calculations using the Pythagorean theorem. But when a student raised an unforeseen (general) question related to the length of the two paths (Fig. 1d), new mathematical connections were constructed: the paths within the rectangle could be compared using the properties of triangles with equal areas and a constant basis or using the properties of the ellipse.

When Einat moved the internal rectangle from the center of the external one, it became clear to her that *the length of the two paths will be different* "because the position of the internal rectangle is asymmetric." This intuitive assumption appeared to be correct for the concrete situation presented in Problem 2, but it was incorrect as a general statement. Einat discovered that not all asymmetrical positions of the internal rectangle resulted in paths of different lengths. When points E and F are on the ellipse (Fig. 1e), the paths are equal in length. Thus, an incorrect intuitive assumption was refuted, and incorrect intuitions were replaced with correct mathematical knowledge. The second critical point for Einat's learning was her intuitive agreement with Opher's conjecture, which was also refuted.

Our additional observation is related to the interrelationship between Einat's mathematical and pedagogical knowledge. It was her pedagogical sensitivity that encouraged her to formulate new problems that led to mathematical discoveries. At the same time her mathematical knowledge allowed her to evaluate the difficulty of the refutations she had produced, and (again) being attentive to her students she designed a new instructional activity using the Dynamic Geometry software.

In sum, in this example, we recognize the development of knowledge in the transformation of intuition into formal knowledge and in the mutual support between pedagogical and subject matter knowledge.

Example 2: Learning from a Student's Mistake: The Case of Lora

Lora, an experienced instructor in a course for pre-service elementary school teachers, taught a lesson on elementary number theory. The following interaction took place:

Lora: Is number 7 a divisor of K, where $K = 3^4 \times 5^6$?

Student: It will be, once you divide by it.

Lora: What do you mean, once you divide? Do you have to divide?

Student: When you go this [points to K] divided by 7 you have 7 as a divisor, this one the dividend, and what you get also has a name, like a product but not a product...

Lora's intention in choosing this example was to alert students to the unique factorization of a composite number to its prime factors, as promised by the Fundamental Theorem of Arithmetic, and to the resulting fact, that no calculation is needed to determine the answer to her question. This later intention is evident in her probing question.

What Did Lora Learn from the Above Interaction?

First, she learned that the term "divisor" is ambiguous, and that a distinction is essential between *divisor of* a number, as a relationship in a number-theoretic sense, and *divisor in* a number sentence, as a role played in a division situation. She learned further that the student assigned meaning based on his prior schooling and not on his recent classroom experience, in which the definition for "divisor" used in Number Theory was given and its usage illustrated. Before this incident, Lora used the term properly in both cases, but was not alert to a possible misinterpretation by learners. The student's confusion helped her make the distinction, increased her awareness of the polysemy (different but related meanings) of the term *divisor* and of the definitions that can be conflicting. This resulted in developing a set of instructional activities in which the terminology is practiced (Zazkis, 1998).

Example 3: Learning from a Student's Solution: The Case of Shelly

Shelly, a teacher with 20 years of experience in secondary school, solved the following problem with her 12th grade students:

Prove that

$$1 + 2t + 3t^2 + 4t^3 + ... + nt^{n-1} = \frac{1 - t^n}{(1 - t)^2} - \frac{nt^n}{1 - t}.$$

She expected her students to prove the equality using mathematical induction, but unexpectedly one of the students suggested the following solution:

$$S(t) = 1 + 2t + 3t^2 + 4t^3 + ... + nt^{n-1} = F'(t), \qquad \text{when}$$

$$F(t) = t + t^2 + t^3 + ... + t^{n-1} + t^n = \frac{t^{n+1} - t}{t - 1}, \qquad \text{thus}$$

$$S(t) = F'(t) = ... = \frac{1 - t^n}{(1 - t)^2} - \frac{nt^n}{1 - t}.$$

Shelly's first reaction was "How could I miss this? Oh well, the problem comes from the mathematical induction topic, and I did not think about derivatives at all."

What Did Shelly Learn in This Episode?

The connection between calculus and mathematical induction was new to Shelly. She was familiar with the use of mathematical induction in geometry, for example, proving a theorem about the sum of interior angles in a polygon. In her experience, mathematical induction was connected with principles of divisibility because many divisibility tests can be proven using induction. Induction was also connected naturally with sequences and series because of the multiple proofs that use induction in these topics. As the following comment indicates, she also was aware that many problems conventionally solved by applying mathematical induction can also be solved in a different way:

Shelly: Even in matriculation exams, they say "prove using induction or in a different way," like 3 is a divisor of $n^3 - n$ because $n^3 - n = (n-1) \times n \times (n+1)$.

But when she was preparing the lesson, Shelley did not consider this solution. Moreover, in more than 20 years of teaching, she has never connected induction with calculus. Her student's solution added a new mathematical connection to her subject matter knowledge, and this problem became part of her repertoire of problems with multiple solutions drawn from different areas of mathematics.

Example 4: Learning from a Student's Question: The Case of Eva

In a geometry lesson, Eva, a teacher with 15 years of experience in secondary school, was proving the following theorem with her 10th grade students:

If *AD* is a bisector of an external angle *CAF* in a triangle *ABC,* then

$$\frac{AB}{AC} = \frac{BD}{CD}$$

(see Fig. 2a).

After the theorem was proven, one of the students asked "What happens if *AD* is parallel to *BC* (Fig. 2b)?" This question led to a classroom discussion in which students arrived at the conclusion that the theorem was correct for *non-isosceles* triangles.

In her reflective analysis of the situation, Eva reported that she "had never thought about whether the theorem was correct for isosceles triangles." Furthermore, when analyzing this situation with the researcher, she unexpectedly connected this geometry problem with the topic of limits:

When AD is parallel to BC, $\lim\limits_{AC \to AB} \frac{BD}{CD} = 1$.
Since BD = BC + CD,
this situation can demonstrate the following rule: $\lim\limits_{x \to \infty} \frac{x-c}{x} = 1$.

What Did Eva Learn from This Lesson?

Eva appeared to be surprised by the connection. First, this lesson led her to develop a "neater formulation of a theorem." She commented that "the theorem was never mentioned in any familiar textbook or mathematics course," and that next time, if students do not consider an isosceles triangle when proving the theorem, she will make them consider this special case. It was a student's question that served as a trigger, but it was the teacher's curiosity and her deep mathematical knowledge that led her to develop new mathematical connections.

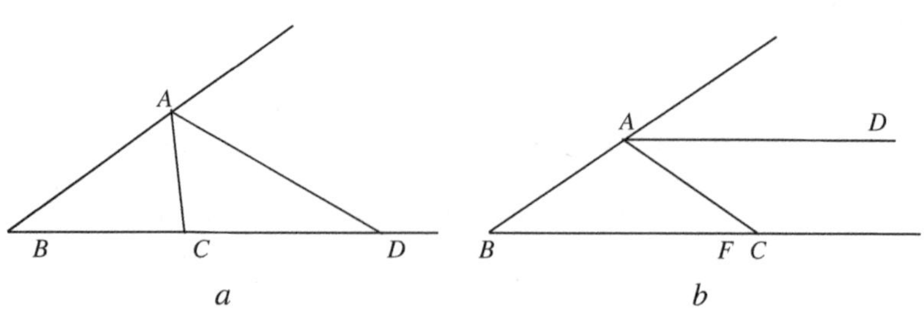

Fig. 2 A bisector of an external angle

Sources of LTT

We acknowledge the crucial importance of instructional interaction and note that abundant interactions with students are necessary but not sufficient conditions for teachers' LTT. The development of teachers' mathematical knowledge depends on their flexibility (Leikin & Dinur, 2007). Einat's episode (Example 1) clearly demonstrated that by opening opportunities for students to initiate interactions and by managing a lesson according to the students' ideas, teachers extend the opportunities for their own learning. Considering instructional interactions, we are particularly interested in the attention teachers pay to student responses, both correct and incorrect.

When observing student conjectures during the interaction, the teachers search for new explanations or clarifications and design activities for students that help verify or refute these conjectures. When attending to the students' mistakes and attempting to correct their understanding, teachers construct new mathematical connections, as illustrated in Lora's case, in Example 2. Another source for teachers' LTT is unexpected correct student ideas, as in Example 3, or surprising questions from students, as in Examples 1 and 4. The teachers' alertness and attention in these matters (Mason, 1998, 2002) play an important role in turning the interaction with students into a learning experience for all concerned.

Is This Knowledge New? Is It Mathematics or Pedagogy?

Teachers learn both mathematics and pedagogy when teaching. In many situations, teachers' pedagogical knowledge develops when they become aware of unforeseen student difficulties. By analyzing the sources of the students' difficulties and misconceptions, teachers gain further awareness of the concepts and greater appreciation of the structure of mathematical thought. Example 2 illustrates a case of developing such awareness: to help students grasp the meaning of the term implied in a given situation, the teacher needed to clarify first the distinction between the different uses of the term "divisor" *for herself.*

In other, less frequent situations, teachers clearly learn mathematics that is new to them. This mathematics then serves them in the instructional design of subsequent lessons. In many situations, it is difficult to differentiate between mathematical and pedagogical learning, and the distinction is blurred because the teachers' knowledge is situated largely in their teaching practice (Leikin & Levav-Waynberg, 2007). Development of the teachers' craft knowledge depends strongly on their systematic knowledge. The teachers' mathematical understanding helps them stay alert to student ideas and develop these ideas further (as shown in Examples 1, 3, and 4). The teachers' pedagogical knowledge and skill are responsible for their awareness of the importance of granting students autonomy in the mathematics classroom and of being open to student ideas. Finally, we found that teachers with more profound mathematical understanding (in the sense used by Ma, 1999) feel safer allowing students to present their mathematical ideas and ask questions.

The Complexity of LTT: Supporting and Impeding Factors

We have already mentioned that everyone learns, or at least has an opportunity to learn, from experience. As such, teachers learn from their teaching experience, but as Watson and Mason (2005) noted, "one thing we do not seem to learn from experience is that we seldom learn from experience alone. Something more is required" (p. 199). We attempted to understand what this "more" entails.

To identify the factors that support teachers' LTT, we must first consider what it is that hinders teachers' LTT. Simon (2007) pointed to the "contrast between the opportunity for learning inherent in teaching and the often-limited knowledge gleaned by teachers" (p. 137). He described the limiting factors as *perception-based perspective* and *empirical learning processes*. From a perception-based (rather than conception-based) perspective, mathematical relationships are external and are perceived similarly by all learners as a result of engagement with particular tasks or representations. A related issue is empirical learning, in contrast to reflective abstraction, where according to Piaget (2001), reflective abstraction is a fundamental necessity in constructing mathematical concepts. To create active learning opportunities for students, a mathematics classroom can operate by collecting data and deriving patterns from sets of data, an activity Hewitt (1992) referred to as "train spotting." This approach results in learning facts rather than the logical necessity of relationships and encourages teachers to focus on whether or not a particular empirically verified relationship is perceived by students. This limits the teachers' opportunity to learn about their students' conceptions, thinking processes, and the obstacles before them.

Although Simon focuses on the teachers' learning about their students' learning rather than on their learning of mathematics, identifying the lack of reflective abstraction as an obstacle for learning applies also to the present discussion. This is consistent with Berliner's (1987) view that bringing experience to the level of expertise depends on the teachers' reflective abilities, and with abundant research in mathematics education identifying reflective ability as the key feature in learning.

In a way, the focus on reflection indicates what "more" is required in order to learn from experience. We can now reformulate our question from "What supports teachers' LTT?" to "What factors are necessary for reflective abstraction to occur in teaching?" Without attempting to provide an exhaustive list, we identified several illustrative components based on the examples above. Considering instructional interaction to be a precondition for LTT, the necessary but not sufficient factors include an open mind, curiosity, and attentiveness to students. But to identify opportunities for learning and take advantage of them, these factors must be supported by the teacher's personal mathematical understanding and profound systematic knowledge of the subject matter.

Finally, we note that teachers are not always aware that they have learned through their teaching, and sometimes they are hesitant to admit it. And even when they are aware of learning, they are not sure that they learned mathematics. They often make statements like "I knew this but have never thought about it." However, we consider anew "thinking about it" – when an instructional situation presents such an opportunity – as an indication of learning. In this case, LTT can be thought of as transferring existing knowledge from the teachers' passive repertoire to their active one. As such, clear criteria that point to teachers' learning of mathematics in LTT need to be further refined.

References

Artzt, A. F., & Armour-Thomas, E. (2002). *Becoming a reflective mathematics teacher*. Mahwah, NJ: Erlbaum.

Atkinson, T., & Claxton, G. (Eds.) (2000). *The intuitive practitioner: On the value of not always knowing what one is doing*. Philadelphia, PA: Open University Press.

Ball, D. L., Hill, H. C., & Bass, H. (2005). Knowing mathematics for teaching: Who knows mathematics well enough to teach third grade, and how can we decide? *American Educator*, 14–17, 20–22, 43–46.

Berliner, D. C. (1987). Ways of thinking about students and classrooms by more and less experienced teachers. In J. Calderhead (Ed.), *Exploring teachers' thinking* (pp. 60–83). Great Britain: Cassell Educational Limited.

Berliner, D. C. (1994). Teacher expertise. In B. Moon & A. S. Mayes (Eds.), *Teaching and learning in the secondary school* (pp. 107–113). Routledge.

Brousseau, G. (1997). *Theory of didactical situations in mathematics*. The Netherlands: Kluwer.

Chazan, D. (2000). *Beyond formulas in mathematics teaching: Dynamics of the high school algebra classroom*. New York, NY: Teachers College.

Chazan, D., Yerushalmy, M., & Leikin, R. (2008). An analytic conception of equation and teachers' views of school algebra. *Journal of Mathematical Behavior*, 27, 87–100.

Cobb, P. (2000). Conducting teaching experiments in collaboration with teachers. In A. E. Kelly & R. A. Lesh (Eds.), *Research design in mathematics and science education* (pp. 307–333). Hillsdale, NJ: Erlbaum.

Cobb, P., & McClain, K. (2001). An approach for supporting teachers' learning in social context. In F.-L. Lin & T. J. Cooney (Eds.), *Making sense of mathematics teacher education* (pp. 207–231). Dordrecht, The Netherlands: Kluwer.

Dewey, J. (1933). *How we think: A restatement of the relation of reflective thinking to the educative process*. Boston: D.C. Heath and Co.

Fischbein, E. (1984). The interactions between the formal, the algorithmic, and the intuitive components in a mathematical activity. In R. Biehler, R. W. Scholz, R. Sträßer & B. Winkelmann (Eds.), *Didactics of mathematics as a scientific discipline* (pp. 231–245). The Netherlands: Kluwer.

Franke, M. L., Carpenter, T. P., Levi, L., & Fennema, E. (2001). Capturing teachers' generative change: A follow-up study of professional development in mathematics. *American Educational Research Journal*, *38*, 653–689.

Hewitt, D. (1992). Train spotters' paradise. *Mathematics Teaching* 140, 6–8.

Jaworski, B. (1998). Mathematics teacher research: Process, practice, and the development of teaching. *Journal of Mathematics Teacher Education*, 1(1), 3–31.

Kennedy, M. M. (2002). Knowledge and teaching. *Teacher and Teaching: theory and practice*, 8, 355–370.

Lampert, M. (2001). *Teaching problems and the problems of teaching*. New Haven: Yale University Press.

Lampert, M., & Ball, D. L. (1999). Aligning teacher education with contemporary K-12 reform visions. In L. Darling-Hammond & G. Sykes (Eds.), *Teaching as the learning profession: Handbook of policy and practice* (pp. 33–53) San Francisco: Jossey-Bass.

Leikin, R. (2005). Qualities of professional dialogue: Connecting graduate research on teaching and the undergraduate teachers' program. *International Journal of Mathematical Education in Science and Technology*, *36*(1–2), 237–256.

Leikin, R. (2006). Learning by teaching: The case of Sieve of Eratosthenes and one elementary school teacher. In R. Zazkis & S. Campbell (Eds.), *Number theory in mathematics education: perspectives and prospects* (pp. 115–140). Mahwah, NJ: Erlbaum.

Leikin, R., & Dinur, S. (2007). Teacher flexibility in mathematical discussion. *Journal of Mathematical Behavior, 26*, 328–347.

Leikin, R., & Levav-Waynberg, A.(2007). Exploring mathematics teacher knowledge to explain the gap between theory-based recommendations and school practice in the use of connecting tasks. *Educational Studies in Mathematics, 66*, 349–371.

Leikin, R., & Levav-Waynberg, A. (2007). Exploring mathematics teacher knowledge to explain the gap between theory-based recommendations and school practice in the use of connecting tasks. *Educational Studies in Mathematics.*

Leikin, R., & Rota, S. (2006). A case study on the development of teacher's proficiency through teaching. *Mathematics Education Research Journal, 18*(3), 44–68.

Lesh, R., & Kelly, A. E. (1994). Action-theoretic and phenomenological approaches to research in mathematics education: Studies on continually developing experts. In R. Biehler, R. W. Scholz, R. Sträßer, & B. Winkelmann (Eds.), *Didactics of mathematics as a scientific discipline* (pp. 277–286). The Netherlands: Kluwer Academic Publishers.

Lloyd, G. M., & Wilson, M. S. (1998). Supporting innovation: The impact of a teacher's conceptions of functions on his implementation of a reform curriculum. *Journal for Research in Mathematics Education, 29*, 248–274.

Ma L. (1999). *Knowing and teaching elementary mathematics: Teacher's understanding of fundamental mathematics in China and the United States*. Mahwah, NJ: Erlbaum.

Mason, J. (1998). Enabling teachers to be real teachers: necessary levels of awareness and structure of attention. *Journal of Mathematics Teacher Education, 1*(3), 243–267.

Mason, J. (2002). *Researching your own practice: The discipline of noticing*. New York: Falmer.

McNeal, B., & Simon M. A. (2000) Mathematics culture clash: Negotiating new classroom norms with prospective teachers. *Journal of Mathematical Behavior, 18*, 475–509.

Nathan, M. J., & Knuth, E. J. (2003). A study of whole classroom mathematical discourse and teacher change. *Cognition and Instruction*, 21, 175–207.

Perrin-Glorian, M.-J., DeBlois, L., and Robert, A. (2008). Chapter 2: Individual practicing mathematics teachers: Studies on their professional growth. In T. Wood & K. Krainer (Eds.), *International handbook of mathematics teacher eduation. Participants in mathematics teacher education: individuals, teams, communities, and networks* (Vol.3, pp. 35–59). Rotterdam, the Netherlands: Sense Publishers.

Piaget, J. (2001). *Studies in reflecting abstraction*. Sussex, England: Psychology Press.

Pressley, M., & McCormick, C. (1995). *Cognition, teaching and assessment*. New York: Harper Collins College Publishers

Scheffler, I. (1965). *Conditions of knowledge. An introduction to epistemology and education*. Glenview, IL: Scott, Foresman & Company.

Schifter, D. (Ed.) (1996) *What's happening in math class? Envisioning new practices through teacher narratives* (Vols. I and II). New York: Teacher College Press.

Schifter, D. (1998). Learning mathematics for teaching: From a teachers' seminar to the classroom. *Journal of Mathematics Teacher Education*, 1, 55–87.

Schön, D. A. (1983). *The reflective practitioner: How professionals think in action*. New York: Basic Books.

Schwartz, J. L., Yerushalmy, M., & Shternberg, B. (2000). *Geometric Supposer* [Computer software]. Israel: CET.

Shulman, L. S. (1986). Those who understand: Knowledge growth in teaching. *Educational Researcher*, 15 (2), 4–14.

Simon, M. A. (1997). Developing new models of mathematics teaching: An imperative for research on mathematics teacher development. In E. Fennema & B. Scott-Nelson (Eds.), *Mathematics teachers in transition*. (pp. 55–86). Mahwah, New Jersey: Lawrence Erlbaum Associates.

Simon, M. A. (2007). Constrains on what teachers can learn from their practice: Teachers' assimilatory schemes. In J.-H. Woo, H.-C. Lew, K.-S. Park, & D.-Y. Seo (Eds.), *Proceedings of the 31st International Conference for the Psychology of Mathematics Education* (Vol. 1, pp. 137–142). Korea: The Korea Society of Educational Studies in Mathematics.

Steinbring, H. (1998). Elements of epistemological knowledge for mathematics teachers. *Journal of Mathematics Teacher Education*, 1(2), 157–189.

Sternberg, R. J., & Horvath, J. A. (1995). A prototype view of expert teaching. *Educational Researcher*, 24(6), 9–17.

Steffe, L. P., & Thompson, P. W. (2000). Teaching experiment methodology: Underlying principles and essential elements. In R. Lesh & A. E. Kelly (Eds.), *Research design in mathematics and science education*, (pp. 267–307). Hillsdale, NJ: Erlbaum.

Stigler J. W., & Hiebert J. (1999). *The teaching gap: Best ideas from the world's teachers for improving education in the classroom.* New York: The Free Press.

Thompson, P. (1979, March). The constructivist teaching experiment in mathematics education research. Paper presented at the *Annual Meeting of the National Council of Teachers of* Mathematics, Boston.

Tzur, R. (2001). Becoming a mathematics teacher educator: Conceptualizing the terrain through self-reflective analysis. *Journal of Mathematics Teacher Education*, 4, 259–283.

Voigt, J. (1995). Thematic patterns of interactions and sociomathematical norms. In P. Cobb & H. Bauersfeld (Eds.), *Emergence of mathematical meaning: Interactions in classroom culture* (pp. 163–201). Hillsdale, NJ: Erlbaum.

Wilson, S., Shulman, L., & Richert, A. E. (1987). "150" Different ways of knowledge in teaching. Representations of knowledge in teaching. In J. Calderhead (Ed.), *Exploring teachers' thinking*. (pp. 1–37). London: Cassell.

Watson, A., & Mason, J. (2005). *Mathematics as a constructive activity: Learners generating examples*. Mahwah, NJ: Lawrence Erlbaum.

Zazkis, R. (1998). Divisors and quotients: Acknowledging polysemy. *For the Learning of Mathematics*, 18(3), 27–30.

Attention and Intention in Learning About Teaching Through Teaching

John Mason

Introduction

Teaching is a marvelous action in which to engage, because it is possible to be learning on many levels simultaneously: About the content being taught, about the students who are learning, about teaching itself, about learning itself, and perhaps most importantly, about yourself. But learning from experience is highly problematic. The case is put forward that most human interaction is based on reacting to others in specific situations with habitual behaviors, and that if significant learning is to take place from and through experience, it is necessary to shift from reacting to responding thoughtfully, even mindfully (Claxton, 1984; Langer, 1997). This is the role of attention directed by intention. Furthermore, when faced with a novel situation, response rather than reaction is vital. The distinction between *reacting* and *responding* applies equally to students and to teachers, but is mainly developed here in relation to teachers learning about their practices while teaching.

In this chapter, I endeavor to show how it is possible to work on sensitizing oneself to notice what previously passed by unnoticed, and how that can be used to inform actions in the future, that is, to learn about teaching through teaching. The process calls upon the Discipline of Noticing (Mason, 2002) which can be used to learn intentionally from experience, by becoming aware of how attention shifts. For example, each technical term in mathematics signals the fact that those who developed the term needed a label to refer to particular features on which to focus attention or to a particular way of attending. Thus *angle* signals attention to turning between two limiting arms, and away from the space between them. It is a reasonable conjecture that learners need to experience a similar shift in their attention if they are to appreciate and understand the term and use it productively and effectively for themselves. Exactly the same applies to pedagogical strategies and didactic tactics. Labels signal shifts in attention, and similar shifts in the form and focus of

J. Mason (✉)
The Open University (emeritus), Milton Keynes, Buckinghamshire; University of Oxford, Oxford, UK

R. Leikin, R. Zazkis (eds.), *Learning Through Teaching Mathematics*, Mathematics Teacher Education 5, DOI 10.1007/978-90-481-3990-3_2,

attention are necessary in order to learn about teaching and to work at developing practices so as to make teaching more effective and pleasurable.

Teachers are subject to numerous forces: Legislation, inspection, practices in their school, expectations of parents and children, and, often submerged beneath these, their own ideals and intentions. Despite conclusions to the contrary, none of these forces preclude working on and developing one's teaching through enquiring into that teaching. What is required is a vision of what might be possible, a source of energy and support, and personal discipline. Discipline is amplified whenever, and to the extent that, appropriate support is present. The vision itself is most effective when dynamic, being renewed and modified periodically, rather than fixed. John Donne's famous meditation (1624 XVII) that "no man is an island, entire of itself" is an excellent reminder that even to work on one's own teaching by oneself, outside connections are necessary in order to avoid descent into solipsism.

Theoretical Underpinnings

I take learning to be the transformation of attention. That is, learning has taken place when people discern details, recognize relationships, and perceive properties not previously discerned, through attending in fresh or distinct ways, and when they have fresh possibilities for action from which to choose. Learning necessarily involves shifts in the form as well as the focus of attention.

I take as given that there is a complex interaction between student(s), content, and teacher within an environment of social, institutional, and political forces and practices. This triad of impulses within its environment is a dynamic system. Each component is constantly adjusting to maintain its Piagetian equilibrium. Each impulse acts at different times and in different ways upon one or other of the other two and is mediated in that action by the third (Bhagavad Gita 18:18 see Mascaró, 1962; see also Bennett, 1956–1966; Mason, 1979, 2008). In relation to this chapter, the environment within which I am working includes a conjecturing atmosphere, in which everything said (by me, and by readers in response) is said in order to consider it more carefully with the intention of altering or modifying it as necessary (Mason, Burton, & Stacey, 1982; Mason & Johnston-Wilder, 2006). Because of the one-way nature of printed text, this means that everything must be tested in your experience. To this end, my brief summary of theoretical underpinnings is followed by succinct descriptions of some phenomena in order to try to ground the subsequent analysis and proposals in experience.

Biologically, a dynamic equilibrium is sustained until there is a significant but not overwhelming disturbance. In an educational context, this means that no learning, no development takes place until there is some disruption of the working out of internalized actions (habits, automaticities), until some dissonance is experienced, whether cognitive (Festinger, 1957), affective (e.g. surprise: See Movshovitz-Hadar, 1988), enactive (Skinner, 1954), or some combination of all three. Following Piaget (1971, 1985), change (assimilation and accommodation) is a reaction or a response to experienced variation, including spontaneous "mutation" in the organism or the environment. For the discipline of teaching (Mason, 2009) this means that it is only when some significant but not overwhelming challenge occurs (change of curriculum or other institutional demands, change in student behavior and attention, change in teacher behavior and attention) that significant development can occur. This applies just as much to student learning as to teacher learning. A student who is "going through the motions" of completing tasks as set by the teacher is not actually learning in any deep sense (such as described by the onion model of understanding see Pirie & Kieren, 1994). A teacher who is going through the motions of curricular or pedagogic reform is not actually disturbing current practices significantly and so is not learning about teaching.

There are dangers, however. When change is sought explicitly and overtly, reaction sets in (think of the last time someone else wanted you to change your behavior). What results is often the opposite of what was intended due to the multitude of possible unintended consequences. Rather, *change* is a partial description of a lived experience of dynamic and constant disequilibrium and equilibration. It comes about through the desire to make sense (note the body-based metaphor of sensation).

The key question in teaching is whether an action is generative or merely repetitive, that is, whether something fresh arises from the action, or whether it is a reworking of habit and internalized automaticities. To make sense of this, it is necessary to consider the commonplace (St. Maurice, 1991) of "learning from experience" and the role of disturbance, which will lead into the structure of the human psyche for inter-relating mindfulness and habit. This then leads me to how it is possible to learn from experience of disturbance. Throughout the chapter, I try to maintain a parallel between student learning and teacher development, because both are instances of learning. Sometimes it is easier to make observations of others learning, and sometimes easier to observe in oneself.

I take an unabashedly phenomenological or experiential stance. I am interested in lived experience and how that can be enriched. My theoretical frame draws eclectically on a range of pedagogical, psychological, and social constructs. The pedagogic constructs of *didactic contract, didactic tension, didactic transposition*, the psycho-social construct *zone of proximal development*, and the psychological constructs of *attention* and *intention*, combined with the Discipline of Noticing and *variation theory*, form the background theory for how teachers can endeavor to learn from their experience of teaching and learning, and can endeavor to promote a parallel action in their students so that they too have the best opportunities to learn. Each construct draws and labels one or more distinctions which remind me to discern details that might otherwise go unnoticed, thus enabling me to recognize relationships in situations, and even to perceive these as instantiations of properties. These, in turn, permit reasoning on the basis of agreed properties which characterizes a fully fledged discipline. Mathematics education still has some way to go to reach this agreement.

The notion of *didactic transposition* introduced by Yves Chevallard (1985) alerts me to the almost inevitable transformation that occurs when someone tries to give or reproduce their experience for someone else. In mathematics education, it usually happens that an expert experiences an insight or surprise, or recognizes the potential pedagogic value of some mathematical exploration or insight, and then tries to reproduce that experience for students by constructing a worksheet or other form of task. Points at which the expert made use of their own powers to investigate and explore are naturally replaced by instructions telling students what to do. Similarly, when teachers try to re-enact a pedagogic strategy or a pre-planned teaching sequence, they tend to miss out on the opportunities to choose to respond sensitively and mathematically to particular students. I summarize the transposition as expert awareness is transposed into instruction in behavior and it applies just as much to policy makers specifying curricula and to teachers trying out a pedagogic strategy, didactic tactic, or rearranged curriculum, as it does to teachers setting tasks for students.

The construct of *didactic situation* introduced by Guy Brousseau (1997) acts as a reminder that the ancient triad of student–teacher–content takes place within a complex environment or *milieu* of social and institutional influences. The associated construct of *didactic tension* is a powerful reminder that the more precisely the teacher indicates the behavior being sought, the easier it is for students to display that behavior without generating it from themselves; when applied to educational reform it translates into the more precisely the policy indicates the specific behaviors (practices) desired, the easier it is for teachers to display that behavior, to go through the motions, without generating it from themselves and without making real contact with students.

It is possible to carry out actions mechanically, either because they have been internalized this way (the result of training of behavior) or because they are being reproduced through mimicry, through adopting the outer, visible form (which is the content of the didactic tension). Again this applies to teachers enacting prepared teaching as much as to students making their way through work sheets and exercise sets.

There are evident resonances with Vygotsky's *Zone of Proximal Development* as elaborated by (van der Veer & Valsiner, 1991, see also Mason, Drury & Bills, 2007). The intention of teaching is that students initiate actions or respond to situations from themselves for themselves rather than having to be triggered or cued by inputs from a worksheet or a more experienced 'other'. The "zone" refers to a state of potential for that transition from a dependent state based on assenting to an independent assertive stance (Mason, 2009). The whole point about learning is that what once had to be pointed out by a teacher becomes available to the learner as a choice, as an option that comes to mind. Attention is then restructured, giving rise to the possibility of a concomitant restructuring of intention. The difference between teachers teaching and students being taught is that whereas the students have a more experienced other (the teacher) on hand, teachers rarely have in-the-moment access to a mentor or a more experienced colleague. So in order to learn from teaching, it is necessary to develop an inner witness who observes without passing judgment or justifications. Put another way, progress in learning is an internalization of practices encountered in others (Vygotsky, 1978), but this internalization requires a transformatory action.

Following Bennett (1956–1966), I take actions to require three impulses: One that initiates or "acts upon," one that responds or is acted upon, and one that mediates between the two, bringing and perhaps holding them together, and so making the action possible. In educational settings these are usually instantiated by the elements of the traditional trio: Student, teacher, and content. By analyzing the combinatorial possibilities, six fundamental modes of interaction can be distinguished: Expounding, expressing, explaining, examining, exploring, and exercising. Here I shall make use of only the first three.

Metaphors

Finding a satisfactory way to articulate the role of the environment on actions is a real challenge. There are several metaphors which direct attention unhelpfully.

Cause and Effect

Cause and effect is a popular and deeply embedded metaphor in the discourse of education, but one which I find particularly misleading. It draws upon the natural sciences, particularly physics, and lies at the heart of why, despite experience, we try to cause learning by initiating teacherly actions, when it is patently obvious that the most we can do is provide opportunities, however sophisticated, for students to use their intention to engage their attention productively. No matter how precisely people try to specify the tasks, the conditions, and the ways of working in a classroom, it is impossible to predict the outcome, precisely because unlike physics which deals with inanimate objects, education involves human beings who can direct their attention by exercising intention, and who are profoundly influenced by the individual and collective energies of peers, and their own imagined assumptions about themselves and their situation.

Social Influence as Forces

I also find myself unconvinced by the physics-based metaphor of forces acting on people which is deeply embedded in sociological discourse. Physical forces are additive whereas the social milieu is a complex mix of influences whose combined effect is not merely the sum of all the parts. Various components interact with, interfere with, and serve to amplify or diminish each other. A slightly more appropriate metaphor would draw upon chemistry, because of the transformations that take place among the various components analogous to chemical reactions. However, in an ideal educational setting, even chemical metaphors are inadequate because, as Maturana and Varela (1972, 1988) suggested, student, teacher, and content form an auto-poetic system (self-constructing in both senses) within an environment which is much more analogous to a biological organism than even a chemical mixture, much less a collection of physical forces.

Sometimes people react automatically without affective or cognitive control; sometimes affect dominates reaction; sometimes cognition exercises control, enabling a response. Under certain conditions the student–teacher–content collective also takes on many of the qualities of a living organism (Davis & Sumara, 2007). The teacher can sometimes be seen as acting as the "consciousness of the collective" just as Bruner (1986, pp. 75–76) saw a tutor being able to act as "consciousness for two." This, of course, brings me back to Piaget (op cit.) and his biological metaphors of *equilibration* through *assimilation* and *accommodation* and to the necessity for some extra impulse in order to "learn from experience."

Learning as Change of State

A metaphor from physics that *is* useful in education is the way change of state comes about: Energy is added to a system and change is observed (for example, temperature rises), then further addition of energy produces no further change (the temperature stops rising while the material changes state), and then, subsequently, the change continues (temperature continues to rise). Often it seems as though learners act in a similar manner: Further instruction or advice produces no visible effect, and then suddenly students appear to have internalized something. They become more informed, more knowledgeable, more flexible, and more dextrous. Classic examples include accommodation of the conjugation of irregular verbs (Brown, 1973), and the gaining of independence with respect to the use of some mathematical technique. The role a teacher can play in this internalization process was elaborated as *scaffolding* (introduced by Wood, Bruner & Ross, 1976 inspired by Vygotsky 1978) then extended to include *fading* (Brown, Collins & Duguid, 1989) and independently expressed in Open University materials as a spectrum of teacher intervention moving through *directed–prompted–spontaneous* use of some action or probe (James in Floyd, Burton, James, & Mason, 1981; Love & Mason, 1992; Mason & Johnston-Wilder, 2004).

Human Psyche as Chariot

One way to see why training behavior alone is ineffective when trying to promote the development of students' powers of thinking mathematically is to make use of a framework developed in the Upanishads (Rhadakrishnan, 1953, p. 263), made use of in various places in Eastern and Western psychology, and underpinning the traditional terms *enaction*, *affect*, and *cognition* which pervade Western psychology.

The human psyche can be seen through the metaphor of a chariot. The chariot is connected by shafts to the harnesses of the horses, and the driver of the chariot uses reins connected to the harnesses in order to direct the horses. The driver attempts to carry out the instructions of the owner. The chariot can be seen as representing the body, or in more modern terms, enaction. It needs to be maintained in order to function properly. The horses are traditionally seen as the senses, or more loosely perhaps, as affect. They are the motive (cf. emotive) power that pulls the chariot, and if uncontrolled will pull the chariot in unintended directions. Thus emotions have to be harnessed, as well as fed and cared for. The shafts can be taken to represent habits and internalized automaticities, the direct connection between chariot and action. The reins are imagination, the means by which we direct our energies through our intentions. The driver is consciousness, cognition, or as I prefer, awareness. The owner is will.

Caleb Gattegno's adage that

> Only awareness is educable

is entirely consonant with and resonant with this chariot metaphor, which also invites extensions:

> Only behaviour is trainable
> Only emotion is harnessable.

To discuss the significance and force of the "only" would take me too far astray, but it has to do with the disturbance these assertions might initiate if they act as *protases* (initial statements of syllogisms) when juxtaposed with specific incidents from your own experience, leading to some sort of conclusion or insight (the syllogistic action) (see Mason, 1998; Mason & Johnston-Wilder, 2006). This is an example of how an action can be initiated in a teacher or among a group of teachers, and which may lead to a transformation of perspective and/or a change of practice or way of articulating both ideals and practices (Mason, 1998).

The three "onlys" can be used on several levels: As constructs which promote discernment of details contributing to the psychologizing of subject matter (as Dewey so cogently described it in 1933) prior to teaching (Mason & Johnston-Wilder, 2006); as reminder when interacting and responding to learners to inform pedagogic and content-based choices; as stimuli to initiate disturbance as part of professional development; and as a way of thinking about the role of attention and intention in learning about teaching through teaching.

Phenomena

Because of my phenomenological stance, the data I have to offer consists of your response to stimuli. These are phenomena abstracted from specific situations that I have experienced directly myself, or that I have recognized in other people's descriptions. Specific situations are turned into phenomena by recognizing something similar in some respects to other situations. Phenomena are then interrogated through a combination of relationship to theoretical constructs and probing of my own experience in similar situations. Often these similar situations are contrived so as to be similar. At the same time I develop and refine tasks, which are challenges to participants in workshops with the aim of seeing whether or to what extent they recognize the phenomena and the associated proposals or actions which might be useful in the future. Participants are then responsible for actualizing the imagined actions and for developing and modifying them to suit their own conditions, ways of working, perspective, etc.

Phenomena A (Own World)

> In the last ten minutes of a problem solving class I suddenly become sharply aware of a 'fog of non-understanding' rolling forward towards me from the audience. In the moment, I choose to keep going rather than stop and regroup.

For me, this is an instance of being so caught up in expressing current insights or mathematical material clearly as to be unaware of what it is like to be in the audience. Similarly, a teacher can be so caught up in orchestrating lesson activities as to be unaware of what students might be experiencing, or so caught up in some agenda such as "making the work interesting" (i.e. relevant, engaging, sufficiently easy) that the mathematical content is submerged. Keeping students on task and occupied may keep them busy and give the impression of engagement, but it may not significantly contribute to any learning. I chose to keep going because of the timing of the session, but then began the next session in a retrospectively reflective mode.

Phenomenon B (Expressing to Others)

> I had a sense of how a proof could be constructed, but each time I tried to write it down, I encountered a difficulty. I found it difficult to make it fully convincing and at the same time clear enough to follow.

For me this is an instance of a general phenomenon where I think I understand something, but it is only when I try to explain it to someone else that I really get to grips with it; it is only when I plan to expound or to guide exploration that I need to be sure about details, variations, ramifications. When struggling with expressing justification for a conjecture, or even with clarifying a conjecture, it can be so helpful to find a colleague who listens politely but need say nothing. This familiar phenomenon may be the source of resonance which makes group discussion and collaboration so popular in socio-cultural and social-constructivist stances, even though it depends for effectiveness on students experiencing a desire to express. It is closely related to the notion of *self-talk* identified by Goffman (1959). Put another way, in a sense, I really learn mathematics through "teaching," that is, expounding or explaining to others. This makes me wonder if one of the ways I learn about teaching is by expressing my thoughts about teaching to others.

Phenomenon C (Training Behavior)

Training learners in the behaviors required to pass specific examinations appears to be helping them achieve their goals, but may be setting up deep trouble for the future. As Herbert Spencer put it,

> ... what with the mental confusion produced by teaching subjects before they can be understood, and in each of them giving generalizations before the facts of which they are generalizations – what with making the pupil a mere passive recipient of others' ideas, and

> not in the least leading him to be an active inquirer or self-instructor .... Examinations being once passed, books are laid aside; the greater part of what has been acquired, being unorganized, soon drops out of recollection; what remains is mostly inert – the art of applying knowledge not having been cultivated; and there is little power either of accurate observation or independent thinking (Spencer, 1878).

The phenomenon of students having learned less than expected has probably been with us since people congregated in groups. I have drawn on Spencer because he was a major influence on Edward Thorndike and John Dewey in the USA and on Alfred Whitehead in the UK, particularly in paying attention to the experience of the learner. Whitehead (1932) used the notion of *inert* knowledge in his own critique of education, in order to account for observed "deficiencies" in student performance. It may be that attempts to teach have been ineffective by failing to appeal to students' logical reasoning, or more generally, by failing to engage students' interests through using and developing their natural powers of specializing and generalizing, conjecturing and convincing, stressing and ignoring, characterizing and classifying, etc. (Mason, 2008a). It may also be that, as John Dewey put it, the effective teacher *psychologizes the subject matter* (Dewey, 1933) by taking into account the ebb and flow of student energies (Dewey, 1913).

Phenomenon D (Hidden Assumptions)

I offered some people the following task:

> I am thinking of two numbers, whose sum is one. I square the larger and add the smaller, then I square the smaller and add the larger. Which of my two answers is likely to be the largest?

After some discussion I took a vote, and almost everyone conjectured that the results would be the same. I assumed that they had done the algebra or perhaps quickly tried a few special cases. A short while later a perplexed voice said "It works for 0 and 1 and those are the only numbers that sum to one!" They had interpreted *number* to mean non-negative integer!

Note: I am deliberately not using a diagram, even though it is part of my way of working on this task in workshops, because my concern here is with the hidden assumptions, not the resolution!

Interpretation or meaning-making is an automatic action. In particular, students interpret tasks so as to make sense of them and so as to be able to do something, and this may mean making implicit assumptions based on a perceived range of permissible change (Watson & Mason, 2005), as here. Teachers also interpret subconsciously when they consider a task for possible use in a lesson, or when they read curricular requirements and reform statements. They too "construe so as to make it possible to act."

The impact of the incident had emotional, cognitive, and behavioral ramifications for me. I noticed, marked, and here am remarking upon that incident (Mason, 2002). I quickly formed a mental image of myself working on this task in the future and guarding against this whole-number assumption. The next time I used the task I found myself sufficiently informed by the past to work with participants on interpretation of *number* before presenting the task itself. I have intentionally learned from experience.

General Comment

Phenomena *A* (*Own World*) refers to examples of momentary awarenesses that I have had, and I expect others have had, usually with a fleeting sense of "I won't do that again" which then fades, leaving the way open to doing it again and again, itself a phenomenon that could be labeled *Never Again*. Phenomenon *B* (*Expressing to Others*) is a common experience amongst mathematicians and can be observed in lessons in which working practices are based on a conjecturing atmosphere. Phenomenon *C* (*Training Behavior*) accounts for what is commonly observed in classrooms, justified by the pressures teachers experience in the struggle to hold onto ideals while retaining student trust and respect. Many theories and frameworks in mathematics education attempt to explain and provide ways to counteract the tendency to emphasize short-term goals over long-term aims; the procedural over the conceptual (Hiebert, 1986); rote learning over understanding. Most curricular reform movements begin with long-term goals based on *understanding*, but over time become rigid through the development of habits and routines. Phenomenon *D* (*Hidden Assumptions*) is a simple example of actually learning from experience.

Never Again is the heart of the matter of learning to teach through the act of teaching. It is not easy to "learn from experience." For example, despite massive evidence from experience, it is hard to learn that acts of teaching neither cause nor guarantee learning. Yet somehow we keep on trying.

Experience

It is commonly said that *we learn by (or from) experience*. My own experience suggests that *one thing we rarely learn from experience is that we do not usually learn from experience alone*. Put another way,

> a succession of experiences does not add up to an experience of that succession

which resonates well with the proposal that

> a succession of feelings is not, in itself, a feeling of succession (James, 1890, p. 628 paraphrasing Immanuel Kant).

Just because I engage in mathematical activity, it does not follow that I am aware of the activity itself as a whole. As many teacher educators have found, some people are disposed to reflect on their experience, to pick out moments and ponder them, and others seem not to be so disposed. That does not mean that they do not process their experience, but if they do, they do it in some subtle manner which is hard to detect.

The sentiment is backed up by a plethora of wise utterances through the ages.

> What experience and history teach is this – that nations and governments have never learned anything from history, or acted upon any lessons they might have drawn from it. (Hegel, 1830/1975)
>
> If men could learn from history, what lessons it might teach us! But passion and party blind our eyes, and the light which experience gives is a lantern on the stern, which shines only on the waves behind us. (Coleridge, 1835, 20 Jan 1834)

An almost lone voice can be found in Roger Ascham, a medieval schoolmaster:

> Learning teacheth more in one year than experience in twenty: Ascham (1570).

We might rephrase this as *real learning* integrates experience and includes making sense of it, which is more informative than a lifetime of experience alone. Most generally, Oscar Wilde beautifully captures the whole point:

> Experience is the name [everyone] gives to their mistakes: Wilde (1893 act 3).

Phenomena *A* (*Own World*) summarizes examples of situations in which it is erroneously assumed that engagement or immersion in activity will produce a transformatory action which, in turn, will result in what is commonly referred to as learning, evidenced through changed behavior in the future. Something more is needed. In order for experience in the past to inform actions in the future, some transformation of attention, intention, and disposition is required.

Reflection

In mathematics education, one transformatory action has been variously referred to as *reflective abstraction* (Piaget, 1970, 2001), *reification* (Sfard, 1991, 1994), *compression* (Thurston, 1994), *retrospective learning* (Freudenthal, 1991, p. 118), *looking back* (Pólya, 1945), *reflection-on-action* (Schön, 1983), and so on and traced back to Locke (1710, Chapter VI) by von Glasersfeld (1991). But before all this retrospective work can be done, there has to be some data. Something has to be discerned and noticed in order to be recalled "in tranquillity." Tranquillity, or at least a change of state from local goal-oriented activity to a more global perspective is at least advantageous and often necessary for there to be any lasting impression. In order to provide energy for this transformation, a disposition to reflect or otherwise process past experience also seems necessary. Attention has to draw back from immersion in immediate action.

Distanciation

Distanciation appears to be a term coined by Bertolt Brecht (1948), which has crept into biblical exegesis (Carson, 1996) and which is currently infiltrating educational discourse. The essence, as the word suggests, is a pulling back from immersion in action in the moment, what Schön (1983) referred to as *reflection-in-action* as distinct from retrospective *reflection-on-action*. In mathematics education it seems to capture the state of awareness-in-the-moment when something triggers the awakening of attention and you participate in a mathematical or pedagogical choice. The best way to prepare for distanciation in the midst of a teaching situation is through retrospective distanciation. An articulation of a systematic approach to this can be found in Mason (2002).

Construal Through Story Telling

One form of transformatory action which supports learning from experience is re-expressing things for yourself, as in *phenomenon B* (*Expressing to Others*). There is a commonplace adage that you only really learn when you try to teach, which I take as an articulation of the insight that it is when trying to express things for and from yourself that you really make contact with the content and clarify your own thinking. This fits with two of the six modes of interaction: *Expressing* and *expounding*. In expressing, there is something about the content (and your affective state) which simply has to be told to someone. Here the content can be seen as taking the initiative, the teacher as respected and experienced other mediates through being the reason as respected audience, and the student comes to appreciate more clearly. In expounding, you try to draw your audience into your own world of perceptions and conceptions. The fact of the audience (students or colleagues) amplifies the need to get things sorted out for yourself as teacher, so their presence (virtual, imminent, or actual) helps to bring you in contact with the subject matter, hopefully in a fresh way. When students take on an expository role, they are taking the role of teacher. Note that the usual meaning of *explaining* (to make plain) overlooks an important feature of this form of interaction, because it involves the teacher endeavoring to enter the world of the student by means of common concern about the content (Leinhardt, 2001). In effective explanation, the content mediates an interaction between student and teacher in which the teacher is drawn into the world of the student.

Two sets of distinctions (frameworks) which can inform pedagogical choices concerning learning from experience through story telling are *Do–Talk–Record* and *Manipulating–Getting-a-sense-of–Articulating* (Floyd et al., 1981; Mason & Johnston-Wilder, 2004). The first can act as a reminder that doing is amplified by talking about what you are doing (whether to yourself or to others) and talking clarifies so that recording becomes easier. Rushing to written records can actually obstruct meaning-making. The second can act as a reminder that the purpose of manipulating material objects, diagrams symbols, or virtual objects is to get a sense of relationships which may be instantiations of general properties. The goal is not to achieve the manipulation but to make contact with underlying generalities in the form of mathematical structure. Another way of approaching this is through the slogans *going with the grain* and *going across the grain* (Watson, 2000), which similarly act as a reminder that recognizing a pattern (going with the grain) is only a precursor to sense-making by examining what structure is revealed (going across the grain).

It is one thing to see students' manipulation of various objects as part of activity which may act as fodder for sense-making through constructing personal stories to account for what happens. It is perhaps another to see acts of teaching as fodder for teacher sense-making through trying to bring to expression, to articulate what is observed about student behavior, mathematical topics, mathematics more generally, and the management of effective interactions with students. Careful observation reveals that there can be moments of collective consciousness (Davis op cit.) as well as individual insights. The latter are of course notoriously unstable over time.

Construal through story telling is an age-old phenomenon. Egan (1986) used it explicitly to propose an alternative stance to teaching in primary schools; Bruner (1990, 1991) put forward the case that human beings are *narrative animals* and Norretranders (op cit.) exploits it in his critique of our illusions about the role of consciousness. We tell stories to ourselves and to other people. Most human narratives involve *accounting-for* actions (Mason, 2002) and are intended to reinforce the coherence of personal identity through the justification of acts. As Norretranders suggests, consciousness as the director of actions and the maker of choices is (often) an illusion. Consciousness lags behind enaction and affect in reacting to events. It takes explicit and significant work to bring consciousness or awareness into play in the moment when a choice is possible. The term *accounts-of* is intended to act as a reminder that what is most valuable is description which is sufficiently brief-but-vivid that others can recognize similar situations in their own experience. Explanations and theory-based *accounting-for* can be indulged in later once colleagues recognize the data they are being offered (accounts-of). As Husserl intimated "describe don't explain" is a basis for phenomenological enquiry. Perhaps the struggle to develop a scientific method within mathematics education could be seen as an attempt to disentangle *accounts-of* observations and *accounting-for* them.

Attention

At the heart of the matter of learning from experience is the person's attention. What is attended to, in what ways, and with what intention and disposition? (Mason 1982) If I am attending to the specifics of some sequence of actions, I may not be in a position to be aware of the fact of those actions or of that sequence, especially when I am being directed by a worksheet or following a prepared lesson plan. Here the *didactic transposition* beautifully captures part of the gap between teaching and learning, and the *didactic tension* describes how attention is directed towards accomplishing tasks as set, rather than using tasks to engage in activity which may promote experience from which it is possible to learn, whether as student or as teacher. Phenomena *A* (*Own World*) and *D* (*Hidden Assumptions*) provide access to specific examples. Teachers whose attention is fully occupied by organizing resources and managing classroom behavior are in a similar position. In order to learn about teaching through their teaching, they need to extract some of their attention to form an inner witness.

Most teachers will have experienced frustration that students don't seem to learn from experience, or that any such learning is short lived, as Spencer (*op cit*) observed in *phenomenon C* (*Training Behavior*). One way to account for this failure to learn from experience is to blame institutional factors (assessment, organization, curriculum pressures, etc.) or even to blame others (students, parents, policy makers). A more productive approach is to attempt to change what can be changed. What can be changed is yourself, because all other components in the auto-poetic system are even more problematic.

One issue is what to do about frustration. In parallel with students, it is most natural for teachers to be caught up in the details of getting through the day, setting tasks, sustaining activity, evaluating and assessing and keeping discipline, as in phenomena *A* (*Own World*). The result is that they too end the day with only a succession of experiences rather than an experience of that succession. Alerted to issues in student learning, and informed by distinctions such as those mentioned here, augmented by many others available in the literature it is always possible to work on changing the conditions in which students are learning. But it takes more than further instances of *Never Again*!

An intelligently designed educational system would recognize that the community as a whole, and the people involved in various roles, all need both time and stimulus to reflect, to reconstruct, and re-enter a succession of experiences in order to locate where their attention could most effectively be concentrated so as to break out of the many stable cycles which constitute what we currently call teaching.

This is where the human being as narrative animal is so important. The negative aspect is that because our psyche is fragmented, and because awareness is the slow mover compared to enaction and affect, we spin stories to account for the functioning of our multiple selves so as to make "our-selves" feel better, feel coherent, and unified. One explanation for the *Never Again* phenomenon is that the self who formulates the intention is not the self in charge later when the change is needed. A positive aspect is that it is possible to learn from experience by engaging all aspects of the psyche: Enaction, affect, awareness, attention, and intention. This requires work on self-integration, including the development of an inner witness or monitor who observes without acting, yet is able to divert attention out of immersion in activity. Again the roots go back to ancient times, such as this stanza from the Rg Veda:

> Two birds, close-yoked companions, both clasp the self-same tree.
> One eats of the sweet fruit; the other looks on without eating (Bennett, 1964, p. 108).

While some interpretations see the tree as the tree of immortality with people ignoring the potential by failing to eat of the sweet fruit, the tree can also be seen as the human psyche with immersion in action (eating of the sweet fruit) accompanied by an inner witness or monitor who merely comments on the action (the bird looking on). As with most Eastern teaching stories, multiple and conflicting interpretations are the source of significance and richness of meaning.

Story telling is one of the best ways to coordinate experiences, to draw back from immersion in activity, to reflect upon rather than simply engage in. By trying to articulate what has been noticed, by trying to re-enter salient moments, by trying to express connections, similarities, and differences between otherwise apparently disparate objects (mathematical concepts, objects, etc.; classroom incidents, mathematical obstacles, pedagogical and didactic choices, etc.), in short by assembling narratives, we can indeed withdraw from the action sufficiently in order to learn from experience. However, it is vital to avoid the negative side of narratives: Self justification, self-calming, and self-promotion. Explaining why some action was impossible or inappropriate so easily degenerates into draining away the energy that comes from noticing, from having attention suddenly sharpened through making distinctions, recognizing relationships and perceiving properties which are instantiated as relationships among particulars.

Reacting and Responding; Habit and Choice

Since learning through teaching involves experiencing some sort of disturbance to the flow of internalized and habitual actions, it is useful, as indicated in the introduction, to distinguish between *reacting* and *responding*. To respond is to make an intentional, conscious, considered choice of action. It is very rare. Usually we react. Something in a situation may trigger a metonymic association, and there may be resonance with some past experiences. Together these produce action. The body acts before consciousness is even aware. Evidence for this can be found in the most ancient of psychological studies such as the Upanishads, through personal investigation (see, for example, Mason, 2002, Chapter 12), or through neurological studies (Norretranders, 1998). Response requires awareness in addition to action.

Habits are the repetition of choices made previously, often a long time previously. The more deeply ingrained or internalized, the harder they are to counteract. But they only need to be counteracted if they are working against desire and intention. Indeed, sometimes they take place because desire and intention are in conflict. For example, the automaticities of arithmetic facts (e.g. single digit additions, multiplication tables, use of associativity, commutativity and distributivity, factoring) and algebraic manipulation are essential for making progress in mathematics. So too is the flexibility to move with facility between process and object (as in $\frac{3}{4}$ as a division and as the answer to a division). The same applies to acts of teaching such as using placement in the classroom to calm or quieten potentially noisy students, or waiting after asking a question. Habits and automaticities are essential, which is why Skinnerian stimulus-response (I prefer stimulus-reaction) works so well for training behavior.

Teaching which focuses on internalizing actions, so as to create automaticities is essentially training behavior. The trouble with training behavior alone is that it tends not to be flexible when conditions change. For example, training children to react to specific language patterns in word problems is more likely to blinker them than to enrich their problem solving flexibility and creativity. Training behavior in concert with educating awareness, including "integration through subordination" (in Gattegno's memorable phrase, 1970), is much more powerful than either dimension alone. The procedural-conceptual dichotomy (Hiebert op cit.) arises from imbalanced evocation of the human psyche.

Various constructs have arisen as a result of failure to train behavior successfully. For example, the rise of constructs such as *transfer* (Detterman & Sternberg, 1993), *situated cognition* (Lave & Wenger, 1991; Watson & Winbourne, 2008), and *situated abstraction* (Noss & Hoyles, 1996) are attempts to account for why what seems to be learned in one context is not called upon or made use of in another (Marton, 2006).

Complication

Any impression that habits are necessarily bad and that flexibility is necessarily good need to be challenged. We cannot function without habits because the effort required to process every stimulus consciously and freshly would overwhelm even our complex brains. It is absolutely vital to internalize responses in order to be able to react quickly and effectively to common situations. If you have to pause and divert attention in order to carry out a simple computation or to simplify an algebraic expression, you are unlikely to plumb the depths of subtle structural relationships. However it is equally important to be able to question and modify some of these internalized automaticities, especially when they begin to obstruct rather than facilitate. One of the weaknesses of our current educational system is that we do not support adolescents in developing techniques for and dispositions to interrogate habits and reactions.

In mathematics, for example, it is vital when manipulating arithmetic or algebra to have a monitor awake to slips and ready to ask "Why are we doing this?" when the going gets tough and perhaps something has gone wrong. The same applies to trying out some new pedagogic strategy: Just because everything does not go swimmingly (notice the metaphor of immersion), there is no reason to abandon ship and conclude that the strategy cannot work, only that it did not work fully in that instance. This, in turn, parallels the way students label themselves and their efforts: "Did not" is much more helpful that "could not," which all too readily turns into "will not," creating mathematics-refusal behavior and affect (Dweck, 2000). Teachers working on their own teaching can also fall into this cycle of decline, by allowing "could not" to displace "did not," leading to "will not" and disaffection with trying to push boundaries and to respond sensitively to students.

Intention and Will

Attention is often described as the manifestation of will. Whatever we are attending to is "where we are" in that moment (James, 1890). It is what "we will it to be," though the experience is very often of being at the mercy of some outside influences which attract our attention and tempt our horses. Intention is a major component of the psyche and a major factor in the effectiveness of teaching and learning, yet curiously difficult to get a handle on (Anscome, 1957). Seen as the means through which the will directs attention, intention is the guide behind the making of choices and the changing of habits, propensities, and dispositions to act, which are components of affect. In the metaphor of the human psyche as a chariot, will is the owner who sets goals for the driver (cognition) who manifests intention through the use of the reins to direct the horses, the sources of motive power.

Intention can usefully be seen as related to affect in the same way that awareness is related to enaction: Awareness guides enaction and enables actions to come to mind in response to problems when they arise; intention guides or flavors affect as energy arising from emotions so as to be directed productively. On the other hand, intention provides a cognitive component to accompany or to challenge desire, which is how energy is accessed and harnessed to make things happen. In that sense, intention is a cognitive dimension of will. Intention is often weaker than the flow of energies of desire, however. The best of intentions may be submerged or diverted by perceived exigencies; ideals may be compromised by pragmatic response to perceived conditions.

For example, everyone recognizes how resolutions (such as those made at New Year) are so quickly forgotten or over-ridden. This can be accounted for by seeing the self that made the resolution as being superseded by other selves with no such commitment. This is probably the most controversial dimension because it challenges the very notion of identity. Deeply embedded in ancient and modern psychology is the notion that the psyche of human beings is not single but composite. Versions can be found in Minsky (1975) and Hudson (1968), as well as in more esoteric sources such as Bennett (1964), Shah (1978), and many others. The idea is that in response to different local environments, we each develop distinct persona or selves. These selves then compete for control of the whole organism. For example, the self who is in charge when chairing a meeting at work is not the self who cooks dinner for the family: There are different dispositions and propensities, different sensitivities as to what is noticed and attended to, different actions skilfully executed, even different epistemological stances. Each self can be thought of as a network of energy flows, directing energy through different channels, thereby activating different collections of habits and actions (see also De Geest, 2006). In Plato's version of the chariot, the servants of a mansion compete to play the role of butler, or even try to usurp the role of the owner (will) who is away; the owner cannot return until the organization of the house is in order and functioning properly.

In order to strengthen the guidance of intention, people have long realized that it is necessary to provide some discipline, whether externally (and not usually effective in the long run) or internally.

The Rg Veda metaphor of the two birds suggests that one of the ways that the will–intention link can be strengthened is through the growth of an inner witness or monitor who, as it were, sits on your shoulder and asks "Why are we doing this?" or "What are we doing?" This, in turn, is fostered by intentional preparation by *prospective* planning based on *retrospective* reflection. The Discipline of

Noticing (Mason, 2002) is an articulation of techniques for enhancing the possibility of noticing opportunities in the moment and participating in a choice to act freshly.

Disciplined Development

The core issue in learning from any experience, and particularly in learning about teaching through teaching, is how to participate in actual choices rather than being dragged by the horses in whatever direction attracts them. Put more sharply, how can one be present in the moment when a choice is being made, an action initiated? Applied to teachers, an answer would shed light on why it is that despite espousing some stance towards teaching and learning, what is observed being enacted is often in considerable contrast. Applied to learners, it would shed light on why it is that despite having been shown or taught some technique or concept, students often do not make use of it when appropriate. This is the age-old problem of transfer: Just because I see the possibility of using technique T or concept C in some situation, what is it that could enable students to be similarly aware (Detterman & Sternberg, 1993, Marton, 2006)?

The chariot metaphor suggests that mental imagery (the reins) is important, together with maintenance of the chariot and the harness and contact between will and awareness. Put another way, attention and intention are crucial elements, the whole of which is encompassed in the Discipline of Noticing. As George Bernard Shaw put it:

> We are made wise not by the recollection of our past, but by the responsibility for our future. (Shaw, 1921, p. 250).

It is the desire to act differently "next time" that drives personal and professional development, but that desire (part of affect) has to be amplified in order to influence intention beyond the particular "self" experiencing the desire in the moment.

In brief, what is required is sufficient discipline to engage in three actions in order to facilitate the important action:

> Collecting pedagogic strategies, didactic tactics and other awarenesses that could inform practice if only they came to mind when needed.
> Engaging in retro-spective distanciation/reflection in order to amplify the energy released by noticing a missed opportunity, including intentionally re-entering moments as fully as possible without judgement or explaining away what happened, in order to locate actions that you wish in retrospect you had tried, or to locate specific relationships that seem problematic (Tripp, 1993);
> Engaging in pro-spective preparation by intensely imaging yourself in some future situation acting in some desired or intended manner.

The latter two actions call upon that fundamental human power of mental imagery: The ability to place ourselves in a remembered past and in an imagined future. The third is how will power is developed, slowly building on fragments of energy released through noticing. Together all three actions contribute to enriching the possibility of having something "come to mind" through a combination of metonymic triggering and metaphoric resonance.

Externalizing and labeling salient moments in retrospect is part of bringing to expression, of story telling. Labels act as axes or foci around which experiences can gather (Mason, 1999). They can come to be associated with distinctions that might be informative, actions that might be relevant, and stances that might be fruitful. The value-laden terms being used here are of course relative to the value system of the individual within the encompassing institutional and hence socio-political milieu.

The combination of these three actions to produce the fourth is the essence of the Discipline of Noticing (Mason, 2002) which provides a detailed structure, method, and philosophical justification.

Maintenance

Following the chariot metaphor, the chariot or carriage, harness, shafts, and reins all require maintenance in order to function efficiently, and, of course, the horses themselves need looking after. Much could be made of this aspect of the metaphor, but suffice it to say here that work on dispositions and propensities (the horses) arises from work on developing the inner witness associated with awareness in conjunction with the strengthening of mental imagery (the reins). The components of the psyche are complexly inter-related beyond what is revealed by the carriage metaphor.

Skills need to be rehearsed because they may atrophy through lack of use. Habits need to be inspected and renewed or replaced as conditions change. This applies to dominantly cognitive skills such as specialized techniques in mathematics and topic specific didactic tactics, to dominantly affective skills such as holding still or poker-facing when a student makes a good conjecture and asks for validation, and to dominantly enactive skills such as classroom behavior management.

When pedagogic strategies and didactic tactics are tried, modified, and found to have potential, they need to be integrated into functioning. In the words of Gattegno (1970) "to integrate through subordination" is achieved through drawing attention out of and away from the carrying out of the skill. In this way, it truly becomes habit. If awareness is not extended and enriched (educated) then training provides a habit on which to call, but may not be sufficiently flexible to cope with changed conditions such as non-routine problems.

One message from this is that a regular practice of working on mathematics at your own level for yourself and with colleagues is an excellent way to rehearse skills, refresh awarenesses, and maintain the chariot! In a sense this is the companion to learning about teaching from and through teaching, because it provides a library of recent personal experience on which to draw and through which to be reminded about what students may be experiencing at their own level.

What Can Teachers Do?

Teachers are expected to do many things, but what is the core of what they can hope to do? I consider this question in relation to students and then in relation to themselves and their teaching apropos of learning about teaching from and through teaching.

What Can Teachers Do for Students?

Teachers can edit and amplify what students do and say (Hewitt, 1996). They do this by how they respond to students through reformulating and rephrasing, through reflecting specific elements, phrases, and objects back to a group of learners. They can pick out some expressions while ignoring others, and by using the same expressions over a period of time in order to enculturate students into using the discourse to express their own awareness. A conjecturing atmosphere, an ethos of enquiry and collegiality is one way to support and amplify this editorial role of a teacher.

By stressing and consequently ignoring teachers can direct student attention in ways which are indicative or representative of the way that mathematicians use their attention. It is a reasonable conjecture that each technical term, each definition, each technique, or method arose originally through a shift in what people attended to and how they attended to it (otherwise, there would neither have been nor be any struggle, any problematicity). Students usually have to experience some similar shift in attention in order to internalize and exploit the new concept or approach. So what a teacher offers is a more sophisticated awareness, more discerning distinctions, wider recognition of relationships, and more insightful perception of properties which are instantiated as relationships. They also display the kind of attention and the kind of reasoning that marks out mathematical thinking from other types of thinking.

Put another way, teachers display higher psychological processes (sometimes called "modeling behavior" but this is problematic because of what people attend to) which over time may be internalized by students, and there are various pedagogic devices for fostering and promoting internalization, including scaffolding and fading and the use of a range of pedagogic constructs (see Mason & Johnston-Wilder, 2006). One popular device for drawing student attention out of immersion in activity, and useful as a contribution to the fading aspect of scaffolding, is meta-questioning such as asking "What question am I going to ask you?" or "What am I going to suggest you do?" If, after a period of using the same or similar prompt, one of these meta-questions is used, student attention can be diverted to experience of the prompt rather than immediately reacting to it.

To learn about teaching through teaching is in large part to work on becoming ever more aware of what as a teacher it is possible to do for students, and to withdraw from trying to do things for students that actually block their learning. Examples include trying to do for students what they can already do for themselves, and usurping their powers in the name of efficiency by specializing and generalizing, conjecturing and reasoning for them. This is not to say that the teacher never does these things, but rather, that when the teacher does these things, not to assume that the students can consequently now repeat them for themselves. Learning about teaching from teaching is a lifetime process of refining sensitivities to students and to the conditions in which learning is fostered and sustained.

What Can Teachers Do for Themselves?

There are parallel actions that teachers can do for themselves. First and foremost, they can maintain their interest and pleasure in mathematics by engaging in mathematics for themselves at their own level. The purpose is not to "learn more mathematics" but to sensitize themselves to the struggles that students experience. Working on mathematics for themselves is an instance of a more general program of engaging in retro-spection and pro-spection in order to support *spection*. This is both a process of alerting oneself to issues that may need probing and actions to take in order to promote responding freshly and more sensitively to situations that emerge when teaching, whether when planning or when engaging with students.

What Can Teachers Do for Each Other?

By collaborating in their enquiries, teachers can display higher psychological processes to and with each other so that as a collective they grow in and into community. They can, for example, edit and amplify descriptions of incidents so as to enrich labels that can serve as triggers to awaken their inner monitor and enable participation in choices in the moment. They can reflect specific elements back to each other or to themselves, drawing them out of the specific action and into states in which mental imagery can be used to prepare for future action. They can direct each other's attention to salient features so that finer distinctions can be made. The power and value of distinctions needs to be tested again and again, especially those that have become so ingrained that they activate even when not appropriate. It is easy to make distinctions, but distinctions which inform future effective action are not so readily located.

Conclusion

For me, Alfred Tennyson (1842) beautifully sums up the role and importance of experience:

> I am a part of all that I have met;
> Yet all experience is an arch wherethro'
> Gleams that untravell'd world whose margin fades
> Forever and forever when I move. [*Ulysses* lines 18–21]

To learn from or attend to experience in this way is to strive to move beyond that arch, an intention which re-inspires:

And this gray spirit yearning in desire
To follow knowledge like a sinking star,
Beyond the utmost bound of human thought. [Lines 30–33]

By working together on mathematics and on pedagogical and didactic choices in a conjecturing and collegial atmosphere, teachers can indeed learn about teaching from and through teaching. But no one says that it is easy!

References

Anscome, G. (1957). *Intention*. London, UK: Harvard University Press.

Ascham R. (1570). *The scholemaster. Or plaine and perfite way of teachyng children, to vnderstand, write and speake, the Latin tong*. Printed for W. Innis at the west end of St. Paul's street and S. Birt in Ave-Maria Lane.

Bennett, J. (1956–1966). *The dramatic universe* (four volumes). London, UK: Hodder & Stoughton.

Bennett, J. (1964). *Energies: Material, vital, cosmic*. London, UK: Coombe Springs Press.

Brecht, B. (1948). *Petit Organon pour le theatre*. Paris, France: Éd. De L'Arche.

Brousseau, G. (1997). *Theory of didactical situations in mathematics: didactiques des mathematiques, 1970–1990*. In N. Balacheff, M. Cooper, R. Sutherland, & V. Warfield, (Trans.), Dordrecht, The Netherlands: Kluwer.

Brown, R. (1973). *A first language*. Cambridge, MA: MIT Press.

Brown S., Collins A., & Duguid P. (1989). Situated cognition and the culture of learning. *Educational Researcher*, *18*(1), 32–41.

Bruner, J. (1986). *Actual minds, possible worlds*. Cambridge, MA: Harvard University Press.

Bruner, J. (1990). *Acts of meaning*. Cambridge, MA: Harvard University Press.

Bruner, J. (1991). The narrative construction of reality. In M. Ammanti & D. Stern (Eds.), *Rappresentazioni e narrazioni* (pp. 17–42). Roma-Bari: Laterze.

Carson, D. (1996). *Exegetical fallacies*. Carlisle, UK: Paternoster.

Chevallard, Y. (1985). *La transposition didactique*. Grenoble, France: La Pensée Sauvage.

Claxton, G. (1984). *Live and learn: An introduction to the psychology of growth and change in everyday life*. London, UK: Harper and Row.

Coleridge, S. (1835). *Specimens of the table talk of the late Samuel Taylor Coleridge*. London: John Murray.

Davis, B., & Sumara, D. (2007). Complexity science and education: Reconceptualizing the teacher's role in learning. *Interchange*, *38*, 53–67.

De Geest, E. (2006). *Energy transactions in the learning of mathematics*. Unpublished PhD Thesis, Milton Keynes, UK: Open University.

Detterman, D., & Sternberg, R. (Eds.) (1993). *Transfer on trial: Intelligence, cognition, and instruction*. Norwood, NJ: Ablex.

Dewey, J. (1933). *How we think*. London: D.C. Heath & Co.

Dewey, J. (1913). *Interest and effort in education*. Boston, MA: Houghton Mifflin. Reprinted by Kessinger Publishing – Rare Reprints of Hard to Find Books.

Donne, J. (1624). *Deuotions vpon emergent occasions, and seuerall steps in my sicknes*. London: Thomas Iones.

Dweck, C. (2000). *Self-theories: Their role in motivation, personality and development*. Philadelphia, PA: Psychology Press.

Egan, K. (1986). *Teaching as story telling: An alternative approach to teaching and curriculum in the elementary school*. Chicago, IL: University of Chicago press.

Festinger, L. (1957). *A theory of cognitive dissonance*. Stanford, CA: Stanford University Press.

Floyd, A., Burton, L., James, N., & Mason, J. (1981). *EM235: Developing mathematical thinking*. Milton Keynes, UK: Open University.

Freudenthal, H. (1991). *Revisiting mathematics education: China lectures*, Dordrecht, The Netherlands: Kluwer.

Gattegno, C. (1970). *What we owe children: The subordination of teaching to learning*. London, UK: Routledge & Kegan Paul.

Goffman, E. (1959). *The presentation of self in everyday life*. Garden City, NY: Anchor Books.

Hegel, G. (1830/1975) H. Nisbet (Trans.) *Lectures on philosophy of world history: Introduction*. Cambridge, UK: University of Cambridge Press.

Hewitt, D. (1996). Mathematical fluency: The nature of practice and the role of subordination, *For the Learning of Mathematics, 16*, 28–35.

Hiebert, J. (1986). *Conceptual and procedural knowledge: The case of mathematics*. Hillsdale, NJ: Erlbaum.

Hudson, L. (1968). *Frames of mind*. London, UK: Methuen.

James, W. (1890 reprinted 1950). *Principles of psychology* (Vol. 1). New York, NY: Dover.

Langer, E. (1997). *The power of mindful learning*. Reading: Addison Wesley.

Lave, J., & Wenger, E. (1991). *Situated learning: Legitimate peripheral participation*. Cambridge: Cambridge University Press.

Leinhardt, G. (2001). Instructional explanations: A commonplace for teaching and location for contrast. In V. Richardson (Ed.). *Handbook of research on teaching* (4th edn., pp. 333–357). Washington, DC: American Educational Research Association.

Locke, J. (1710). In P. Nidditch (Ed.), *Essay on human understanding* (1975). Oxford, UK: Clarendon Press,

Love, E., & Mason, J. (1992). *Teaching mathematics: Action and awareness*. Milton Keynes, UK: Open University.

Marton, F. (2006). Sameness and difference in transfer. *Journal of the Learning Sciences, 15*, 499–535.

Mascaró, J. (Trans.) (1962). *The Bhagavad Gita*. Harmondsworth, UK: Penguin.

Mason, J. (1979). Which medium, which message. *Visual Education, 2*, 29–33.

Mason, J. (1982). Attention. *For the Learning of Mathematics, 2*(3), 21–23.

Mason, J. (1998). Protasis: a technique for promoting professional development, in C. Kanes, M. Goos, & E. Warren (Eds.) *Teaching mathematics in new times: Proceedings of MERGA 21* (Vol. 1, pp. 334–341). Australasia: Mathematics Education Research Group of Australasia.

Mason, J. (1999). The role of labels for experience in promoting learning from experience among teachers and students. In L. Burton (Ed.), *Learning mathematics: From hierarchies to networks* (pp. 187–208). London: Falmer.

Mason, J. (2002). *Researching your own practice: The discipline of noticing*. London, UK: Routledge Falmer.

Mason, J. (2008). Making use of children's powers to produce algebraic thinking. In J. Kaput, D. Carraher, & M. Blanton (Eds.), *Algebra in the early grades* (pp. 57–94). New York, NY: Lawrence Erlbaum.

Mason, J. (2008a). From concept images to pedagogic structure for a mathematical topic. In C. Rasmussen & M. Carlson (Eds.), *Making the connection: research to practice in undergraduate mathematics education*. MAA Notes (pp. 253–272). Washington, DC: Mathematical Association of America.

Mason, J. (2009). From assenting to asserting. In O. Skvovemose, P. Valero & O. Christensen (Eds.), *University science and mathematics education in transition* (pp. 17–40). Berlin, Germany: Springer.

Mason, J. (2009). Teaching as disciplined enquiry. *Teachers and Teaching: Theory and Practice, 15*(2–3), 205–223.

Mason, J., Burton, L., & Stacey, K. (1982). *Thinking mathematically*. London, UK: Addison Wesley.

Mason, J. Drury, H., & Bills, E. (2007). Explorations in the zone of proximal awareness. In J. Watson & K. Beswick (Eds.), *Mathematics: Essential research, essential practice. Proceedings of the 30th annual conference of the Mathematics Education Research Group of Australasia.* (vol. 1, pp. 42–58). Adelaide SA, Australia: MERGA.

Mason, J., & Johnston-Wilder, S. (2004). *Fundamental constructs in mathematics education*. London, UK: Routledge Falmer.

Mason, J., & Johnston-Wilder, S. (2006). *Designing and using mathematical tasks* (2nd edn.). St. Albans, UK: Tarquin.

Maturana, H., & Varela, F. (1972). *Autopoesis and cognition: The realization of the living*. Dordrecht, The Netherlands: Reidel.

Maturana, H., & Varela, F. (1988). *The tree of knowledge: The biological roots of human understanding*. Boston, MA: Shambala.

Minsky, M. (1975). A framework for representing knowledge. In P. Winston, (Ed.) *The psychology of computer vision* (pp. 211–280). New York, NY: McGraw Hill.

Movshovitz-Hadar, N. (1988). School mathematics theorems – an endless source of surprise. *For The Learning of Mathematics*, *8*(3), 34–40.

Norretranders, T. (1998). (J. Sydenham Trans.). *The user illusion: Cutting consciousness down to size:* London, UK: Allen Lane.

Noss, R., & Hoyles, C. (1996). *Windows on mathematical meanings: Learning cultures and computers*. Mathematics Education Library, Dordrecht, The Netherlands: Kluwer.

Piaget, J. (1970). Genetic Epistemology. New York: Norton.

Piaget, J. (1971). *Biology and knowledge*. Chicago, IL: University of Chicago Press.

Piaget, J. (1985). *The equilibration of cognitive structures*. Chicago, IL: University of Chicago Press.

Piaget, J. (2001). *Studies in reflective abstraction*. Sussex, UK: Psychology Press.

Pirie, S., and Kieren, T. (1994). Growth in mathematical understanding: How can we characterize it and how can we represent it? *Educational Studies in Mathematics*, *26*(2–3), 165–190.

Pólya, G. (1945). *How to solve it: A new aspect of mathematical method.* Princeton: Princeton University Press.

Rhadakrishnan S. (1953). *The principal upanishads*. London, UK: George Allen & Unwin.

Schön, D. (1983). *The reflective practitioner: How professionals think in action*. London, UK: Temple Smith.

Sfard, A. (1991). On the dual nature of mathematical conceptions: Reflections on processes and objects as different sides of the same coin. *Educational Studies in Mathematics. 22*, 1–36.

Sfard, A. (1994). Reification as the Birth of Metaphor, *For the Learning of Mathematics*, *14*(1), 44–55.

Shah, I. (1978). *Learning how to learn*. London, UK: Octagon.

Shaw, B. (1921). *Back to Methuselah*. London, UK: Constable.

Skinner, B. F. (1954). The science of learning and the art of teaching. *Harvard Educational Review*, *24*(2), 86–97.

St. Maurice H. (1991). A Guide to Commonplaces: On the use of loci in educator's discourse. *Journal of Curriculum Studies*, *2–3*(1), 41–54.

Tennyson, A. (1842). *Poems, In two volumes*. London, UK: Moxon.

Thurston, W. (1994). Proof and progress in mathematics. *Bulletin of the American Mathematical Society*, *30*(7), 161–177. Reprinted in 1995 by *For the learning of mathematics*, *15*(1), 29–37.

Tripp, D. (1993). *Critical Incidents in Teaching: Developing professional judgement*. London, UK: Routledge.

van der Veer, R., & Valsiner, J. (1991). *Understanding Vygotsky*. London, UK: Blackwell.

Vygotsky L. (1978). *Mind in Society: The development of the higher psychological processes*. London, UK: Harvard University Press.

Von Glasersfeld, E. (1991). Abstraction, re-presentation, and reflection. In L. Steffe (Ed.), *Epistemological foundations of mathematical experience* (pp. 45–67). New York, NY: Springer.

Watson, A., & Mason, J. (2005). *Mathematics as a constructive activity: Learners generating examples*. Mahwah, NJ: Erlbaum.
Watson, A., & Winbourne, P. (2008). *New directions for situated cognition in mathematics education*. New York, NY: Springer.
Watson, A. (2000). Going across the grain: Mathematical generalization in a group of low attainers, *Nordisk Matematikk Didaktikk*, *8*, 7–22.
Whitehead, A. (1932). *The aims of education and other essays*, London: Williams and Norgate.
Wilde, O. (1893). *Lady Windermere's fan: A play about a good woman*. London, UK: Elkin Mathews and John Lane.
Wood, D., Bruner, J., & Ross, G. (1976). The role of tutoring in problem solving, *Child Psychology*, *17*, 89–100.

How and What Might Teachers Learn Through Teaching Mathematics: Contributions to Closing an Unspoken Gap

Ron Tzur

Introduction

My three-fold thesis in this chapter is rather simple:

1. Mathematics teachers' practice is a strategic site for learning to teach in ways that will substantially improve students' mathematical understandings (Leikin & Zazkis, 2007; Perrin-Glorian, Deblos, & Robert, 2008; Zaslavsky & Leikin, 2004);
2. By and large, such LTT is an unrealized potential (Simon, 2007);
3. Articulating *how* and *what* LTT is prerequisite to realizing the potential; such articulation can greatly benefit from the reflection on activity–effect relationship (*Ref*AER*) framework (Simon, Tzur, Heinz, & Kinzel, 2004; Tzur & Simon, 2004).

Accordingly, I first provide rationale for the importance of LTT. Then, focusing on psychological and epistemological shifts in teachers' perspectives, I present an approach to how and what teachers may learn via their practice. Finally, I discuss questions for further research on LTT.

Rationale

Asian countries' practices of teacher professional development indicate that they have long noticed, and acted upon, the underlying premise of the above threefold thesis: No matter how much preparation teachers gain prior to becoming practitioners, their profession *requires* much more on-the-job learning. The Japanese Lesson

R. Tzur (✉)
University of Colorado Denver, Denver, CO, USA
e-mail: ron.tzur@ucdenver.edu

This chapter extends the discussion of Teacher Learning Through Teaching Research Forum, PME-31, Seoul.

R. Leikin, R. Zazkis (eds.), *Learning Through Teaching Mathematics*, Mathematics Teacher Education 5, DOI 10.1007/978-90-481-3990-3_3,

Plan Study (Lewis & Tsuchida, 1998; Yoshida, 2008) and the Chinese teacher collaboration (Ma, Zhao, & Tuo, 2006; Paine & Ma, 1993) approaches are but two clear examples of this insight. Throughout mathematics teachers' professional careers in Asian countries, they continually learn through figuring out the particular mathematics they intend their students to grasp, planning sequences of tasks that seem useful to promote students' robust understandings, and reflecting on their own and others' teaching activities in terms of impact on students' progress. I see these approaches to LTT as insightful for three major reasons.

First, LTT is *a must* because the time it takes to develop as an effective mathematics teacher goes well beyond the typical, 3–5 years of pre-service teacher education programs. In these programs, prospective mathematics teachers need to not only understand a vast body of highly abstract concepts, approaches, and theories from several domains, but also begin interconnecting those into a coherent "mathematics teaching" praxis. These domains include profoundly understanding mathematics (Ma, 1999) not merely as a user but also for teaching it to others (Adler & Huillet, 2008; Ball & Bass, 2000; Hill, Rowan, & Ball, 2005; Mason, 2001; Schifter, 1998). They include theories of learning and development along with their epistemological underpinnings (Steinbring, 1998) and psychological attributes of different ages (cf. Piaget, 1964). They include issues of organizing and managing classrooms, schooling and assessment, and dealing with diverse populations of constituencies—students, parents, and administrators. Moreover, all these novel understandings are to be developed by graduates of an educational system that does not equip them with good enough baseline knowledge, yet they would eventually have to both join and reform this system (Cohen & Ball, 1990). And they are supposed to learn while experiencing rather diverse methods of teaching—from prevalent traditional lectures to most advanced but pretty rare, reflection-oriented, laboratory-like approaches. The novice-expert literature (Leinhardt, 1989; Livingston & Borko, 1990) indicates that understanding and interconnecting these domains demand a very difficult and long process.

Second, except for the single, semester-long and stressful experience of student teaching, the above overwhelming amount of knowledge that teachers need to synthesize lacks an imperative, *qualitatively different* ingredient of effective practice, namely, noticing the impact (or lack thereof) of one's teaching on students' learning. At best, prospective teachers can learn, in the absence of real action, to plan lessons/units and use available curricula to this end. More often than not, these plans are traditional in nature and/or assimilated into prospective teachers' deeply engrained conceptions of traditional teaching. To novices, what experienced teachers have been doing and are observed doing in schools, day-in-and-day-out, seems much more intuitive than what their reform-oriented mathematics educators profess (Tzur, 2008b). To shift teachers' attention to and develop pedagogies that revolve around student learning, teachers must experience and reflectively link between their plans (anticipation of activities that engender learning) and the effects of those planned activities (Mason, 1998, 2008; Yoshida, 2008). That is, due to immersion in action, LTT is suitable for what and how is needed to reform the current, insufficient teaching practices.

The third reason is implied by the second—LTT is embodied in experiences that are more conducive to a meaningful and long-lasting shift in one's practice. In their practice, teachers continually carry out the planning–implementing–reflecting cycle, with reflective processes occurring both *in* and *on* action (Schön, 1987). Such cycles endow teachers with a wealth of opportunities to be perturbed, that is, to identify gaps between what they meant their teaching activities to engender and what students actually learned. Being perturbed by the lack of efficiency of *their own* goal-directed (planned) activities (Krainer, 2008a) can contribute to teachers' attentively listening to students (Davis, 1996; Empson & Jacobs, 2008) and *problematizing* learning (and teaching), that is, to making the reforming of practice *their own problem.* Consequently, teachers are more likely to maintain and adhere to practice adjustments that will grow out of such problematizing, because these adjustments represent new, empowering links *they* make between their activities and the effects of their activities in terms of student learning.

One then must ask why, at least in the Western world, in spite of teachers' ongoing experience of the plan–implement–reflect cycle, by and large their current practices do not change and often show strong resistance to change (Leikin, 2008)? Along with the various reasons usually provided, such as time stress, low salaries, and lack of instructional support (Ingersoll, 2001), I see three deep-seated reasons. First, when teacher learning is encouraged, it is usually via means outside the organic plan–implement–reflect cycle of their practice. Teachers receive from mathematics teacher educators messages that may range from "you are doing great and just need some extra tips" through "you fail to accomplish the desired student outcomes." This entire continuum, however, seems rooted in the *teacher educators'* problematizing and reforming of learning. As long as (problematizing) their own teaching cycle is not the source of teacher learning, suggested improvements will, at best, be adopted superficially and bound for quick decay.

The second reason for minimal LTT seems to be the disproportion in emphasis on two key principles. Most reform efforts seem to focus on the psychological principle of *active learning*, which is common to social–cultural perspectives (Lerman, 2006; Lompscher, 2002; Vygotsky, 1978; Wertsch & Toma, 1995) and constructivism (Dewey, 1933; Piaget, 1970, 1971, 1985). They seem to pay little or no attention to the epistemological constructivist principle of assimilation, a disproportion to which von Glasersfeld (1995) referred as *trivial constructivism.* Active learning is evident in the discourse in which reform-oriented mathematics educators engage teachers—solving challenging, realistic tasks in small groups, with concrete and/or digital manipulatives, and discussing their solutions with the whole group (Markovits & Smith, 2008; Watson & Sullivan, 2008). Teachers are then also engaged in discussing the advantages of doing the same with their students, in hope that the teachers would follow suit to emulate their own positive learning experiences. That is, teachers are engaged in active, hands-on experiences so they come to "see" the intended mathematics and appreciate reform-oriented ways of teaching it.

The assimilation principle requires teachers to understand students' mathematics as qualitatively different from the teachers' understanding and, thus, as the conceptual force that constrains and affords the mathematics students can "see" in the world. Steffe (1995) referred to this as the distinction between first order models (one's own math, be it a student or a teacher) and second order models (teachers' models of students' math). For example, assimilation implies that one would see a base-ten, place-value number system in the so-called base ten blocks only if she has already constructed necessary mental structures and operations with which to "see" (assimilate) the relationship among numbers that the system was made to symbolize (Gravemeijer, 1994). That is, assimilation entails a learning paradox (Bereiter, 1985; Pascual-Leone, 1976), which is a key constituent of the problematizing that mathematics teachers (and their educators) must make their own in order to deeply appreciate the difficulties their students face when asked to "see" what the teachers came to "see" via reform-oriented professional development (Tzur, 2008b). The disproportion between active learning and assimilation was captured in the categorization of three qualitatively distinct perspectives (further discussed below) that my colleagues and I have postulated to underlie teaching: Traditional, perception-based, and conception-based perspectives (Simon, Tzur, Heinz, Kinzel, & Smith, 2000; Tzur, Simon, Heinz, & Kinzel, 2001).

The third reason for the ever-present gap between potential and actual LTT seems to be the lack of articulation, from a conception-based perspective, as to *how* and *what* can teachers learn (Sánchez & García, 2008). At issue, I contend, is the balanced application of *both* principles (assimilation, active learning) to the *teachers' learning* as a means for them to understand and practice the same approach. In Tzur (2008b), I noted that accomplishing this balance requires development and facile implementation of conception-based pedagogies by mathematics teacher educators. What follows is such an articulation (theoretical account), which focuses on how teachers' own plan–implement–reflect cycle can be used to promote their understanding and adoption of both principles. I argue for using the recently elaborated framework of learning a new mathematical conception through reflection on activity–effect relationship (*Ref*AER*) (Simon et al., 2004). I believe that this framework provides a good basis for the needed theoretical account, though adaptations to the complexities of teacher learning will most likely be needed.

Accounting for How and What Can Teachers Learn Through Teaching

In this section, I first briefly present key constructs of the *Ref*AER* account. Then, I address the two leading questions: *How* might LTT and *what* might teachers learn that is worthwhile learning (i.e., likely to benefit student learning)?

Activity–Effect Relationship: Assimilation, Anticipation, and Reflection

The theoretical account of forming and having a mathematical conception, which I (Tzur, 1996, 2004, 2007) and colleagues (Simon & Tzur, 2004; Simon et al.,

2004; Tzur & Simon, 2004) have postulated, draws on three core constructivist notions—*assimilation, anticipation*, and *reflection* (Dewey, 1933; Piaget, 1985) and on the three-part model of a scheme introduced by von Glasersfeld (1995):—(a) situation-goal, (b) activity, and (c) result. We postulated that, upon assimilation, a person anticipates and reflects on a single mental relationship—a "compound" the brain forms between a mental activity and the effect(s) of that activity (*AER*). We termed the mechanism for abstracting a new conception *reflection on activity–effect relationship* (*Ref***AER*).

The *Ref***AER* mechanism commences with a person's assimilation of "information" (e.g., a mathematical task) into the situation part of an available scheme, which also sets the person's goal in that situation. To accomplish this goal, the scheme's second part is called up and executed while being regulated by the person's goal from within the mental system (Piaget, 1985). As the activity progresses the person may notice effects that the activity produces, including discrepancies between those effects and the goal. Through reflection on solutions to similar problems and reasoning about them, the learner may abstract a new invariant—a relationship between an activity and its anticipated effect(s). This view is consistent with cognitive neuroscience researchers' assertion that learning is essentially a goal-directed process (Baars, 2007) and is echoed in Krainer's (2008b) view on teacher education as goal-directed intervention.

The notion of *reflection* refers to two types of comparison continually performed in the human brain—consciously or subconsciously (Simon et al., 2004; Tzur, in press). *Type-I reflection* (comparison) focuses on differences between the effects of the activity and the person's goal, which engenders sorting of *AER* records. *Type-II reflection* focuses on comparisons among stored records of experiences (situations) in which such *AER* were used and engenders abstraction of a new *AER* as an anticipated, reasoned invariant. This invariant involves a reorganization of the situation that called upon the *AER* compound in the first place, that is, of the person's previous assimilatory scheme(s).

The construction of a new scheme via the *Ref***AER* process is postulated to occur in two stages; they are distinguished by the extent to which a learner has access to a newly formed conception (Tzur & Simon, 2004). The *participatory* (first) stage is marked by the person's dependence on being prompted for the activity at issue in order to bring forth and use the invariant *AER* compound. At this stage only a provisional anticipation of an *AER* identical in its *content* to the new conception has been formed. Once prompted and reinstated, this anticipation includes reasoning of why the effects follow the activity. The participatory stage is marked by the well known "oops" experience. For example, upon hearing a student's correct solution to a problem, a teacher who is learning to ask students how they solved a mathematical problem may first continue on to posing the next problem. As she does so, however, she may independently notice and regard this move as a mistake she *could and should* have avoided, with an accompanying "oops" thought and/or utterance.

The *anticipatory* (second) stage is marked by a person's independent calling up and using an anticipated *AER* proper for working on a given problem situation. That is, the person has explicitly linked between a newly formed *AER* and a set of situations—a link that was not yet abstracted in the participatory stage. Thus, the person can intentionally and spontaneously act toward the set goal. The crucial understanding about the stage distinction is that in both stages the essence of the anticipated relationship is the same (and thus observed behaviors are the same); what differs is its availability to the person in a given situation as she or he recognizes it.

How Might Teachers Learn Through Teaching?

Viewed through the *Ref*AER* lens, every activity of teaching is a potential source for teacher learning. The reason is straightforward—a teaching activity, whether planned, adjusted, or in response to unforeseen classroom events—*is* an expression of the teacher's anticipation of desired student learning effects. More often than not, the teacher can notice students for whom the actual effects do not fit with her anticipation. At issue is (a) the extent to which the teacher takes the time and effort to consciously reflect (*Type-I*) on this perturbation, (b) how does she resolve the perturbation (e.g., dismissing the event as "another example that some students can never get this"), and (c) whether or not further comparison (*Type-II*) across relevant records of such experiences is carried out to examine plausible adjustments to the teaching activities. Most importantly, the teacher's perspective on what constitutes "learning" and how teaching may promote such an effect in students affords and constrains what she may notice (Mason, 1998, 2008) and, crucially, what relationships she may or may not (trans)form between her teaching activities and their effects. Put differently, the scope and nature of the teacher's perturbation, as well as how she resolves it, are excellent indicators for her assimilatory conceptions of mathematics learning and teaching. According to this view, one would consider the teacher to have learned if changes in the teacher's anticipation can be identified in the form of an invariant link, which is novel for the teacher, between what she plans to do, what she does, and why she plans/does it.

The above view of LTT, particularly the key role that anticipation and reflection play in this process, is compatible with Clarke's (2008) approach to using teacher-generated curricula as a source for professional development. Likewise, this view was illustrated in all four articles of the PME Research Forum (Borba, 2007; Leikin & Zazkis, 2007; Liljedahl, 2007; Simon, 2007). I contend that each of the authors' examples is a manifestation of teachers' learning as change(s) in anticipation, when students' and/or peers' unanticipated reactions become prompts for the teacher's reflection on pedagogical/math *AER*. To substantiate this contention, I briefly discuss three examples from those papers.

Leikin and Zazkis' (2007) second example reports on a teacher who anticipated that students would solve a problem via induction. The teacher's own mathematical conceptions afforded her assimilation of an *un*anticipated student's solution obtained via calculus. Her pedagogical conceptions afforded acceptance of different solutions to the same problem, which triggered her goal of making sense of the student's unanticipated solution. That is, the student's solution served as a prompt that engendered the teacher's extension of both her mathematical anticipation of proper solutions to such problems and her pedagogy—proactively planning for fostering students' understandings of both solutions. The report indicated that this change in the teacher's anticipation was quickly solidified into an anticipatory stage, as she not only accepted it but also planned on presenting it in future lessons on the same topic.

In Liljedahl's (2007) study, teachers worked on creating tasks for assessing students' numeracy. Initially, he considered the teachers to be using a narrow and rather procedural understanding of numeracy, which triggered his intervention. He interjected a follow-up prompt (What would you, teachers, consider the qualities of a successful (numerate) student?). This prompt fostered further interaction among the teachers that, in turn, oriented their reflection on and transformation in their conception of what "numerate students" should be anticipated to do/reason. Liljedahl's work, which draws on Wenger's (1998) meaning for reification, resonates with Pirie & Kieren's (1994) emphasis on the need to continually promote expressing of one's action-generated ideas as a means for clarifying these ideas to both oneself and others.

Borba (2007) reported on teachers' learning via online exchanges in which they each expressed their actions on the "same" virtual geometrical object. They uploaded their solutions to a common website and reacted to one another's solutions. Through those virtual, distance exchanges, teachers were exposed to their peers' actions, which often did not match one's anticipation of actions she would have carried out in that situation. As indicated by one of Borba's participants ("to cope with math activities for our students we had to revisit our own math"), peers' *un*anticipated actions prompted further reflection, hence LTT.

The above examples of reflection-based changes in teachers' anticipations did not always specify the goals that regulated the teachers' activities and reflections. The *Ref*AER* framework requires such specification in order to provide deeper analyses of LTT. However, those articles provided ample instances of where, as mathematics teacher educators and researchers, we can identify teachers' goals. For example, goals were indicated through experiences a teacher provided with her students that she considered as different from her own school experiences, correction of student mistakes, prediction of student responses, resolutions of her disagreements with peers and/or her own cognitive conflicts, attempts to satisfy school's requirement to use software, desires to improve one's own math, etc. Articulating these goals allows to form a better model of the teachers' rationale (anticipation) for why their teaching activities would engender the anticipated student learning. In turn, this articulation of goals and activities can assist in making sense of what teachers notice anew and link as novel teaching-learning anticipations. In other words, empirically grounded analyses can capitalize on the *Ref*AER* framework, alongside constructs such as awareness and noticing (Mason, 1998) and the dualistic, action-expression nature of understanding (Pirie and Kieren, 1994), for developing powerful explanations of the complex mechanisms, contexts, and stages in teacher change toward productive, reasoned practices.

What Might Teachers Learn Through Teaching?

To address the issue of what might LTT so they teach well and their students learn well (Sullivan, 2008b), I first present the three perspectives (traditional, perception-based, and conception-based) postulated to underlie practices of mathematics teaching (Heinz, Kinzel, Simon, & Tzur, 2000; Simon et al., 2000; Tzur et al., 2001). I consider the latter as an essential goal for mathematics teacher development. I then introduce a three-prong pedagogical approach (a "*Teaching Triad*") that underlies three central contributions of LTT to the improvement of mathematics education for both teachers and students. These three contributions are presented following a developmental (not logical) sequence implied by my own experience, by research literature, and by the *Ref**AER* framework: (a) Mathematics knowing for teaching (MKT), (b) task design and adjustment, and (c) epistemological shift toward profound awareness of the learning paradox (PALP) (Tzur, 2008b).

Teaching Perspectives

The distinction among traditional, perception-, and conception-based perspectives grew out of our work in the context of the NSF-funded Mathematics Teacher Development (MTD) project.[1] A more detailed exposition of this distinction, and empirical studies of teachers whose teaching we identified to manifest the perception-based perspective, can be found in the three papers referenced above. Here, I focus on further organizing those distinctions on the basis of the two principles discussed in the Rationale, namely, the psychological principle of active learning and the epistemological principle of assimilation (see Fig. 1).

A *traditional perspective* is characterized by a passive stance toward learning coupled with a "harvesting" stance toward mathematics—it exists outside and independent of the learner (or knower) who needs to obtain it. Thus, traditional teachers are likely to adhere to and utilize a transmission approach to teaching. They feel responsible for logically organizing and clearly presenting the mathematical content; students' role is to listen attentively to the teacher's directions, complete assigned work, and memorize millennia-old crystallized mathematical procedures

	Psychology	Epistemology
Traditional	Passive learning	Math exists independent of learner
Perception-Based	Active learning	Math exists independent of learner
Conception-Based	Active learning	Math depends on learner's assimilatory conceptions

Fig. 1 Three pedagogical perspectives

[1] The research was conducted as part of the NSF Project No. REC-9600023, Mathematics Teacher Development. All opinions expressed are solely those of the author.

and facts. A *perception-based perspective* (PBP) markedly differs from the traditional in its psychological emphasis on learning as necessarily an active process; however, it is rooted in the same learner-independent epistemological stance toward knowing mathematics. Thus, perception-based teachers are likely to adhere to and utilize a discovery (Platonic) approach. They feel responsible for engaging students in first-hand experiences of solving meaningful problem situations (tasks), via manipulating concrete objects and discussing with others patterns they identify until coming to "see" (discover) for themselves the intended (existing) mathematics (Simon, 2006b). Quite often, this is typical of teachers informed by and oriented toward reforms advocated by documents such as the NCTM *Principles and Standards for School Mathematics* (2000).

Like the PBP, a *conception-based perspective* (CBP) draws on the psychological principle of active learning. However, it differs markedly from PBP in its adherence to the radical constructivist principle of assimilation (von Glasersfeld, 1995). Assimilation implies that knowing and coming to know mathematics depends on the learner's available schemes by which she interprets and solves "external" problem situations. Simply put, one can only "see" mathematical ideas through conceptual lenses that are already established in her mental system; the "existence" of those ideas *is determined* (afforded and constrained) by her assimilatory conceptions. Such a perspective problematizes learning: How can one come to "see" new mathematical ideas while not having available conceptions through which to assimilate those ideas? Pascual-Leone (1976) and Bereiter (1985) referred to this as the learning paradox. Thus, conception-based pedagogy adheres to and utilizes not a discovery but rather a reorganization (conceptual transformation) approach. Teachers feel responsible for (a) engaging learners in realistic tasks that bring forth the learners' available conceptions (goals, activities, effects) to commence learning and (b) orienting learners' reflection on their goal-directed activities so they notice new aspects of those activities and reorganize the previously established schemes. Building on the notion of hypothetical learning trajectory (Simon, 1995; Simon & Tzur, 2004), I introduced the *Ref*AER* framework as a plausible conception-based pedagogy that addresses the learning paradox, along with a seven-step cycle that further specifies the teacher's responsibilities (Tzur, 2008b).

A three-Prong Pedagogical Approach

The two essential questions every teacher of mathematics continually contemplates upon are *what* and *how* to teach her students next. As one would expect, traditional pedagogies typically address both of these questions by letting the crystallized body of knowledge, as a knowing adult conceives of it, predetermine the logical sequence of concepts, algorithms, techniques, and facts to be taught. This is a single-prong approach driven by mathematics; it is independent of the learners who are to passively attain it. This single-prong approach is evident in what I call *one-column curriculum* (intended and implemented) that inversely breaks down the expert's (models of) mathematics, from advanced to basic pieces of knowledge, and then re-orders those pieces in terms of logical (to an expert) prerequisites.

Energized by the outcry to reform the largely failing traditional approaches, perception-based pedagogies typically address the *what and how to teach next* questions by coordinating the vast body of expert mathematical models with tasks and materials that promote learner activities. This is a two-prong approach that tailors to the pieces in the expert-driven prerequisite lists success-proven teaching-learning (inter)activities. This two-prong approach is evident in reform-oriented curricula produced around the world in the last two decades. I refer to it as a *two-column curriculum*, because it consists of (a) the intended mathematics and (b) corresponding activities for endowing students with "seeing" it. Evidently, two-prong approaches take into account students' interest and capacities. However, by and large, teachers' decision to move on and their selection of activities to promote what's to be learned next is still determined by the expert-driven mathematical prerequisite lists rather than by learners' assimilatory conceptions and how they might be reorganized. Lambert's (2008) recent study of traditional and reform-oriented tasks for teaching counting-on to first graders clearly demonstrated that both approaches do not account for the conceptual reorganization needed.

A conception-based pedagogy addresses the *what and how to teach next* questions very differently. It always begins with articulated accounts (second-order models) of students' assimilatory schemes. It then specifies research-based accounts of expert-intended (first-order) mathematical understandings into which student assimilatory schemes may be reorganized, focusing on what Simon (2006a) called key developmental understandings (KDUs). It finally proposes a set of tasks selected for their explicitly reasoned, hypothetical power to engender the transformation from students' available schemes to the intended mathematics. This three-prong approach, illustrated in the *Teaching Triad* diagram below (Fig. 2), tailors what's to be learned next and how to what students know precisely because the mathematics of students afford and constrain goal-directed activities they can bring forth and reflect upon as a means to construct novel (to them) ideas. To the best of my knowledge, a corresponding *three-column curriculum* has not yet been produced commercially, although examples of limited-in-scope units/lessons can be extracted from numerous constructivist studies that focused on students' assimilatory conceptions.

LTT and Mathematics Knowledge for Teaching

From the *Ref*AER* framework and the three-prong approach it implies for mathematics teacher education, LTT seems a strategic site for transforming teachers' subject matter knowledge into the desired mathematical knowledge for teaching *(MKT)* (Hill et al., 2005). I take this as a core of the Asian insightful

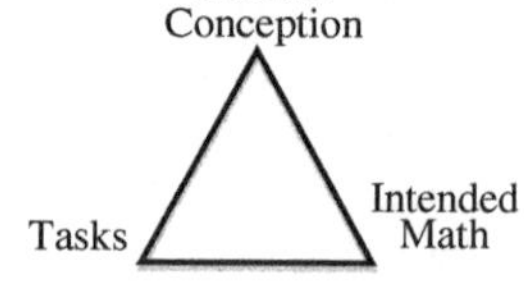

Fig. 2 A three-prong, conception-based pedagogy (Teaching Triad)

LTT endeavors. These endeavors organize teachers' reflection via exchanges among teachers about what a teacher planned (anticipation) and implemented (activity) in her classroom, and what she and her peers considered to have been brought forth (or not) in terms of student learning (effects). The examples from the studies by Leikin and Zazkis (2007), Liljedahl (2007), and Borba (2007) manifested this: Teachers can learn (and re-learn) MKT via reflecting on students' work in response to the teachers' planned/implemented activities. The resulting mathematical schemes of the teachers endow them with understandings of what and how can be learned by their students, that is, empowers the teachers' MKT.

Two aspects of such learning are worthy of noting. First, realizing the potential for LTT seems to depend heavily on a teacher's gradual development of a predisposition toward unexpected situations as an opportunity, not as a threat to be eliminated (Empson & Jacobs, 2008). Welcoming unexpected situations as an opportunity is likely to trigger a productive cycle, because it encourages students to make more contributions, hence more learning opportunities for a teacher. Second, the mathematics a teacher can learn with and without guidance seems rather different. In particular, teacher educators can prompt teachers' noticing in situations that would otherwise go unnoticed (Mason, 2008) and orient teachers' reflection onto relationships the teachers either overlook (Yoshida, 2008) or avoid (Sullivan, 2008b) due to lack of knowledge for coping with what the teachers do not know. On the other hand, mathematical interactions with teacher educators may add to the teachers' sense of threat. Thus, further studies of how to strike a balance between teachers' learning with and without guidance seem of importance.

LTT and Task/Lesson Design/Adjustment

Like Watson and Sullivan (2008), I consider tasks that teachers use for actively engaging their students in mathematics as a natural source for their professional growth. From the *Ref*AER* framework, this is almost self-explanatory. A task is probably the most representative artifact of a teacher's pedagogical scheme—anticipation of relationship between the mathematics her students will learn (effect) and the teaching (activity) that will engender such learning in a designated time/space (situation). For LTT to take place, however, at issue is the extent to which teachers systematically reflect on this anticipated relationship as a means to learn and improve task use—both how to think about the role tasks can play in students' learning of the intended mathematics (task features) and how to implement them properly (task pedagogy). An entire Research Forum carried out during PME-32 (Herbst, 2008; Sullivan, 2008a; Tzur, 2008a; Tzur, Zaslavsky, & Sullivan, 2008; Watson, 2008) focused on how, too often, very effective tasks may be "lost in teachers' translation" (lost in the sense of lowering task demands/impact). This is one area where purposeful intervention on the part of mathematics teacher educators is likely to be needed—to orient teachers' reflection onto different ways of thinking about and using tasks.

In this regard, I think that the most important goal for LTT is the articulation of *why or why not* a task engenders the intended learning. Such an articulation should be made explicit prior to using the task, as much as possible during task use, and afterward (in reflection). For the reflection part, it should consist of both *Type-I* comparisons between the planned and actual effects (student learning) a task engendered, and *Type-II* comparisons across situations in which a particular task did (or did not) engender that learning. My interpretation of Asian discourses suggests that they are powerful precisely because both types of reflection are *systematically* carried out before, during, and after a lesson.

Besides the foundational learning to design, reflect on, and adjust tasks, LTT can and should focus on translating curriculum-given or teacher-generated tasks into lessons . For example, once a core, "platform" task has been selected, variations are needed to enable less advanced and high-flyer students to productively engage in and learn via the task. Sullivan, Mousley, and Zevenbergen (2003) introduced an approach for designing/using enabling and challenging prompts that provide such variation. Integrated with the *Ref*AER* framework, this approach can yield significant LTT by using the design of prompts as a pedagogical task to orient teachers' reflection on the relationship between students' available conceptions and how they interpret and work on a given task. Our experience in the Mathematics Teacher Development project indicated that comparing across different learners' responses could turn into teachers' initial differentiation between the mathematics they could see (first order models) and the mathematics their students could or could not see due to the students' available understandings (second order models). This highly desired shift in teachers' epistemological stance is further discussed below.

LTT and Epistemological Paradigm Shift

Leikin (2006) proposed a three-dimensional model of teachers' knowledge. I consider the third dimension—forms of knowledge—as guidance for identifying a critical goal for LTT. Liljedahl (2007) and Simon (2007) emphasized the need for mathematics teachers to progress from intuitive to formal ways of thinking about teaching. This progression is consistent with Mason's (1998, 2008) emphasis on changes in teachers' awareness. Liljedahl (2007) illustrated this goal in terms of changes in teacher reactions to students; Simon (2007) illustrated it in terms of the need for teachers to question and reflect on their hidden epistemological assumptions. Drawing on the distinction among three pedagogical perspectives and their implied one-prong, two-prong, and three-prong approaches, I propose the shift toward epistemological stance that embraces the assimilation principle as a crucial goal for LTT. That is, I contend that teachers' progress to formal ways of thinking about teaching should involve construction of conception-based perspectives, including familiarity with research-based developmental (conceptual) "maps" and with methods (e.g., interviewing) for figuring out their students' available conceptions (see, for example, Tzur, 2007). As I noted above, promoting this difficult transformation implies that teachers need to construct a mindset (pedagogical scheme) of openness to and acceptance of unexpected situations as a key, self-generated source for their own professional development. That is, teachers need to form an invariant anticipation of actively *listening for* the unexpected (Empson & Jacobs, 2008).

For most teachers, developing a conception-based perspective as the core of intuitive-to-formal (or systematic) transition is not likely to happen without substantial, guided, long-term interventions. The reason is that teachers' existing schemes and perspectives (traditional or perception-based) serve as an assimilatory trap (Stolzenberg, 1984): What they notice and act upon is afforded and constrained by the paradigm they have yet to alter. For example, it seems that a teacher who concluded that "there are different ways to solve a problem" (Leikin & Zazkis, 2007), or teachers who rethought "what serves as evidence for numeracy" (Liljedahl, 2007), were yet to transform their epistemological stance toward the role of students' assimilatory conceptions in what they do or do not notice. I believe that making a conception-based perspective—adopting the principle of assimilation—the explicit goal for teachers' learning could greatly support the mathematics teacher educators' work.

Viewed from the *Ref*AER* framework, perception-based practices have the potential of serving as "material" for teacher reflection on their own epistemological anticipations. One obvious type of such reflection consists of *Type-I* comparisons between anticipated and actual student responses. For example, Leikin and Zazkis (2007) reported on a teacher who learned, through noticing a student's unexpected solution, that students may interpret a mathematical term (e.g., "divisor"), which she precisely defined in the class lecture, differently than the teacher's meaning for the conventional term. This then being the teacher's goal could be capitalized upon by mathematics teacher educators for promoting her abstraction of the epistemological role of assimilating a mathematical expression by the students. In turn, this can lead the teacher to examine how the term "factor" as used in the task could have brought forth in her students the activity–effect relationship relevant for figuring out if *any* natural number, presented as multiplication of primes to some power, is divisible by *any* other natural number. Similarly, teachers' learning of how to use computer software (Borba, 2007) can become the source for reflection on the role of mathematical activity in a medium, as well as on how the goal of a software user regulates what she or he notices (i.e., considers as "machine feedback").

A second, powerful type of such reflection consists of *Type-II* comparisons between tasks the teacher would use for teaching a particular conception to a specific group of students and tasks designed and used by mathematics teacher educators to accomplish the same student learning. In Tzur (2008b), I have discussed the powerful impact of such comparisons. Initially, teachers simply could not fathom how tasks and lesson plans I created would actually lead to the desired student learning (e.g., counting-on, fractions). What, from a conception-based perspective (specifically, *Ref*AER*), seemed perfectly intuitive to me, seemed counter-intuitive to the teachers; my plans contradicted their pedagogical anticipations. In that paper I discussed how teachers' witnessing the accomplishment of their goals for student learning via the counter-intuitive plan could be turned into constructive reflections that began "shaking" their deeply entrenched (traditional or perception-based) anticipations. That is, intentionally contrasting teachers' anticipations for failure of a teacher educator's task/plan and the surprising (to the teacher) success of such a plan proved powerful in the teachers' questioning of their own anticipations. I believe that both types of comparisons are needed to intentionally promote teacher transition from a perception- to a conception-based perspective.

Concluding Remarks

In this chapter, I argued for the need to close the gap between the potential and actual LTT of mathematics. I explained why augmenting LTT is important and suggested three deep-seated reasons why, at least in the Western world, the gap seems rather wide: The learning process not being problematized by teachers, the disproportion between the active learning and the assimilation principles, and the insufficient articulation of *how* and *what* teachers LTT. I introduced the *Ref*AER* framework, developed for mathematics learning and teaching, and proposed it can assist in better articulating both questions. Then, I provided an adapted model of *Ref*AER* for *how* might LTT take place. Finally, I introduced a novel coordination between pedagogical perspectives identified in research (traditional, perception, and conception) and corresponding approaches to curriculum design/use (one-, two-, and three-prong, respectively). I demonstrated why the latter, captured by the Teaching Triad notion that uses tasks as the interface between student assimilatory conceptions and the intended mathematics, provides three core goals for LTT (i.e., *what* they might learn to teach well).

I believe that the different sections of this chapter contribute to closing the largely unspoken gap between what teachers of mathematics actually and could potentially learn through teaching. To boost these contributions, I culminate with a few issues that deserve further theoretical and empirical scholarly attention:

1. How are guided and non-guided teacher learning different? Similar?
2. How does teachers' continual engagement in expressing their ideas to others contribute to LTT?
3. How does openness to students' unexpected reactions, which is a necessary condition for noticing such reactions and treating them as *contribution* to the teacher's own learning, evolve over time in relation to teachers' confidence (in math, in pedagogy)?
4. How might researchers use/measure changes in teachers' anticipatory schemes of teaching actions—schemes of which teachers are quite often unaware?
5. Like in quantum mechanics, it seems that the medium through which researchers interact with and observe teachers' behaviors may change what is observed. What are the methodological and educative implications of such changes?
6. How do teachers' goals and implicit assumptions impact LTT? This question bears both a theoretical elaboration and an articulation of teachers' practical focus (e.g., improve lesson plans, build assessment tasks, etc.)?
7. Derived from #6, what tasks and prompts can teacher educators employ (and why!) to foster LTT, including scaffoldings that can serve the teachers in the

absence of direct guidance, and why would such tasks work, or adjusted when not?

8. As a constructivist, I take for granted that what a mathematics teacher educator or a researcher can consider as lack of coherence in teachers' knowledge/beliefs/practice may be un-problematically coherent for the teacher (Liljedahl, 2007). Thus, in addressing both #6 and #7 above the onus is on the scholarly community to make explicit teachers' conceptions that afford/constrain hypothetical LTT.
9. And, last but certainly not least: What do we mean by "the teacher has learned" and what do we take as evidence for it (i.e., what is the meaning/measure of success in math teacher education)?

References

Adler, J., & Huillet, D. (2008). The social production of mathematics for teaching. In P. Sullivan & T. Wood (Eds.), *The international handbook of mathematics teacher education: Knowledge and beliefs in mathematics teaching and teaching development* (Vol. 1, pp. 195–221). Rotterdam, the Netherlands: Sense.

Baars, B. J. (2007). Mind and brain. In B. J. Baars & N. M. Gage (Eds.), *Cognition, brain, and consciousness: Introduction to cognitive neuroscience* (pp. 3–29). Amsterdam, The Netherlands: Elsevier.

Ball, D. L., & Bass, H. (2000). Interweaving content and pedagogy in teaching and learning to teach: Knowing and using mathematics. In J. Boaler (Ed.), *Multiple perspectives on the teaching and learning of mathematics* (pp. 83–104). Westport, CT: Ablex.

Bereiter, C. (1985). Toward a solution of the learning paradox. *Review of Educational Research, 55*(2), 201–226.

Borba, M. C. (2007). Integrating virtual and face-to-face practice: A model for continuing teacher education. In J.-H. Woo, H.-C. Lew, K.-S. Park, & D.-Y. Seo (Eds.), *Proceedings of the 31st annual meeting of the international group for the psychology of mathematics education* (Vol. 1, pp. 128–131). Seoul, Korea: The Korean Society of Educational Studies in Mathematics.

Clarke, D. (2008). The mathematics teacher as curriculum maker: Developing knowledge for enacting curriculum. In P. Sullivan & T. Wood (Eds.), *The International handbook of mathematics teacher education: Knowledge and beliefs in mathematics teaching and teaching development* (Vol. 1, pp. 133–151). Rotterdam, the Netherlands: Sense.

Cohen, D. K., & Ball, D. L. (1990). Policy and practice: An overview. *Educational Evaluation and Policy Analysis, 12*(3), 233–239.

Davis, B. (1996). *Teaching mathematics: Toward a sound alternative*. New York: Garland.

Dewey, J. (1933).*How we think: A Restatement of the relation of reflective thinking to the educative process*. Lexington, MA: D.C. Heath.

Empson, S. B., & Jacobs, V. R. (2008). Learning to listen to children's mathematics. In D. Tirosh & T. Wood (Eds.), *The international handbook of mathematics teacher education: Tools and processes in mathematics teacher education* (Vol. 2, pp. 257–281). Rotterdam, the Netherlands: Sense.

Gravemeijer, K. (1994). *Developing realistic mathematics education*. Utrecht, The Netherlands: Freudenthal Institute.

Heinz, K., Kinzel, M., Simon, M. A., & Tzur, R. (2000). Moving students through steps of mathematical knowing: An account of the practice of an elementary mathematics teacher in transition. *Journal of Mathematical Behavior, 19*, 83–107.

Herbst, P. (2008). The teacher and the task. In O. Figueras, J. L. Cortina, S. Alatorre, T. Rojano & A. Sepulveda (Eds.), *Proceedings of the 32nd annual conference of the international group of psychology of mathematics education* (Vol. 1, pp. 125–132). Morelia, Michoacán, Mexico: Michoacán University of Saint Nicholas of Hidalgo.

Hill, H. C., Rowan, B., & Ball, D. L. (2005). Effects of teachers' mathematical knowledge for teaching on student achievement. *American Educational Research Journal, 42*(2), 371–406.

Ingersoll, R. M. (2001). Teacher turnover and teacher shortage: An organizational analysis. *American Educational Research Journal, 38*(3), 499–534.

Krainer, K. (2008a). Individuals, teams, communities and networks: Participants and ways of participation in mathematics teacher education – An Introduction. In D. Tirosh & T. Wood (Eds.), *The international handbook of mathematics teacher education: Participants in mathematics teacher education* (Vol. 3, pp. 1–10). Rotterdam, the Netherlands: Sense.

Krainer, K. (2008b). Reflecting the development of a mathematics teacher educator and his discipline. In B. Jaworski & T. Wood (Eds.), *The international handbook of mathematics teacher education: The mathematics teacher educator as a developing professional* (Vol. 4, pp. 177–199). Rotterdam, the Netherlands: Sense.

Lambert, M. A. (2008). *From counting-on to number: Analysis of manipulative-based tasks.* Unpublished Masters Thesis Purdue, West Lafayette, IN.

Leikin, R. (2006). Learning by teaching: The case of Sieve of Eratosthenes and one elementary school teacher. In R. Zazkis & S. Campbell (Eds.), *Number theory in mathematics education: Perspectives and prospects*, (pp. 115–140). Mahwah, NJ: Erlbaum.

Leikin, R. (2008). Teams of prospective mathematics teachers: Multiple problems and multiple solutions. In D. Tirosh & T. Wood (Eds.), *The international handbook of mathematics teacher education: Participants in mathematics teacher education* (Vol. 3, pp. 63–88). Rotterdam, the Netherlands: Sense.

Leikin, R., & Zazkis, R. (2007). *A view on the teachers' opportunities to learn mathematics through teaching.* Paper presented at the Thirty-First Annual Conference of the International Group for the Psychology of Mathematics Education, Seoul, Korea.

Leinhardt, G. (1989). Math lessons: A contrast of novice and expert competence. *Journal for Research in Mathematics Education, 20*(1), 52–75.

Lerman, S. (2006). Socio-cultural research in PME. In A. Gutiérrez & P. Boero (Eds.), *Handbook of research on the psychology of mathematics education: Past, Present, and Future* (pp. 347–366). Rotterdam, The Netherlands: Sense.

Lewis, C., & Tsuchida, I. (1998). A lesson is like a swiftly flowing river: How research lessons improve Japanese education. *American Educator, 22*, 12–52.

Liljedahl, P. (2007). Teachers' learning reified: The professional growth of inservice teachers through numeracy task design. In J.-H. Woo, H.-C. Lew, K.-S. Park, & D.-Y. Seo (Eds.), *Proceedings of the 31st annual meeting of the international group for the psychology of mathematics education* (Vol. 1, pp. 132–137). Seoul, Korea: The Korean Society of Educational Studies in Mathematics.

Livingston, C., & Borko, H. (1990). High school mathematics review lessons: Expert-novice distinctions. *Journal for Research in Mathematics Education, 21*(5), 372–387.

Lompscher, J. (2002). The category of activity as a principal constituent of cultural-historical psychology. In D. Robbins & A. Stetsenko (Eds.), *Voices within Vygotsky's non-classical psychology: Past, Present, Future* (pp. 79–99). New York: Nova Science.

Ma, L. (1999). *Knowing and teaching elementary mathematics: Teachers' Understanding of fundamental mathematics in China and the United States.* Mahwah, NJ: Lawrence Erlbaum.

Ma, Y., Zhao, D., & Tuo, Z. (2006). Difference within communalities: How is mathematics taught in rural and urban regions in Mainland China? In L. Fan, N.-Y. Wong, J. Cai, & L. Shiqi (Eds.), *How chinese learn mathematics: Perspectives from insiders* (pp. 413–442). Singapore: World Scientific Publishing.

Markovits, Z., & Smith, M. (2008). Cases as tools in mathematics teacher education. In D. Tirosh & T. Wood (Eds.), *The international handbook of mathematics teacher education: Tools and processes in mathematics teacher education* (Vol. 2, pp. 39–64). Rotterdam, the Netherlands: Sense.

Mason, J. (1998). Enabling teachers to be real teachers: Necessary levels of awareness and structure of attention. *Journal of Mathematics Teacher Education, 1*(3), 243–267.

Mason, J. (2001). Knowing and learning mathematics for teaching: Proceedings of a workshop. *Journal of Mathematics Teacher Education, 4*(4), 339–348.

Mason, J. (2008). Being mathematical with and in front of learners: Attention, Awareness, and Attitude as sources of differences between teacher educators, teachers, and learners. In B. Jaworski & T. Wood (Eds.), *The international handbook of mathematics teacher education: The mathematics teacher educator's as a developing professional* (Vol. 4, pp. 31–55). Rotterdam, the Netherlands: Sense.

Paine, L. W., & Ma, L. (1993). Teachers working together: A dialogue on organizational and cultural perspectives of Chinese teachers. *International Journal of Educational Research, 19*(8), 675–697.

Pascual-Leone, J. (1976). A view of cognition from a formalist's perspective. In K. F. Riegel & J. A. Meacham (Eds.), *The developing individual in a changing world: Vol. 1 Historical and cultural issues* (pp. 89–110). The Hague, The Netherlands: Mouton.

Perrin-Glorian, M.-J., Deblos, L., & Robert, A. (2008). Individual practising mathematics teachers: Studies on their professional growth. In D. Tirosh & T. Wood (Eds.), *The international handbook of mathematics teacher education: Participants in mathematics teacher education* (Vol. 3, pp. 35–59). Rotterdam, the Netherlands: Sense.

Piaget, J. (1964). Development and learning. In R. E. Ripple & V. N. Rockcastle (Eds.), *Piaget rediscovered: Report of the conference on cognitive studies and curriculum development* (pp. 7–20). Ithaca, NY: Cornell University.

Piaget, J. (1970). *Genetic epistemology* (E. Duckworth, Trans.). New York: Columbia University.

Piaget, J. (1971). *Biology and knowledge* (B. Walsh, Trans.). Chicago: The University of Chicago.

Piaget, J. (1985). *The equilibration of cognitive structures: The central problem of intellectual development* (T. Brown & K. J. Thampy, Trans.). Chicago: The University of Chicago.

Pirie, S. E. B., & Kieren, T. E. (1994). Beyond metaphor: Formalising in mathematical understanding within constructivist environments. *For the Learning of Mathematics, 14*(1), 39–43.

Sánchez, V., & García, M. (2008). What to teach and how to teach it: Dilemmas in primary mathemaics teacher education. In B. Jaworski & T. Wood (Eds.), *The international handbook of mathematics teacher education: The mathematics teacher educator as a developing professional* (Vol. 4, pp. 281–297). Rotterdam, the Netherlands: Sense.

Schifter, D. (1998). Learning mathematics for teaching: From a teachers' seminar to the classroom. *Journal of Mathematics Teacher Education, 1*(1), 55–87.

Schön, D. A. (1987). *Educating the reflective practitioner: Toward a new design for teaching and learning in the professions* (1 edn.). San Francisco: Jossey-Bass.

Simon, M. A. (1995). Reconstructing mathematics pedagogy from a constructivist perspective. *Journal for Research in Mathematics Education, 26*(2), 114–145.

Simon, M. A. (2006a). Key developmental understandings in matheamtics: A direction for investigating and establishing learning goals. *Mathematical Thinking and Learning, 8*(4), 359–371.

Simon, M. A. (2006b). Pedagogical concepts as goals for teacher education: Towards an agenda for research in teacher development. In S. Alatorre, J. L. Cortina, S. Mariana, & A. Méndez (Eds.), *Proceedings of the twenty eighth annual meeting of the north american chapter of the international group for the psychology of mathematics education* (Vol. 2, pp. 730–735). Merida, Yucatán, Mexico: Universidad Pedagógica Nacional.

Simon, M. A. (2007). *Constraints on what teachers can learn from their practice: Teachers' assimilatory schemes.* Paper presented at the Thirty-First Conference of the International Group for the Psychology of Mathematics Education, Seoul, Korea.

Simon, M. A., & Tzur, R. (2004). Explicating the role of mathematical tasks in conceptual learning: An elaboration of the hypothetical learning trajectory. *Mathematical Thinking and Learning, 6*(2), 91–104.

Simon, M. A., Tzur, R., Heinz, K., & Kinzel, M. (2004). Explicating a mechanism for conceptual learning: Elaborating the construct of reflective abstraction. *Journal for Research in Mathematics Education, 35*(3), 305–329.

Simon, M. A., Tzur, R., Heinz, K., Kinzel, M., & Smith, M. S. (2000). Characterizing a perspective underlying the practice of mathematics teachers in transition. *Journal for Research in Mathematics Education, 31*(5), 579–601.

Steffe, L. P. (1995). Alternative epistemologies: An educator's perspective. In L. P. Steffe & J. Gale (Eds.), *Constructivism in education* (pp. 489–523). Hillsdale, NJ: Lawrence Erlbaum.

Steinbring, H. (1998). Elements of epistemological knowledge for mathematics teachers. *Journal of Mathematics Teacher Education, 1*(2), 157–189.

Stolzenberg, G. (1984). Can inquiry into the foundations of mathematics tell us anything interesting about mind? In P. Watzlawick (Ed.), *The invented reality: How do we know what we believe we know* (pp. 257–309). NY: Norton.

Sullivan, P. (2008a). Designing task based mathematics lessons as teacher learning. In O. Figueras, J. L. Cortina, S. Alatorre, T. Rojano, & A. Sepulveda (Eds.), *Proceedings of the 32nd annual conference of the international group of psychology of mathematics education* (Vol. 1, pp. 133–138). Morelia, Michoacán, Mexico: Michoacán University of Saint Nicholas of Hidalgo.

Sullivan, P. (2008b). Knowledge for teaching mathematics: An introduction. In P. Sullivan & T. Wood (Eds.), *The international handbook of mathematics teacher education: Knowledge and beliefs in mathematics teaching and teaching development* (Vol. 1, pp. 1–9). Rotterdam, the Netherlands: Sense.

Sullivan, P., Mousley, J., & Zevenbergen, R. (2003). Being explicit about aspects of mathematics pedagogy. In N. A. Pateman, B. J. Dougherty & J. Zilliox (Eds.), *Proceedings of the 2003 Joint Meeting of PME and PME-NA* (Vol. 4, pp. 267–274). Honolulu, Hawai'i: University of Hawai'i.

Tzur, R. (1996). *Interaction and children's fraction learning*. Ann Arbor, MI: UMI Dissertation Services (Bell & Howell).

Tzur, R. (2004). Teacher and students' joint production of a reversible fraction conception. *Journal of Mathematical Behavior, 23*, 93–114.

Tzur, R. (2007). Fine grain assessment of students' mathematical understanding: Participatory and anticipatory stages in learning a new mathematical conception. *Educational Studies in Mathematics, 66*(3), 273–291.

Tzur, R. (2008a). A researcher perplexity: Why do mathematical tasks undergo metamorphosis in teacher hands? . In O. Figueras, J. L. Cortina, S. Alatorre, T. Rojano, & A. Sepulveda (Eds.), *Proceedings of the 32nd Annual conference of the international group of psychology of mathematics education* (Vol. 1, pp. 139–146). Morelia, Michoacán, Mexico: Michoacán University of Saint Nicholas of Hidalgo.

Tzur, R. (2008b). Profound awareness of the learning paradox (PALP): A journey towards epistemologically regulated pedagogy in mathematics teaching and teacher education. In B. Jaworski & T. Wood (Eds.), *The international handbook of mathematics teacher education: The mathematics teacher educator's as a developing professional* (Vol. 4, pp. 137–156). Rotterdam, the Netherlands: Sense.

Tzur, R. (accepted for publication). Brain-mind theory-driven task design. In E. Kelly, S. Jensen, & J. Baek (Eds.), *Neuro Math Education*. Cambridge, MA: Cambridge University Press.

Tzur, R., & Simon, M. A. (2004). Distinguishing two stages of mathematics conceptual learning. *International Journal of Science and Mathematics Education, 2*, 287–304.

Tzur, R., Simon, M., Heinz, K., & Kinzel, M. (2001). An Account of a teacher's perspective on learning and teaching mathematics: Implications for teacher development. *Journal of Mathematics Teacher Education, 4*(3), 227–254.

Tzur, R., Zaslavsky, O., & Sullivan, P. (2008). Examining teachers' use of (non-routine) mathematical tasks in classrooms from three complementary perspectives: Teacher, teacher educator, researcher. In O. Figueras, J. L. Cortina, S. Alatorre, T. Rojano, & A. Sepulveda (Eds.),

Proceedings of the 32nd annual conference of the international group of psychology of mathematics education (Vol. 1, pp. 123–125). Morelia, Michoacán, Mexico: Michoacán University of Saint Nicholas of Hidalgo.
von Glasersfeld, E. (1995). *Radical constructivism: A way of knowing and learning*. Washington, DC: Falmer.
Vygotsky, L. S. (1978). *Mind in society: The development of higher psychological processes* (Vol. 1). Cambridge, MA: Harvard University Press.
Watson, A. (2008). Task transformation is the teacher's responsibility. In O. Figueras, J. L. Cortina, S. Alatorre, T. Rojano, & A. Sepulveda (Eds.), *Proceedings of the 32nd annual conference of the international group of psychology of mathematics education* (Vol. 1, pp. 146–153). Morelia, Michoacán, Mexico: Michoacán University of Saint Nicholas of Hidalgo.
Watson, A., & Sullivan, P. (2008). Teachers learning about tasks and lessons. In D. Tirosh & T. Wood (Eds.), *The international handbook of mathematics teacher education: Tools and processes in mathematics teacher education* (Vol. 2, pp. 109–134). Rotterdam, the Netherlands: Sense.
Wenger, E. (1998). *Communities of practice: Learning, meaning, and identity*. NY: Cambridge University.
Wertsch, J. V., & Toma, C. (1995). Discourse and learning in the classroom: A sociocultural approach. In L. P. Steffe & J. Gale (Eds.), *Constructivism in education*(Vol. 1, 1 edn., pp. 159–174). Hillsdale, NJ: Lawrence Erlbaum.
Yoshida, M. (2008). Exploring ideas for a mathematics teacher educator's contribution to lesson study. In D. Tirosh & T. Wood (Eds.), *The international handbook of mathematics teacher education: Tools and processes in matheamtics teacher education* (Vol. 2, pp. 85–106). Rotterdam, The Netherlands: Sense.
Zaslavsky, O., & Leikin, R. (2004). Professional development of mathematics teacher educators: Growth through Practice. *Journal of Mathematics Teacher Education*, *7*(1), 5–32.

Learning Through Teaching Through the Lens of Multiple Solution Tasks

Roza Leikin

Introduction

During the past decade my research focused on teachers' learning through teaching, with special attention to the mathematics learned by facilitators of the learning process: teachers, and teacher educators. Although my focus was on the teachers' mathematical knowledge, I realized (very naturally) that the mathematics they learned in the classroom is not separate from the pedagogy they learned. There was a two-way connection between mathematics and pedagogy (Fig. 1): Teachers' advanced pedagogy was manifested in their flexibility (Simon, 1997; Leikin & Dinur, 2007) and improvisation (Sawyer, 2004) in the classroom; their success in creating a-didactic situations (in the sense defined by Brousseau, 1997) resulted in their learning mathematics. At the same time, teachers' connected and advanced mathematical knowledge (Leikin, 2007; Zazkis & Leikin, 2009) displayed in their problem-solving expertise promoted teacher flexibility and the ability to design challenging settings in which students learned autonomously (Steinbring, 1998). These findings were reported, for example, in Leikin (2006) and in Leikin and Dinur (2007).

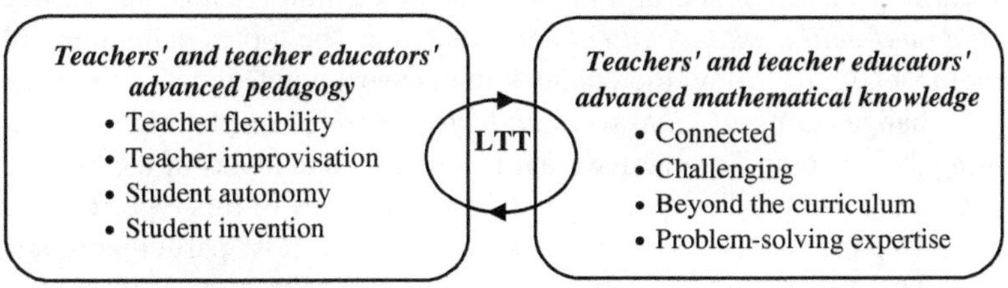

Fig. 1 The dual nature of LTT

R. Leikin (✉)
University of Haifa, Haifa, Israel
e-mail: rozal@construct.haifa.ac.il

R. Leikin, R. Zazkis (eds.), *Learning Through Teaching Mathematics*, Mathematics Teacher Education 5, DOI 10.1007/978-90-481-3990-3_4,

In this chapter I illustrate and explain the phenomenon of learning through teaching (LTT) using multiple solution tasks (MSTs). A "multiple solution task" is one in which learners are explicitly required to solve a mathematical problem in multiple ways. The distinctions between the solutions can be based, for example, on (a) use of different representations of a mathematical concept; (b) use of different properties (definitions or theorems) of a mathematical concept; and (c) use of mathematics tools from different branches of mathematics (Leikin, 2003; Leikin, 2007; Leikin & Levav-Waynberg, 2007, 2008). Systematic use of MSTs both in teacher educational programs and in school mathematics classrooms is an example of advanced interweaving of mathematics and pedagogy. This approach requires facilitators of the learning process to demonstrate advanced mathematical knowledge as well as pedagogical flexibility and improvisation because the use of MSTs intensifies the creation of unpredicted learners' responses.

Below I outline the research methodology I used to study LTT and the way in which MSTs were incorporated in the study. Next, I describe the case of Rachel, a mathematics teacher educator, who used MSTs to learn mathematics in her teacher education courses. I use the story-telling method to present the case. After the story I introduce the notion of solution spaces, a useful construct for analyzing the processes and outcomes associated with teaching MSTs. Then I present a longitudinal exploration of LTT by means of MSTs. Finally, I return to the connection between teachers' mathematics and pedagogy in teaching and learning.

Methodology of LTT Research

I used two main methodologies in studies exploring LTT: multiple cases of (relatively short) teaching experiments and longitudinal investigation of a teacher's development experiment.

Multiple cases of teaching experiments: In the first chapter of this volume, Leikin and Zazkis argued that the Teaching Experiment (TE) is a powerful framework for teacher learning. Based on this argument, I conducted multiple teaching experiments on *teaching unfamiliar mathematical tasks* matching the topics being taught by the individual teacher. To follow the changes in teacher knowledge and the manner in which the change occurred, *data were collected in triads* of planning the teaching, performing the teaching (interactive stage), and critical analysis of the two previous stages. More than 40 teachers participated in experiments of this type. The teaching experience of participants varied from 1 to 20 years; most participants taught in secondary school and some in elementary school. The duration of teaching varied from one to three lessons. The data were recorded in written protocols or videotaped and transcribed. At least two people analyzed the data (for a detailed description of some of the cases see Leikin, 2005a; 2005b; 2006; Leikin & Rota, 2006; Leikin & Dinur, 2007).

Using similar methodology (but less formally), I also explored several cases of development of teacher educators' mathematical knowledge and recorded my own learning as well. One of these cases – the case of Rachel – is presented and analyzed in detail in this chapter.

The *longitudinal study* used the teacher development experiment (TDE) methodology (Simon, 2000; Leikin, 2003). Twelve secondary school mathematics teachers (MTs) participated in the study over a period of three years (2003–2006). In the first year, the teachers volunteered to take part in a 56-hour professional development course focusing on MSTs. During that year the teachers were asked to avoid implementing MSTs in their classrooms, but in the second year they were required to implement them. To follow teachers' LTT, we interviewed them at the end of the first-year course and at the end of the second year of the research intervention. At the end of the second year, the teachers also participated in four meetings focusing on the implementation of MSTs. During these meetings, we also conducted group interviews. Figure 2 shows the main elements of the research methodology used in the two studies. During the third year we followed the effect of the intervention on teachers' work through individual communications with the teachers.

Common to the two studies were *triads of data*, which included interviews and observations, and the *presence of unfamiliar elements in teaching*. The use of unfamiliar approaches or contents was aimed at intensifying the data collection. During the interviews the teachers were consistently asked to solve problems related to the lessons they taught and provide instructional examples. During the pre-interviews teachers were asked to predict student responses, whereas during the post-interviews they were asked to analyze student responses. Comparing the pre- and post-interviews, we analyzed the development of the teachers' mathematical knowledge based on how they solved and exemplified tasks, and the development of their pedagogical knowledge based on their prediction and analysis of student responses. Exemplification tasks often provided additional evidence for the teachers' pedagogical knowledge because teachers tended to reason about their students when providing examples (Leikin & Levav-Waynberg, 2007). Following the

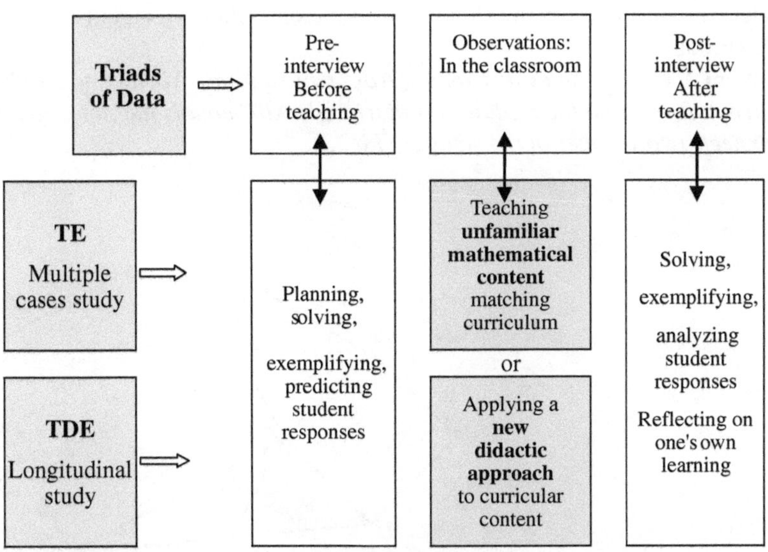

Fig. 2 Research methodology for the investigation of LTT

teaching, teachers were asked to analyze their own learning, and the resulting information was combined and compared with the researchers' perception of teachers' learning. We used MSTs as a research tool to examine the teachers' LTT and found them to be an effective tool for LTT.

The Case of Rachel

Rachel (all the names mentioned in the chapter have been changed) is an experienced teacher educator. I present her case as a story, a methodology that has been acknowledged as a valid means of describing processes of teaching and learning to teach (Krainer, 2001; Chazan, 2000; Lampert, 2001; Schifter, 1996; Zaslavsky & Leikin, 2004). Krainer (2001) suggested three learning levels associated with stories: (1) Stories help provide authentic evidence of typical development in teacher education; (2) stories can extend our theoretical knowledge about the complex processes of teacher education; and (3) stories serve as starting points for reflection and promote insight into our mental processes and challenges (ibid., p. 271). Rachel's story provides authentic evidence of a typical case of LTT and combines all the elements of data collected about her teaching and learning.

Rachel was an instructor at a professional development course on MSTs for secondary school MTs. During the course participating MTs were asked to solve every mathematical problem in at least two ways. At the beginning of the course, MTs had difficulty finding multiple solutions for the problems presented to them. In the middle of the course, Rachel and the MTs were surprised by the change in their problem-solving proficiency. For one of the home tasks Rachel asked MTs to solve Problem 1 (Fig. 3) in at least 3 different ways. Her objective was to make MTs consider typical auxiliary constructions for problems in which a median is one of the main elements in the problem: doubling the median (Proof 1.1) and doubling the triangle (Proof 1.2).

Problem 1 *On two sides of triangle ABC two squares, ABED and AFGC, are constructed. Prove that the median AM of triangle ABC equals half the segment FD that connects two vertices of the squares (Fig. 3).*

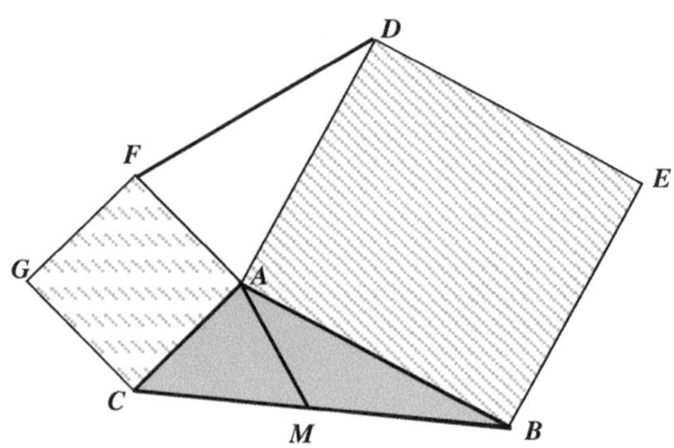

Fig. 3 Problem 1

Five Proofs Presented at the Lesson

Median-Based Proofs

As Rachel expected, at the beginning of the lesson MTs presented two median-based proofs (Fig. 4a,b provide outlines of the proofs). The whole-group discussion about Proofs 1.1 and 1.2 focused on the use of the two constructions, which the teachers considered to be "typical auxiliary constructions for problems where the median of a triangle is given." The teachers proceeded to discuss the connections between the objects obtained by the two auxiliary constructions, e.g., the connections between the equality of the areas of the two small triangles created by the median and the equality of the areas of the four small triangles in the parallelogram created by the two diagonals. When doubling the triangle, the median is also a midline in the bigger triangle. Participants also discussed why the two different constructions produce congruent triangles (ΔFAD ≅ ΔNBA (Proof 1.1) and ΔFAD ≅ ΔPAB (Proof 1.2), so that ΔPAB ≅ ΔNBA). After the discussion of Proofs 1.1 and 1.2, Rachel asked MPTs to present additional solutions to the task. MTs presented three proofs.

Proof 1.1: Doubling the median

Auxiliary construction AN = 2AM
⇒ABNC – parallelogram
⇒ ΔFAD ≅ ΔNBA
⇒ FD = AN ⇒ FD = 2AM

Proof 1.2: Doubling the triangle

Auxiliary construction PB = 2AM,
⇒AM – midline in ΔCPB
⇒ ΔFAD ≅ ΔPAB
⇒ FD = PB ⇒ FD = 2AM.

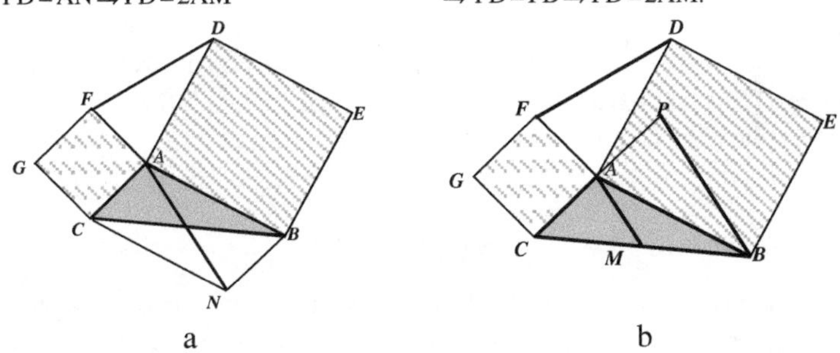

Fig. 4 Median-based proofs

Rotation-Based Proofs

MTs presented three new proofs, two of which were based on rotation. Mali suggested rotating triangle ABC by 90° about vertex A (Fig. 5a), and Soha rotated the "small" square by 90° about Q, which is the center of one of the squares (Fig. 5b).

These two proofs were unusual from the point of view of the school curriculum, but the MTs were experienced in solving problems using symmetry. Rachel was glad that the teachers found these solutions and disappointed that she did not see them herself before the lesson. At this point, the whole-group discussion focused on

Proof 1.3: Rotating the triangle

Rotation of triangle ABC by 90° about vertex A
AC moves to AF; AB moves to AL
$\Rightarrow$ AM – median in ΔABC, moves to AK – median in ΔALF
$\Rightarrow$ KA is the midline in triangle LFD
$\Rightarrow$ FD = 2KA $\Rightarrow$ FD = 2AM.

Proof 1.4: Rotating the "small" square

Rotation of square ACGF by 90° about Q, which is the center of the second square
$\Rightarrow$ AFTD – parallelogram.
ΔABC moves to ΔDAT $\Rightarrow$ TA = CB
FD and TA (diagonals in the parallelogram) intersect at R $\Rightarrow$ FD = 2RD,
AM moves to RD $\Rightarrow$ FD = 2AM.

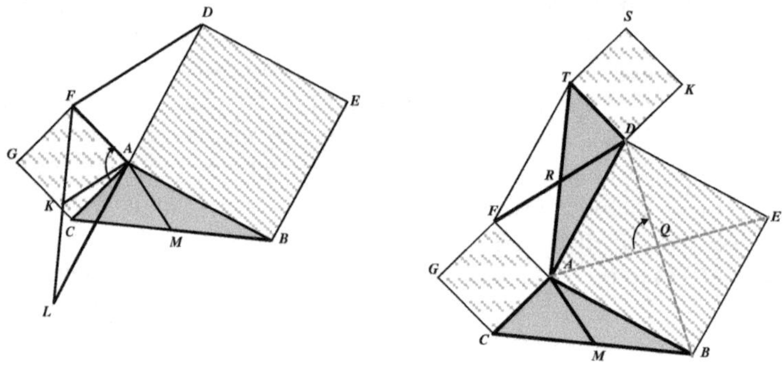

a b

Fig. 5 Rotation-based proofs

the elegance of the proofs, their clarity, and effectiveness. The teachers held diverse opinions and most of them considered that these two proofs were not applicable in their classes. "Students will be confused by these proofs," they said, "They won't be able to explain these solutions mathematically enough."

Proof 1.5: We Obtain One More Square Inside

An additional proof, provided by Ronit, was based on the observation that "we obtain one more square inside." She constructed centers O and Q of the two squares and connected them with points M and R, where R is the midpoint of FD (Fig. 6). She then proved that ORQM is a square, and based on this fact she proved that triangles RDQ and MAQ are congruent. Therefore FD=2AM (because RD=AM).

Unfortunately Ronit's proof that ORQM is a square was very long, based on multiple congruencies of triangles, some of which were superfluous. Rachel asked the MTs: "Can we prove that ORQM is a square in a shorter way?" After a few minutes, Anat reasoned aloud: "If ORQM is a square, its diagonals FB and CD must be perpendicular and equal." Based on this observation, the MTs searched for a proof that triangles CAD and FAB are congruent (Fig. 6). One of the MTs proved this by a comparison of angles and segments. Congruence of the triangles proved the equality of segments FB and CD. She then showed that the segments are perpendicular by calculating the angles.

Proof 1.5: An "inside" square

ORQM is a square
$\Rightarrow \Delta RDQ \cong \Delta MAQ$
$\Rightarrow FD=2AM$

Proof that ORQM is a square
Rotation of triangle CAD by 90°
transforms it into triangle FAB.
$\Rightarrow$ CD and FB are equal and
perpendicular
$\Rightarrow$ ORQM is a square

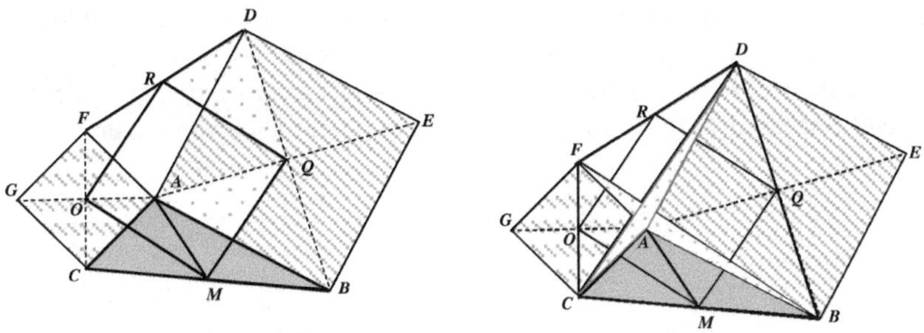

a b

Fig. 6 A square inside

At this point Rachel demonstrated to MTs that rotation of triangle CAD by 90° transforms it into triangle FAB. Thus, CD moves to FB, and they are equal and perpendicular. Rachel was familiar with this proof.

The group discussion again turned to such aspects of the five proofs as elegance, clarity, applicability in the classroom, and the practicality of "spending time on different solutions" in the classroom. As usual, MTs had different preferences about the implementation of MSTs in school, and different opinions about "which proof is the most beautiful."

Not Only the Proofs, Not Only at the Lesson

Note that Rachel learned geometry together with her students. First, three of the five proofs presented at the lesson were unfamiliar to her, so these new proofs enriched her knowledge, and Problem 1 expanded the collection of the multiple-solution tasks with at least three solutions that she was compiling.

Reaching beyond the new proofs that she learned during the lesson, Rachel discovered similarities between the two rotation-based proofs and the median-based proofs. The parallelogram obtained in Proof 1.4 can be constructed using a strategy of *doubling the median* (this time in triangle ADF), which was used in Proof 1.1. Triangle LFD obtained in Proof 1.3 can be constructed using a strategy of *doubling the triangle* (in this case DAF), which was used in Proof 1.2. Thus, the median in triangle FAL is the midline in the doubled triangle LFD. Conversely, triangle APB, created in Proof 2 (Fig. 4b) can be obtained by rotating triangle FAD by 90° about A.

The lesson helped Rachel discover connections between different proving strategies, between Euclidean and transformational geometry.

Furthermore, Rachel understood that "the squares do not depend on the given triangles" and realized that a collection of new statements was proven in the course of the lesson. These statements can be presented as an independent problem, and their collection may be presented as a problem chain (Fig. 7).

Continuing to ponder the issue of the transformation of the given problem into a new one, in which two squares with a common vertex are given (Fig. 7), led Rachel to a new question: *Does the mutual position of the two squares affect the regularities observed?* She used a dynamic geometry environment (Schwartz, Yerushalmy, &

Problem 2

If

ACGF and ABED are two squares with a common vertex A

Then

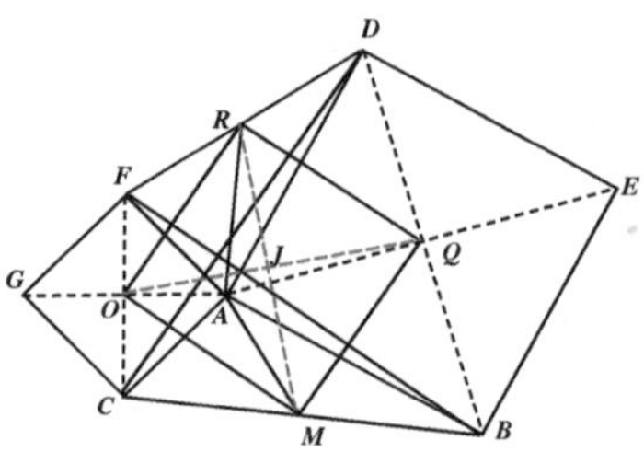

1. FB = CD and FB ⊥ CD
2. FD = 2AM and FD = 2AR
   where M and R are midpoints of the segments CB and FD correspondingly
3. ROMQ is a square
   where O and Q are centers of squares ACGF and ABED correspondingly
4. Points ROMQ are on a circle with the center at J
   where J is the intersection point of OQ and RM

4a. OQ = RM, OQ ⊥ RM
4b. OJ = JQ = RJ = JM

5. Triangles FAD and ABC can be obtained as:
   a. two parts of a bigger triangle divided by its median

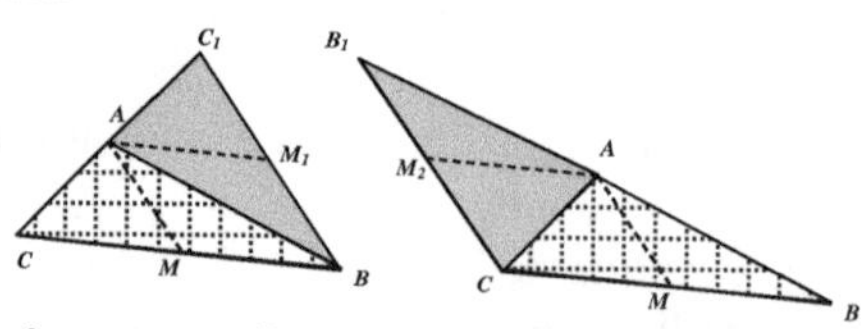

   b. pairs of not necessarily congruent small triangles obtained through the division of a parallelogram in four triangles by the intersecting diagonals
   c. "two big halves" of a parallelogram with a common side.

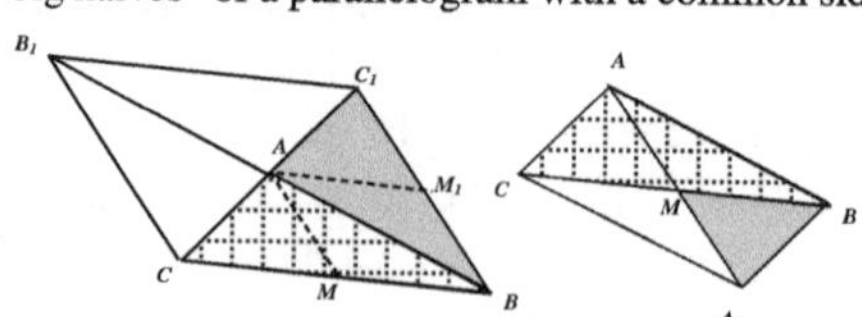

Fig. 7 New facts (theorems)

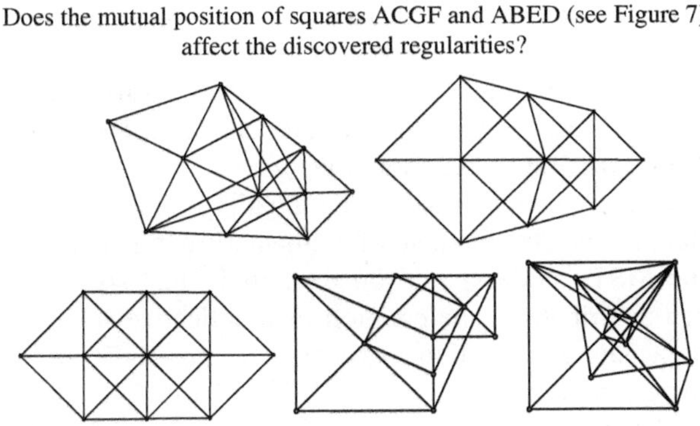

Fig. 8 A new question

Shternberg, 2000) to examine this question (Fig. 8). I leave this question open for the readers to investigate and prove.

LTT as a Transformation of Solution Spaces for MSTs

To outline the LTT mechanisms associated with teaching MSTs, I use the construct of *solution*[1] *spaces* for MSTs (Leikin, 2007; Leikin & Levav-Waynberg, 2008; Leikin, 2009). In Leikin (2007) I suggested considering solution spaces for MSTs analogously with the metaphor of *example spaces* introduced by Watson and Mason (2005) to describe example-generation processes. The following types of solution spaces for MSTs are defined with respect to individuals or groups of individuals solving a task:

An individual solution space is a set of solutions to a problem that an individual provides. Depending on a person's ability to produce solutions to an MST with or without prompts, individual solution spaces are of two kinds: An *available personal solution space* consists of solutions that a solver produces without help from others; a *potential personal solution space* consists of solutions produced with the help of others (cf. the concept of ZPD defined by Vygotsky, 1978). When the MTs solved Problem 1 at home, Proofs 1.1 and 1.2 belonged to available individual solution spaces for most of the teachers, whereas Proofs 1.3, 1.4, and 1.5 belonged to the individual solution space of one teacher only.

An *expert solution space* of a problem is the most complete set of solutions to a problem known at a given time. *Collective solution spaces* characterize solutions produced by groups of participants, and they are manifest in the whole-group discussions and written tests.

[1]Here a proof is a concrete type of a problem solution when the problem requires proving (like Problem 1 in this paper).

Solution spaces can also be characterized by their conventionality: *Conventional solutions* appear in curriculum-based instructional materials (Proofs 1.1. and 1.2 are conventional proofs for Problem 1); *unconventional solutions* either are not included in curriculum-based instructional materials (e.g., Proofs 1.3, 1.4, and 1.5 for Problem 1) or curriculum-based solutions applied in an unusual situation (not typical for geometry problems). The conventionality of solution spaces is a useful measure in evaluating the development of mathematical creativity.

When solving MSTs students rely on their individual solution spaces (indSS in Fig. 9), which contribute to the creation of a collective solution space for the MST. The dynamic nature of solution spaces (i.e., the possibility of transforming the spaces) is a basic element of learning with MSTs (Fig. 9). The interaction between individual and collective solution spaces enriches and expands individual solution spaces to include more solutions, to involve more concepts and properties, and to construct more mathematical connections.

Usually, at the moment when a teacher or teacher educator presents a MST to her students the teacher solution space becomes one of the individual solution spaces of the classroom community. Through interactions with students, during problem solving, individual solution spaces (including the teacher's solution space) can be expanded.

I suggest that *for a student* whose individual solution space of a MST is completed by the collective solution space generated by the whole group of students, some of the new solutions added to the student's solution spaces belong to his/her

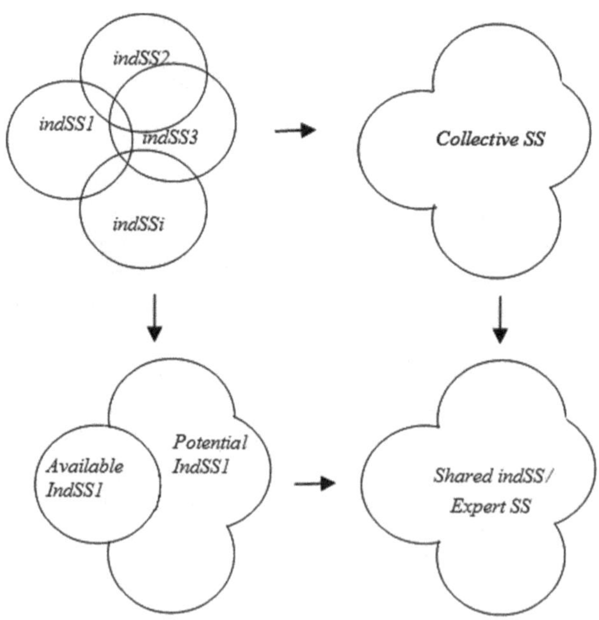

Fig. 9 The dynamic nature of solution spaces

new *potential solution space* of this MST. In other words, additional solutions presented in the classroom belong to the student's ZPD.

At the same time, *for the teacher*, this shared solution space becomes the available individual solution space, serving in future as an *expert solution space*. Rachel's expert solution space for the task, described in the chapter, at the end of the lesson was richer and more complete than at the beginning of the lesson. Her expert solution space, constructed in the classroom as a shared individual solution space, was enriched both by the learners' solutions and by her own unplanned solutions generated during the lesson.

Using Connecting Tasks to Promote Teachers' Learning

In this section I present a study of a different kind. In this study we compared teacher development through learning (in systematic mode) and through teaching (in craft mode) (Leikin, 2003; Leikin & Levav-Waynberg, 2007, 2008). In the first year of the study, the teachers learned to solve problems in different ways to become familiar with MSTs and learn how to use them. In the second year teachers implemented MSTs in their classrooms. The tailed methodology of this longitudinal investigation carried out with a group of 12 secondary school MTs is described earlier in the Methodology of LTT research section.

The teachers participated in three interviews: before the learning state (int-A), after the learning and before the teaching stage (int-B), and after the teaching stage (int-C). All interviews were identical in structure. The teachers were asked to solve all the problems in as many different ways as possible. All the problems in the interviews were borrowed from algebra, geometry, and calculus textbooks. In the various interviews, similar tasks were of the same level of difficulty and had an approximately equal number of solutions in the expert solution spaces. The interviews included problems of two types:

Conventional MSTs (CMSTs), problems with multiple solutions commonly taught in school: a system of linear equations (int-A), a quadratic inequality (int-B), and an absolute value inequality (int-C) (Fig. 10). All tasks required a comparable number of solution stages of similar complexity. The school curriculum usually requires multiple solutions to these problems (several algebraic and graphic solutions). At the same time, all the tasks can be solved using symmetry considerations, which are unconventional for both teachers and students.

Unconventional MSTs (UMSTs), problems that are usually solved according to the school curriculum in a particular way. In each of the three interviews we included a maxima-minima problem (Fig. 11), a word problem, and a geometry problem. Maxima-minima problems were used as UMSTs since in school they were usually assigned to calculus and solved using derivatives, and therefore all other solutions for maxima-minima problems were unconventional. The word problems did not fit any of the conventional categories usually introduced in schools, and were therefore considered UMSTs. Geometry problems in regular classes are usually associated

	Int-A	Int-B	Int-C
Solve in as many ways as possible	$x^2 < 25$	$(x-2)^2 \neq x^2-4$	$\mid 2x-3 \mid > 7$
Solution 1	$\lvert x \rvert < 5$ $(x<0, x>0)$	$x^2-4x+4 \neq x^2-4$ $x \neq 2$	$2x-3>7$ *or* $2x-3<-7$
2	$(x-5)(x+5)<0$	$(x-2-x)(x-2+x) \neq 4$	$\lvert x-1.5 \rvert > 3.5$ 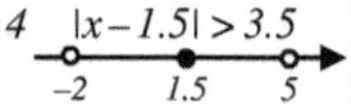
3	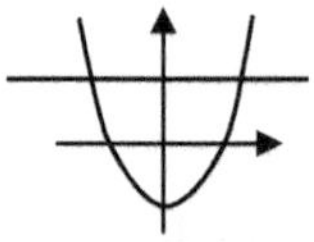	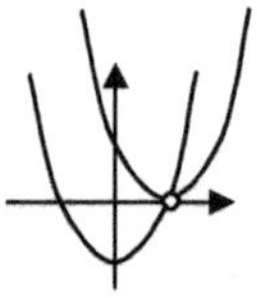	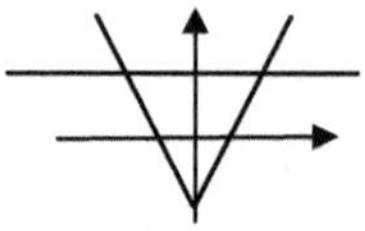

Fig. 10 Conventional MSTs assigned at the three interviews

Among all the isosceles triangles with side b, which one has the maximal area?

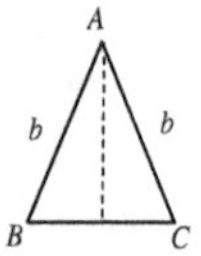

Solution 1: $f(x) = \frac{1}{2} b^2 \sin(\angle BAC)$, $f(x) - \max \Leftrightarrow \sin(\angle BAC) = 1$

Solution 2: $f(x) = \frac{1}{2} b^2 \sin(\angle BAC)$, $f(x) - \max \Leftrightarrow f'(x) = 0$

Solution 3:

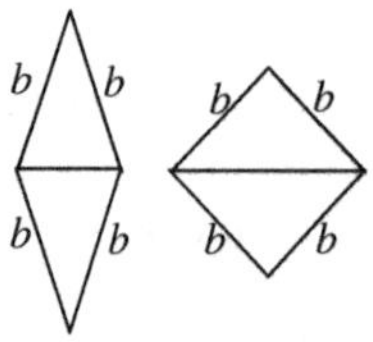

The square has the maximal area among rhombuses with a given side

Solution 4:

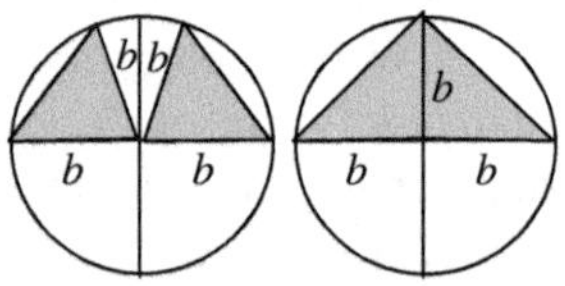

The sum of the areas of the two triangles is maximal when they combine in a triangle

Fig. 11 Unconventional MST: maxima-minima problem

with a particular theorem used to solve the problem. Because teachers are usually aware of this assignment, any additional solution to the problem can be considered unconventional.

At the beginning of the intervention, when the teachers were asked to solve problems, they reproduced solutions prescribed by the curriculum, and the teachers' solution spaces were mostly conventional. Naturally, they produced more solutions for CMSTs than for UMSTs. In the course of teaching both individual and collective solution spaces expanded, and half the individual spaces included unconventional solutions. The growing numbers of solutions in both individual and collective solution spaces were due not only to the reproduction of solutions offered during the course but also to the production of new ones. These findings demonstrate the importance of the teachers' systematic learning for the development of their mathematical knowledge.

LTT Depended on the Conventionality of the Tasks

Teaching experience was clearly reflected in the teachers' individual and collective solution spaces (see Fig. 12, and for a detailed description of the study see Leikin & Levav-Waynberg, 2008). When MSTs belonged to *topics that the teachers did not teach* during the year, their individual solution spaces were smaller than at the end of the course in which teachers participated beforehand. When teachers taught topics related to the MSTs, changes in teachers' solution spaces during the third interview depended on the conventionality of the tasks.

The number of solutions in the individual solution spaces for *CMSTs* either remained unchanged or were reduced, whereas unconventional solutions disappeared from the teachers' individual solution spaces. Analysis of the teachers' lessons in which they implemented MSTs shows that topics related to the conventional tasks were taught only by implementing conventional solution. Thus, teachers preserved in their individual solution spaces only conventional solutions.

LTT related to *UMSTs* showed several tendencies depending on whether MSTs were incorporated deliberately, belonged to topics being taught without deliberate use of MSTs, or belonged to topics that were not being taught. For topics in which MSTs were incorporated deliberately, solution spaces expanded. LTT began with teachers searching for multiple solutions to the problems when planning the lessons. For topics taught without incorporating connection tasks we did not observe changes in the number of individual solution spaces, but as in the case of the conventional

No. of solutions / Type of Task		Individual solution spaces		Collective solution spaces	
		Total	Unconventional	Total	Unconventional
Conventional task taught according to the curriculum		Unchanged or decreased	Disappeared	Decreased	Disappeared
Min-Max problem	Not taught	Decreased	Decreased	Decreased	Decreased
	Taught according to curriculum	Unchanged	Decreased	Unchanged	Decreased
	Taught as MST	Increased	Increased	Increased	Increased
Geometry problem		Increased		Increased	

Fig. 12 LTT – Changes in solution spaces in the year of teaching MSTs

MSTs, all the solution spaces become conventional. In these cases, the teachers did not plan MSTs for the lessons and did not include multiple solutions as an integral part of their teaching; unconventional solutions were never presented according to students' initiative. At the same time, as a result of awareness developed during the systematic stage of intervention, teachers were attentive to student responses and willing to discuss all the solutions that students provided during the lesson. Consequently, the number of solutions in the teachers' individual solution spaces remained unchanged, but the spaces became conventional. At the same time, the number of solutions in the collective spaces grew for all the unconventional MSTs, as the individual solution spaces of these tasks were grounded in the teachers' individual practice and differed from each other.

The number of solutions suggested by all teachers for geometry problems increased when the teachers implemented MSTs. We relate this phenomena to our finding that teachers became more attentive to student solutions, started collecting them, and allowed students to "always present all the solutions they found" without saying "this is good but we don't have enough time for it." We hypothesize that this combination of awareness (Mason, 2002) and pedagogical flexibility (Leikin & Dinur, 2007) enabled teachers to learn multiple solutions in geometry from their students. The teachers' practice in solving geometry problems in different ways made their mathematical reasoning more flexible, freed them from "being locked" on the first solution they found, and enabled them to search for different solutions to the problems.

Finally, based on the comparison of changes in the solution spaces for the different types of teaching experience (Fig. 9), I speculate that purposeful incorporation of MSTs in teaching practice developed the teachers' creativity because it was only for tasks of this type that teachers provided unconventional solutions during int-C.

Conclusions

MSTs, LTT, and Didactic Situations

The teachers' main task in the teaching process is the devolution of good tasks to their students (Brousseau, 1997; Steinbring, 1998). To achieve this objective, teachers create didactic situations in which students construct knowledge according to the teachers' aims. Among all didactic situations designed by teachers to encourage the learners' knowledge construction according to the teacher's plan, Brousseau (1997) distinguishes one that meets the conditions needed to promote student learning regardless of the teacher, with the actions performed by students depending entirely on the problem they must solve. In an a-didactic situation, students are responsible for their learning progress. Based on this definition, MSTs may be considered an effective tool for creating didactic and a-didactic situations. When solving problems in different ways is an explicit objective of the mathematical activity, the situation is didactic with respect to the MST and a-didactic with respect to the construction of mathematical connections by comparing different solutions to the problems. The teachers' choice of MSTs is usually directed at the construction of the students' mathematical knowledge in particular fields and of mathematical connections between the fields. At the same time, an explicit requirement to produce multiple solutions encourages the students' learning autonomy. Any a-didactic situation has great potential for teacher LTT because it includes a variety of unpredicted situations. Systematic implementation of MSTs by teachers usually allows the creation of a-didactic (and sometimes non-didactic) situations, which form an effective environment for teacher learning.

Leikin and Levav-Waynberg (2007) showed the complexity of implementing MSTs in school and the gap that exists between theory-based recommendations and school practice in the use of MSTs. Despite this gap, however, even when not planning MSTs, teachers encounter various student solutions in their classrooms and learn from them. This tendency was apparent when teachers were asked to present examples of problems that admit multiple solutions: We realized that about half the examples of MSTs provided by teachers were student-generated. The teachers' ability to connect with student responses during the individual interviews served as an additional indication of the LTT process that takes place in mathematics classrooms. I suggest that systematic implementation of MSTs intensifies both student and teacher learning. Rachel's case demonstrates that when MSTs are part of the didactic contract, teachers learn in the classroom with their students.

Leikin and Levav-Waynberg (2007) showed that teachers' understanding of mathematics and pedagogy within the community of practice is bounded by socially constructed webs of beliefs that determine the teachers' perception of what needs to be done (Roth, 1998; Brown, Collins, & Duguid, 1989). Using MSTs in the classroom is not a simple matter, and sometimes it is difficult because school mathematics is generally result-oriented and topic-centered (Schoenfeld, 1991). The portion of the study presented in this chapter supports the claim that systematic sources are a necessary but not sufficient component of teacher knowledge development when incorporating changes in the school mathematics curriculum. To be able to use didactic tools effectively (e.g., MSTs), teachers must communicate with the communities of practitioners who use and value these tools and implement them in their own practice (Brown et al., 1989).

Relationships Between Pedagogy and Mathematics in LTT

MSTs play a dual role in the studies described in this chapter. They are shown to be an effective didactic tool that develop the connectedness of learners' mathematical knowledge, their flexibility (and hence creativity), and problem-solving expertise. At the same time MSTs are an effective research tool that enables the tracing of the development of teachers' knowledge through implementation of MSTs.

As a research tool, MSTs demonstrated the mutual relationship between teachers' mathematical and pedagogical knowledge (Fig. 1).

Implementation of MSTs requires teacher flexibility, i.e., lesson management that respects student responses. The case of Rachel demonstrated her willingness to provide learners with autonomy in creating proofs for given MSTs, her ability to listen to learners, discuss their proofs, and connect between them. These pedagogical skills created opportunities for Rachel to learn new proofs and new connections, in other words, to learn mathematics. The relationship between the teachers' pedagogical expertise and the development of their mathematical knowledge was also clear in the second study. Teachers who purposefully implemented MSTs by requiring students to provide their own solutions developed problem-solving expertise more than teachers who did not implement MSTs purposefully.

At the same time, Rachel's mathematical knowledge served as the basis for the mathematical activity that she initiated. Her mathematical knowledge provided her with the confidence to direct her lessons toward a variety of unpredicted mathematical directions based on ideas raised by the students. Her mathematical knowledge also allowed her to identify MSTs with broad and rich (i.e., those that include unconventional solutions) solution spaces, reflect on learner solutions, and evaluate their correctness and elegance. When proof 1.5 was presented during the lesson, Rachel intuitively realized that a more elegant proof existed, asked the learners to provide a shorter proof, and added her own to the classroom discussion. This proof was based on her previous problem-solving experience and on the connection between the current task and another mathematical problem solved earlier. As often, her learning continued after the lesson. This learning was based on her mathematical knowledge, her personal mathematical curiosity, and her attentiveness to students' solutions and ideas.

References

Brousseau, G. (1997). *Theory of didactical situations in mathematics*. The Netherlands: Kluwer.

Brown, J. S., Collins, A., & Diguid, P. (1989). Situated cognition and the culture of learning. *Educational Researcher*, *1*, 32–41.

Chazan, D. (2000). *Beyond formulas in mathematics teaching: Dynamics of the high school algebra classroom*. New York, NY: Teachers College.

Krainer, K. (2001). Teachers' growth is more than the growth of individual teachers: The case of Gisela. In F.-L. Lin & T. J. Cooney (Eds.), *Making sense of mathematics teacher education* (pp. 271–293). Dordrecht, The Netherlands: Kluwer Academic Publishers.

Lampert, M. (2001). *Teaching problems and the problems of teaching*. New Haven: Yale University Press.

Leikin, R. (2003). Problem-solving preferences of mathematics teachers. *Journal of Mathematics Teacher Education*, *6*, 297–329.

Leikin, R. (2005a). Qualities of professional dialog: Connecting graduate research on teaching and the undergraduate teachers' program. *International Journal of Mathematical Education in Science and Technology*, 36(1–2), 237–256.

Leikin, R. (2005b). Teachers' learning in teaching: Developing teachers' mathematical knowledge through instructional interactions. The paper presented at the 15th *ICMI*

Study: The Professional Education and Development of Teachers of Mathematics. http://stwww.weizmann.ac.il/G-math/ICMI/log_in.html

Leikin, R. (2006). Learning by teaching: The case of the Sieve of Eratosthenes and one elementary school teacher. In R. Zazkis & S. Campbell (Eds.), *Number theory in mathematics education: Perspectives and prospects* (pp. 115–140). Mahwah, NJ: Erlbaum.

Leikin, R. (2007). Habits of mind associated with advanced mathematical thinking and solution spaces of mathematical tasks. *The fifth conference of the european society for research in mathematics education – CERME-5.* (pp. 2330–2339) (CD-ROM and On-line). Available: http://ermeweb.free.fr/Cerme5.pdf

Leikin, R. (2009). Multiple proof tasks: Teacher practice and teacher education. In F-L. Lin, F-J. Hsieh, G. Hana & M. De Villiers (Eds.), *The proceeding of the 19th ICMI Study conference: Proofs and proving in mathematics education* (vol. 2, pp. 31–36). National Taipei University, Taiwan.

Leikin, R., & Dinur, S. (2007). Teacher flexibility in mathematical discussion. *Journal of Mathematical Behavior, 26*, 328–347.

Leikin, R., & Levav-Waynberg, A. (2007). Exploring mathematics teacher knowledge to explain the gap between theory-based recommendations and school practice in the use of connecting tasks. *Educational Studies in Mathematics, 66*, 349–371.

Leikin, R., & Levav-Waynberg, A. (2008). Solution spaces of multiple-solution connecting tasks as a mirror of the development of mathematics teachers' knowledge. *Canadian Journal of Science, Mathematics and Technology Education, 8*(3), 233–251.

Leikin, R., & Rota, S. (2006). A case study on the development of teacher's proficiency through teaching. *Mathematics Education Research Journal, 18*(3), 44–68.

Mason, J. (2002). *Researching your own practice: The discipline of noticing*. New York: Falmer.

Roth, W.-M. (1998). *Designing communities*. Boston: Kluwer Academic Publication.

Sawyer, R. K. (2004). Creative teaching: collaborative discussion as disciplined improvisation. *Educational researcher, 33*, 12–20.

Schifter, D. (Ed.) (1996) *What's happening in math class? Envisioning new practices through teacher narratives*. Volumes one and two. New York: Teacher College Press.

Schoenfeld, A. H. (1991). On mathematics as sense-making: An informal attack on the unfortunate divorce of formal and informal mathematics. In J. F. Voss, D. N. Perkins, and J. W. Segal (Eds.). *Informal reasoning and education* (pp. 311–343). Hillsdale, NJ: Lawrence Erlbaum.

Simon, A. M. (1997) Developing new models of mathematics teaching: An imperative for research on mathematics teacher development. In E. Fennema & B. Scott-Nelson (Eds.), *Mathematics teachers in transition* (pp. 55–86). Mahwah, NJ: Erlbaum.

Vygotsky, L. S. (1978). *Mind in Society*. Cambridge, MA: Harvard University Press.

Steinbring, H. (1998). Elements of epistemological knowledge for mathematics teachers. *Journal of Mathematics Teacher Education, 1*(2), 157–189.

Schwartz, J. L., Yerushalmy, M & Shternberg, B. (2000). *Geometric Supposer* [Computer software]. CET, Israel.

Simon, M. A. (2000). Research on the development of mathematics teachers: The Teacher Development Experiment. In A. E. Kelly & R. A. Lesh (Eds.), *Research design in mathematics and science education* (pp. 335–359). New Jersey, NJ: Lawrence Erlbaum Associates.

Watson, A., & Mason, J. (2005). *Mathematics as a constructive activity: Learners generating examples*. Mahwah, NJ: Lawrence Erlbaum.

Zaslavsky, O., & Leikin, R. (2004). Professional development of mathematics teacher-educators: Growth through practice. *Journal of Mathematics Teacher Education, 7*, 5–32.

Zazkis, R., & Leikin, R. (2009). Advanced mathematical Knowledge: How is it used in teaching? In the *Proceedings of CERME-6.*

Part II
Examples of Learning Through Teaching: Pedagogical Mathematics

Interlude 1

Part I of this book provided a theoretical background on teachers' learning through teaching (LTT). However, it was not devoid of examples. In fact, the chapters by Leikin, Mason, and Leikin and Zazkis, despite their focus on theoretical or methodological issues, provided numerous examples of teachers' learning. Similarly, the authors of chapters in Parts II and III, despite their focus on particular instances of LTT, provide further theoretical and methodological considerations. They employ a variety of theoretical perspectives as a lens for describing and analyzing examples of learning through teaching.

As mentioned in the introduction to this volume, we found that distinguishing between the mathematics and pedagogy that have been learned by a teacher through teaching was extremely complex. We decided to address this complexity by using the notions of *mathematical pedagogy* and *pedagogical mathematics* as introduced by Mason (2007). In the chapter that opens Part II of this volume, Zazkis explicitly analyses the interrelationship between mathematics and pedagogy as used by the teacher educator with the purpose of developing the mathematics and pedagogy of prospective mathematics teachers. Zazkis demonstrates how the examination of this interrelationship led to mathematical discoveries and didactical insights. The theoretical framework employed in this chapter exemplifies the complexity of distinctions between mathematics and pedagogy that teachers learn and endorses the use of Mason's constructs – *mathematical pedagogy* and *pedagogical mathematics* – in structuring our book.

The chapters in Part II focus on pedagogical mathematics; they describe particular cases of LTT and analyze the ways in which learning occurs.

Pedagogical Mathematics

Let us consider the particular examples presented by the authors.

For several teachers featured in these chapters, learning included solving a new (for them) mathematical problem, or learning a new solution to a known problem, while connecting several mathematical ideas. Several examples of this kind are presented in Part I of this volume. Rachel, the teacher in Leikin's chapter (in Part I), learned a number of new solutions for a given problem in geometry. Consequently, she extended her *solution space* – a notion introduced by Leikin to explain some mechanisms of LTT and is exemplified by Rachel's case. Moreover, Rachel continued her mathematical explorations, based on her lesson with prospective teachers, and discovered new for herself mathematical facts. Shelly and Einat, the teachers in Leikin and Zazkis' chapter, connected calculus with mathematical induction, and a conic section with maximum-minimum problems, respectively. In this same chapter, based on a student's inquiry, Eva extended a given "familiar" theorem to include a special case of an isosceles triangle.

Rina (in Zazkis' chapter) learned a new theorem related to invariances in affine transformations. Michael, the teacher in Kieran and Guzman's study, acquired a new student-generated solution to the task of proving that $(x+1)$ is always a factor of x^n-1. Students' proofs also expanded Michael's solution space. Marcelo and Rubia (in Borba and Zulatto's chapter) developed a new explanation that distinguishes the "look-alike" conic sections of parabola and half-hyperbola. They also designed a technology-based illustration of their explanation. Ms. Alley and Ms. Lewis (in Markus and Chazan's study) enhanced their own knowledge of equations in two variables as they explored this mathematical content for teaching.

Jackiw and Sinclair provide a very unique perspective on learning mathematics. In their study, learning mathematics involves learning mathematical discourse, where the computer plays the role of the traditional student and the teacher's role is given to the hypothetical user-learner of the computer software, which can be either a teacher or a student.

In all these examples of enhanced mathematical knowledge, learning mathematics followed critical pedagogical events: a repeated mistake of students (Zazkis), an unexpected feedback from software (Jackiw & Sinclair), learners' questions or suggestions (Leikin & Zazkis, Leikin, and Borba & Zulatto), and acknowledgement of students' difficulty and a search for ways to help students build their mathematical understanding of the algebra (Markus & Chazan). Implementation of technological tools that allowed learners to engage in mathematical explorations (Kieran & Guzman and Borba & Zulatto) seems to intensify teachers' learning of mathematics. Moreover, we see in researchers' reports (e.g., Kieran & Guzman and Marcus & Chazan) and self-reports (e.g., Zazkis and Borba & Zulatto) that teachers' learning not only followed but also resulted in new pedagogical approaches, activities, or explanations. That is to say, the newly-learned mathematical content became a part of these teachers' pedagogical repertoire.

By considering the examples mentioned above as examples of pedagogical mathematics, we are extending Mason's definition. (Recall: "Pedagogical mathematics involves mathematical explorations useful for, and arising from, pedagogical considerations"). These examples indeed arise from pedagogical considerations, even when these considerations are not mentioned explicitly. They are triggered by interactions with students, the desire to accommodate students' ideas and queries, as well as the flexibility in doing so. However, we see "explorations" as only one of several indicators of pedagogical mathematics. Other indicators include mathematical problem posing, enriching solution spaces of a mathematical problem, extending the repertoire of explanations, reinforcing mathematical discourse, and fostering mathematical connections. These examples also demonstrate strength in teachers' prior mathematical knowledge. This strength is essential in developing and accommodating new ideas and, as such, is essential in developing pedagogical mathematics.

To summarize, in our view, pedagogical mathematics involves a broad range of mathematical objects, actions, activities, and tools constructed in a pedagogical context and useful in teaching. The chapters in Part II instantiate teachers' learning of mathematics through their own teaching, where all notions – teachers, mathematics, and teaching – are interpreted broadly.

References

Mason, J. (2007, June). *Mathematical pedagogy and pedagogical mathematics. Two talks at the pacific institute for mathematical studies 10th anniversary*. Retrieved in March 2009 from http://mcs.open.ac.uk/jhm3/Presentations/%20Presentations%20Page.htm

What Have I Learned: Mathematical Insights and Pedagogical Implications

Rina Zazkis

I cherish the rare opportunity, presented to me in writing this chapter, to reflect on some of my personal learning. As teachers, we often learn by trial and error or by mimicking good examples, and seldom are we explicitly aware of a personal learning that took place. Successful experiences become parts of our instructional repertoire and after years it is impossible to determine how a specific mathematical issue or a specific pedagogical strategy was acquired. We learn to anticipate students' questions, their difficulties and their errors, and then teaching, or navigating learning, resembles a walk on a familiar trail, where sharp turns or other obstacles are either anticipated or avoided.

However, there are some memorable experiences in teaching that shake the routine. Those are the encounters that were not anticipated. Those are experiences that compel us to reconsider or extend our practice, develop a new method or a new assignment, redesign instructional sequence or approach, or, in short, help us learn.

For example, several years ago, following a student's question on why the divisibility test for 7 'works', I have not only proved it, but also developed similar tests for (almost) any number (Zazkis, 1999). Of course, these tests are part of the shared knowledge of the community, and I could have learned them in other ways, but it was the student's curiosity that triggered my learning. In what follows, I share with the reader several such "critical incidents" in teaching that equipped me with new knowledge. As will be clear from the descriptions below, I treat "teaching" rather broadly, where it includes "formal" presentation of the material, as well as issues surrounding in-class interaction with students, attending to students' written work, preparing a lesson or an activity, responding to students' questions, choosing and assigning tasks, among others.

R. Zazkis (✉)
Simon Fraser University, Burnaby, BC, Canada

R. Leikin, R. Zazkis (eds.), *Learning Through Teaching Mathematics*, Mathematics Teacher Education 5, DOI 10.1007/978-90-481-3990-3_5,

Disaggregated Perspective on Learning

A large part of my teaching is within a teacher education program, where we pursue a dual task: To enhance students' mathematical powers and to highlight and exemplify issues pertaining to the pedagogy of mathematics. In Liljedahl, Chernoff, and Zazkis (2007) we illustrated the way of examining the use of tasks in teacher education with a 2×2 array, presented in Fig. 1. We suggested reading the content of the four cells as the "The use of ***x*** to promote understanding of ***Y***."

We acknowledged that "we use our knowledge of mathematics and pedagogy to produce understanding of mathematics and pedagogy within our prospective teachers" (p. 240). The suggested array served to disaggregate "our knowledge of mathematics and use of pedagogy from the mathematical and pedagogical understandings we wish to instill within our students" (ibid.).

I find this array helpful in organizing my personal reflection on learning through teaching and sharing my stories. Because the issue here is that of a personal learning, rather than task development, the interpretation of the cells should be slightly changed to read "I learned ***x*** and I use it to promote students' understanding of ***Y***" or "I learned how to use ***x*** to promote students' understanding of ***Y***."

pM: As a result of instructional interaction I developed a pedagogical approach and I use it to enhance students' mathematics. In particular, I report on classroom conversations pertaining to a horizontal translation of parabola that inspired formal research and development of an untraditional instructional approach, unraveling some of the mystifying nature of transformations.

mM: As a result of teaching, I learned some new (to me) mathematics and I use it to develop students' mathematics. In particular, I share my puzzlement with some of the solutions that students provided when asked to use Affine coordinates to prove geometric statements, my making sense of their incorrect approaches that resulted in correct conclusions, and my attempt to turn this experience into an instructional task.

mP: As a result of teaching and facing unexpected students' difficulties, I developed a mathematical approach that can be seen as exemplification of a general pedagogy. In particular, I remind the reader of a classical puzzle of "the missing dollar" and suggest a mathematical variation that not only demystifies the situation, but also serves to equip students with a powerful strategy for some of their future endeavors.

USAGE \ GOALS	Mathematics (M)	Pedagogy (p)
mathematics (m)	**mM**	**mP**
pedagogy (p)	**pM**	**pP**

Fig. 1 Goals and usage grid

pP – This cell overshadows almost all the activities in teacher education. However, since it leaves mathematics in the background, I leave it out of scope of this chapter.

Story 1: Counterintuitive Translation of Parabola (pM)

In teaching a "methods" course I frequently attend to questions that ask "How would you explain this or that to a student?" For example, how would you explain division by zero? How would you explain to students that $a^0 = 1$? How do you explain to students the multiplication of negative numbers?

Rather often, these and similar questions are presented by teachers, both practicing and prospective, and I use them as an opportunity to provoke conversation. I often turn these questions back to the class in order to gather different ideas and a variety of possible explanations, either before or instead of presenting my own. Teachers know the facts, or the so-called rules, but they do not necessarily understand the logic behind the facts. As such, prospective teachers' questions of "how to explain to students" are, in fact, a disguised question of "how do you make sense of" and a masqueraded plea, "help me make sense."

In one of those "how to explain to students" conversations the following question was posed: "How do you explain that parabola $y=(x–3)^2$ moves right rather than left?" What was meant by "parabola's move" was the location of $y=(x–3)^2$ with respect to the canonical graph of the parabola $y=x^2$. As a habit, before offering any explanation of my own, I referred the question to the class, gathering possible explanations that teachers can offer.

Different explanations were offered and various sources of discomfort were voiced by teachers. What appeared confusing was that the vertical translation of quadratics behaved "as expected," that is, the function $y=x^2–3$ "moved down," relative to $y=x^2$. This further was in accord with the expected behavior of linear function, where $y=x–3$ also moved 3 units down. On the other hand, the behavior of horizontal translation was described as "strange," "unexpected," and "counterintuitive." It was further noted that "unexpected" was only the initial reaction for many teachers, but there was a sufficient amount of combined experience in the room to "know what to expect."

Among various explanations to the phenomenon that were offered none was really convincing. The "problem" of horizontal translation is known in mathematics education research (Baker, Hemenway, & Trigueros, 2000; Borba & Confrey, 1996; Eisenberg & Dreyfus, 1994), however, it did not attract my interest until I experienced personal dissatisfaction with the explanations provided by teachers. As such, I initiated a more formal data collection, exploring a range of explanations provided by high school students and teachers, both practicing and prospective. A detailed account of these data is found in Zazkis, Liljedahl, and Gadowsky (2003). To provoke a conversation about horizontal translation of a parabola, participants were asked first to sketch the graphs of $y=x^2$ and $y=(x–3)^2$ on the same coordinate system, and then to explain the transformation.

All the teachers participating in this study have sketched the graph of $y=(x-3)^2$ correctly. For the practicing teachers it was an immediate and effortless recall from memory, the way one would recall, rather than derive, a basic multiplication fact. However, the prospective teachers needed a few minutes of thinking and checking. It was evident that for some prospective teachers the horizontal translation of a parabola was not in their immediate repertoire of knowledge, but the location of the graph was derived correctly and without major effort.

Teachers' Explanations

While students responding to the same task had a similar tendency to rely on memorized rules, there was a considerable variety in teachers' responses to the interviewer's follow-up request to explain the movement of the parabola. The explanations of prospective teachers did not differ in content or in assortment from those provided by practicing teachers. However, when prompted by the interviewer, the majority of practicing teachers were able to provide more than one explanation. A brief summary of their responses, organized by the themes emerging in teachers' explanations, is presented below.

Citing Rules

A majority of the interviewed teachers referred to the "rule of horizontal translation." According to this "rule," $y=(x-3)^2$ has the same shape as $y=x^2$ but is located 3 units to the right.

To reinforce memorization and to explain function translation to their students several practicing teachers have formulated "the law of opposites." They indicated that having "the law" helped their students in "getting it right." This of course raises issues on the purposes and values of mathematics education. If the purpose is to "get it right" on an exam, then introducing such a law has its merit. However, if the purpose is to teach mathematical thinking, then the creation of such a law jeopardizes the consistency of mathematical structure and directs students' attention to memorization rather than to explanation.

Pointwise Approach

Plugging numbers into the equation, creating a table of values, and then plotting the points seemed more convincing for some teachers than simply accepting what the computer or graphing calculator was showing. These teachers explained that they saw advantage in using the point-by-point creation of $y=(x-3)^2$ as an explanatory tool for their students. Interestingly though, a pointwise approach was not mentioned by any of the students. It appears that the utility and availability of graphing calculators and the lack of extensive experience with creating graphs manually, point by point, influences students' perception of graphs and suggests that the perceived convincing power of teachers' explanations needs to be reexamined.

Attending to Zero and "Making Up"

Another common explanation, suggested by the participating teachers was to find the zero (x=3), or the vertex, of the parabola and imply that the rest of the points are "symmetrically determined" around it. Those teachers were prompted to explain in what way the location of zero would determine the location of the rest of the points. In most cases, preservation of shape and symmetry were put forward as justifications.

Transforming Axes

Additional explanation provided by a few teachers considered transformation of axes rather than transformation of a graph. These participants convinced themselves that (x–3) actually meant moving the Y-axis to the left, and, therefore, the parabola "looked as if" it moved to the right.

Search for Consistency

The standard form in which the parabola is discussed in the local curriculum guide and conventional textbooks is $(y–k) = a(x–h)^2$. It is stated that the value of k determines the size of the vertical translation, while the value of h determines the size of the horizontal translation. However, the direction of the translation is omitted.

A repeated tendency of practicing teachers, likely influenced by this general form, was to have their students consider the parabola $y=x^2-3$ as $y+3=x^2$. In this representation "adding 3 to y" results in a vertical translation in the negative (downward) direction. This view helps in achieving consistency with the horizontal translation, but does not provide an adequate explanation for the move of the graph in the "opposite" (that is, inconsistent with the initial expectation) direction.

In summary, teachers provided a variety of explanations, the most common of which were citing the rules, considering the function pointwise, and attending to the zero of the function. Most teachers were not completely satisfied with their explanations, but claimed to "have never seen a better one." As previously mentioned, there was no significant difference between practicing and preservice teachers in their explanations of the translation. However, preservice teachers' responses differed from the responses of practicing teachers on two accounts. The first is the ease of recall, acknowledged early in this section. Automatic and fluent retrieval is considered to be one of the indicators of expert knowledge (Bransford, Brown, & Cocking, 2000), and therefore, it is not surprising that practicing teachers exhibited better expertise in the subject matter they taught than preservice teachers. Second, as expected by assuming expertise not only in content but also in pedagogical content knowledge, practicing teachers had a better understanding of the perceived inconsistency and of the problematics that a horizontal translation presents to a learner. As a result, they had developed a larger repertoire of explanations as to why the horizontal translation of a function "behaves" as it does.

Pedagogical Approach: Rerouting

While analyzing teachers' explanations, I came up with the scary realization that I did not have myself an explanation with which I was totally satisfied. In an attempt to develop such an explanation, I concluded that the difficulty learners have in understanding horizontal transformations of a function is a consequence of the instructional sequence in which the treatment of transformations of functions is presented. That is, transformations of functions, in general, and of quadratics, in particular, are usually presented in the curricula in the context of exploring functions, rather than in the context of exploring transformations. I suggest that presenting the discussion of translations of functions in the context of transformations may prevent the problem of perceived inconsistency and counterintuitive behavior of functions. In fact, the context for transformations is present in many curricular sequences. For example, to "apply transformations and use symmetry to analyze mathematical situations" is one of the Geometry standards identified by the NCTM for K-12 (NCTM, 2000). Following a progressive sequence for students in different grades, NCTM outlines the following expectation for upper grades:

> understand and represent translations, reflections, rotations and dilations of objects in the plane by using sketches, coordinates, vectors, *function notation* and matrices (Grades 9–12), (p. 308).

It is the function notation briefly mentioned by the NCTM in the Geometry standard for Grades 9–12 that I wish to focus on here. In what follows my main focus is on translations, however, a similar approach can be extended to other transformations as well.

Translation on a Coordinate Plane

Translation on a plane is defined by a vector (or directed segment) that specifies the direction of the motion and the distance. Breaking this motion into its horizontal and vertical components leads to the natural introduction of the formal notation for translation

$$(x, y) \Rightarrow (x + a, y + b) \text{ or } T((x, y)) = (x + a, y + b),$$

where a and b are horizontal and vertical components of the motion, respectively. This function notation for a transformation is often referred to as a mapping rule. Figure 2 illustrates the effect of $T((x,y)) = (x+5, y-3)$ on a set of points in a triangle; A'B'C' (on the right) is the image of ABC.

Once function notation is introduced, students should be given ample opportunity to connect the visual image of translation to the mapping rule. This is achieved by carrying out transformations according to given mappings as well as by identifying mappings according to given visual images. Specifically, students will identify

Fig. 2 $T((x,y)) = (x+5, y-3)$ applied on a triangle ABC

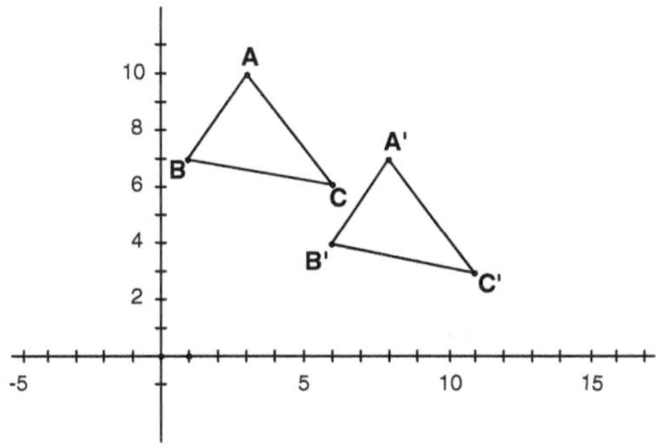

positive and negative values of "a" with motion to the right or to the left, respectively; positive and negative values of "b" with motion up and down, respectively. Furthermore, they will associate the strict horizontal motion with $b=0$ and strict vertical motion with $a=0$.

Since any set of points can be translated according to the mapping rule, this set of points can be a parabola. This is the focus of the next section.

Translating a Parabola

As mentioned above, I wish to situate the discussion of transformations of functions in the context of transformations. Let us start by exploring a translation of a parabola $y=x^2$. Given the experiences described in the previous section, students understand that $T((x,y)) = (x+3, y)$ represents a horizontal translation by 3 units to the right. Now restrict the set of points (x, y) to a canonical parabola and apply T to it. Figure 3 illustrates the effect of $T((x,y)) = (x+3, y)$ on a set of points of the canonical parabola; the image appears to the right of the pre-image.

The task now becomes connecting the translation image to its algebraic representation. Recall that the set of points of the source is described by $y=x^2$. Without loss of generality, focus on a point (a, b) of the source set that was translated to the point (c, d) of the image set. According to the specific translation performed, $d=b$ and $c=a+3$. We wish to connect c and d in an equation. Relating c to d we obtain the following: $d=b$; $c=a+3$ which implies $a=c-3$. However, $b= a^2$, as (a, b) is a point on the source parabola. Substitution leads to $d=(c-3)^2$. Since the above is true for every point of the image set, the image of the translation is described by the equation $y=(x-3)^2$. (Of course, one can work directly with x's and y's, but switching to a's, b's, c's and d's , can be easier for students). This explains the "unexpected" appearance of "–3" in the horizontal translation to the right.

This method can be also used for transformations that are not "counterintuitive" or "problematic." Consider the same approach applied to a vertical translation $T((x,y)) = (x, y+3)$. Focus on a point (a, b) of the source that is translated to point

Fig. 3 $T((x,y)) = (x+3, y)$ applied on the canonical parabola

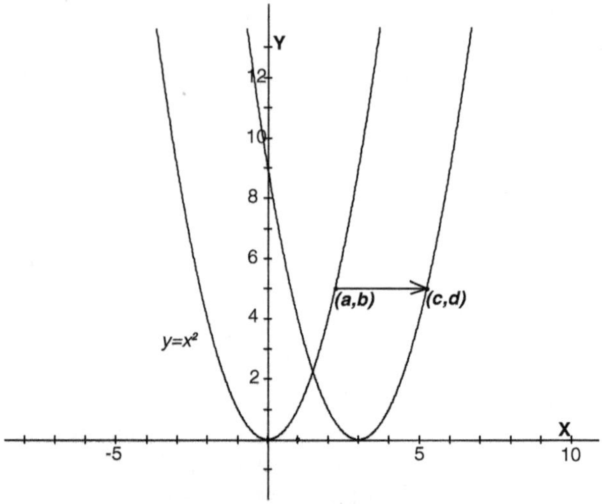

(c, d) of the image (Fig. 4). In this case $a=c$ and $d=b+3$. Points of the source satisfy $b=a^2$. Therefore, $d = b+3 = a^2 + 3 = c^2 + 3$. Since this is true for any point on the image, the set of image points is described by $y = x^2 + 3$, as expected. Figure 4 illustrates the effect of $T((x,y)) = (x, y+3)$ on a set of points of the canonical parabola; the image appears above the pre-image.

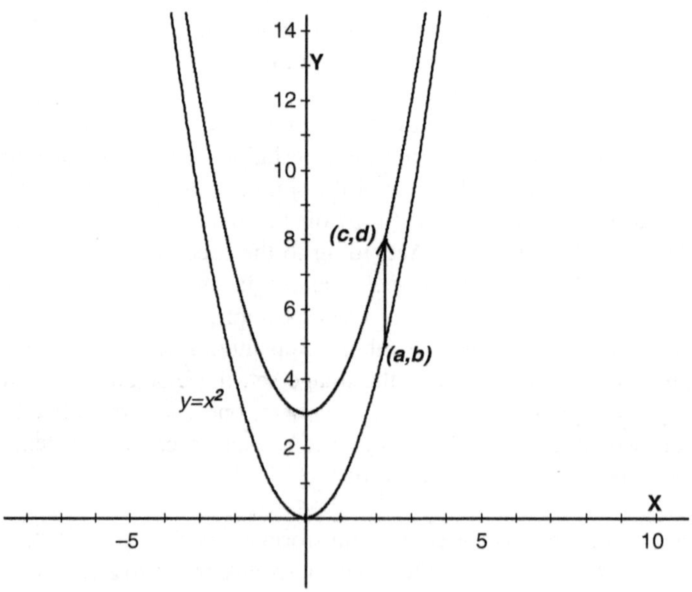

Fig. 4 T((x,y)) = (x, y+3) applied on the canonical parabola

"Waw" – is the usual reaction of teachers once this approach is presented. This exclamation is accompanied with reactions like, "finally this makes sense" or "Why didn't they explain this before?" Who the mysterious "they" are is not elaborated upon.

Returning to the discussion on learning through teaching, and the 4-cell matrix, the above example illustrates the didactical approach I developed in order to reinforce teachers' mathematical connections (pM). Of course, consideration of a geometric transformation as a starting point is not limited to translations. It is desirable for students and teachers alike to explore a variety of transformations, making a connection between the conventional definition of a transformation and its effect on specific sets of points on the plane. Such sets could be simple geometric shapes in the initial stages of exploration and graphs of specific functions or relationships in later stages.

Searching for Consistency Again: Illusion in a Linear Transformation

It was mentioned above that by attending to the general form of parabola and seeing $y=x^2-3$ as $y+3=x^2$ some learners achieved consistency within initially counterintuitive transformations. That is, "adding 3" to y results in a downward transformation, while "adding 3" to x results on a horizontal transformation "left," while the direction associated with addition is a positive one, upward, and right.

The question remains, what about linear functions? Everyone was indoctrinated to the idea that $y=x-3$ "moves down" in relation to $y=x$. But does it? A serious scrutiny of this idea will detect an inconsistency on two counts. First, a change in "x" is expected to result in a horizontal translation, not a vertical one. Second, following the detailed discussion connecting translations to the function notation, there is a new expectation that "subtracting" should result in a move in the positive direction.

The consistency can be achieved in two ways. First, $y=x-3$ can be seen as $y+3 = x$. With this view the consistency is achieved: The translation is vertical, as expected with manipulating "y" and the direction is negative, as expected with "+3." The second view requires imagination. In fact, what we are used to seeing as a move down can also be seen as a move to the right, as shown in Fig. 5. The consideration of the graph of a line as a whole, rather than pointwise, disguises the possibility of such an interpretation. However, this ensures a complete consistency with the perceived counterintuitive translation of a parabola: Substituting "$x-3$" in place of "x" results in a horizontal move in the positive direction.

What have I learned in this experience? I have developed an answer to "How do you explain that $(x-3)^2$ moves right rather than left?" To address this question I learned to explicate a mathematical connection by the means of a new instructional sequence. This mathematical connection resulted in a pedagogical strategy and a didactical sequence designed to support students' and teachers' mathematics (pM).

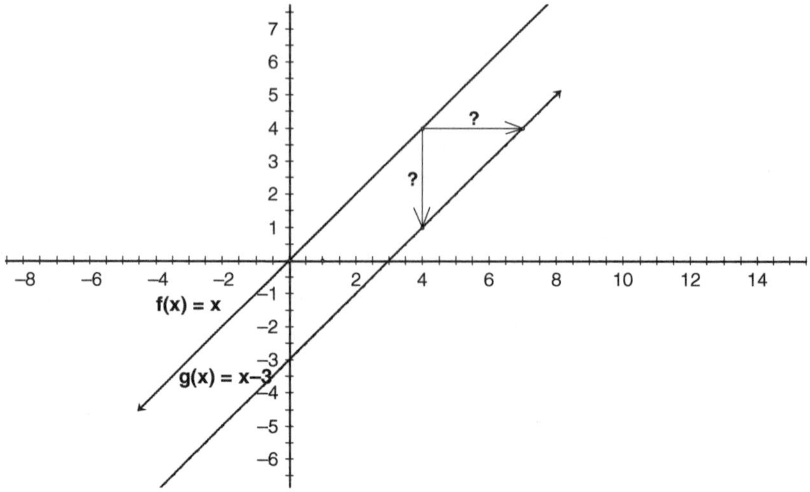

Fig. 5 Considering $g(x) = x{-}3$ with respect to $f(x) = x$

Story 2: Geometry with Affine Coordinates (mM)

In a problem solving class I often introduce students to Affine coordinates, as a means to extend their notion of coordinate system and also as a tool for some proofs in geometry.

Affine Coordinates: An Introduction

The familiar Cartesian coordinate system is an orthogonal homogeneous system, that is, on a plane it is defined by a choice of two perpendicular lines, of a positive direction on each line and of a unit that is equal on both lines. This coordinate system induces one-to-one correspondence between ordered pairs of real numbers and points on the plane. However, the Cartesian system of coordinates is only one particular example of Affine coordinates, which can be neither orthogonal nor homogeneous. Affine coordinate system is defined on a plane by any three non-collinear points (O, I, J). Lines OI and OJ establish two axes intersecting at the point of origin O, the directed segments OI and OJ determine the positive direction and a unit on each one of the axes (Fehr, Fey, & Hill, 1973). Coordinates of a point on a plane are found by parallel projection, that is, by drawing parallel lines to the axes through this point, and noting points of intersection of these lines with the axes (Fig. 6).

Among the first tasks presented to teachers when introducing an Affine coordinate system is to assign a correspondence between points on a plane and their coordinates. But even this simple task is challenging at first to individuals "stuck"

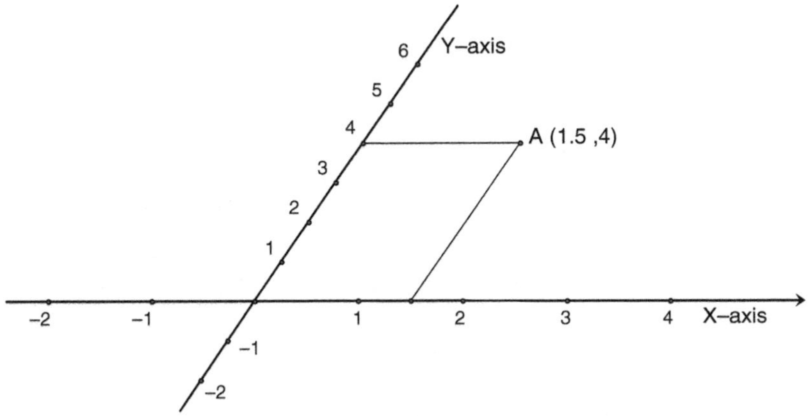

Fig. 6 Determining coordinates of a point in an Affine plane

in the Cartesian plane. For example, in describing how to find the coordinates of a point on a plane, many teachers attempt to "draw a perpendicular from the point to the axes." Of course, this description is valid only when the axes are orthogonal. Facing the limitations of this description, and striving for an appropriate one, can be a starting point for a proof that within an Affine system of coordinates, every point on a plane has a unique representation as an ordered pair of numbers and vice versa.

After becoming familiar with plotting points and assigning coordinates, teachers attend to graphs of polynomial functions of the first degree, or so-called linear equations. This may be a good opportunity to reconsider their knowledge of linear equations plotted with the Cartesian coordinates and to explore what holds for any Affine coordinate system. Possible questions to be considered are the following: What is defined by parameters m and b in $y=mx+b$? Do two lines have a unique point of intersection? Does the graph of $y=x$ form a 45 degrees angle with the positive direction of the X-axis? If not, how can this angle be determined? What predicts how steep the slope is?

Exploring Medians

One basic notion that invites re-examination with the help of Affine coordinates is that of length. In my teaching I asked teachers to prove the well-known theorem about the medians of a triangle by using the sides of a triangle to assign the system of Affine coordinates (see Fig. 7). The theorem states that (a) the three medians intersect at one point and (b) the ratio of the two segments on the median determined by this point is 2:1.

Embedding the triangle in an Affine coordinate system, as shown in Fig. 7, the equations of the three lines are given by the following:

Fig. 7 Medians in a triangle *ABC*

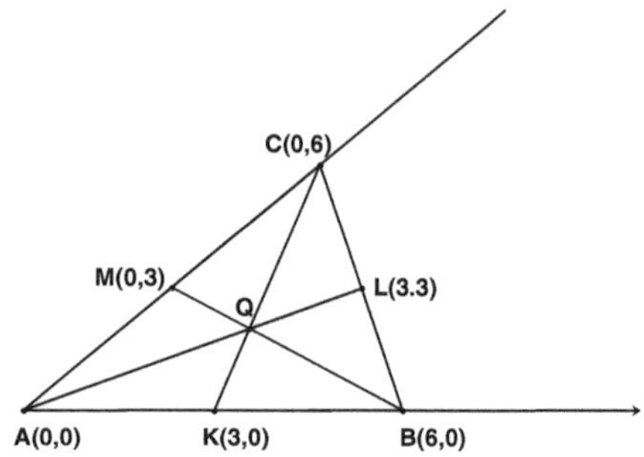

$$AL\colon y = x,$$
$$BM\colon y = -{}^{1}\!/_{2}\,x + 3,$$
$$CK\colon y = -2x + 6.$$

And the intersection point of each pair of these lines is calculated to be (2, 2). This completes the proof of (a).

Considering (b), after calculating the coordinates of the intersection Q, teachers had to prove that $AQ{:}QL = BQ{:}QM = CQ{:}QK = 2{:}1$. Surprisingly or not, most teachers used the "distance formula" $\left(d = \sqrt{(x_1 - x_2)^2 + (y_1 - y_2)^2}\right)$ to determine the lengths of the segments.

For example,

$$CQ = \sqrt{(0-2)^2 + (6-2)^2} = \sqrt{4+16} = \sqrt{20} = 2\sqrt{5},$$
$$QK = \sqrt{(2-3)^2 + (2-0)^2} = \sqrt{1+4} = \sqrt{5},$$
$$CQ : \mathrm{QK} = 2{:}1$$

or

$$BQ = \sqrt{(6-2)^2 + (0-2)^2} = \sqrt{16+4} = \sqrt{20} = 2\sqrt{5},$$
$$QM = \sqrt{(2-0)^2 + (2-3)^2} = \sqrt{4+1} = \sqrt{5},$$
$$BQ : \mathrm{QM} = 2{:}1.$$

These calculations led to an exciting conversation of the origins of the formula, its scope of applicability, and of the notion of "length," in general. In this particular case the unit of length is determined only with reference to the particular axis, and, therefore, there cannot be a numerical assignment to the length of a given segment. Nevertheless, comparison of lengths in terms of ratios is possible by attending to the difference in *X*- and *Y*- coordinates, respectively. In students' solutions a correct statement was proven using an incorrect method. However, why did this "work"? – this question puzzled me.

I was aware that the proof presented above related only to a particular case of an isosceles right angle triangle in which the equal sides are located on the *X* and *Y* axes. Why was this relationship of lengths preserved? I was further perplexed facing an unexpected solution to a more difficult problem, presented in the next section.

Exploring Tridians

Problem *Similarly to a median, we define a* tridian *as a segment that connects a vertex of a triangle to a point on the opposite side that marks 1/3 of the side's length. As shown in Fig. 8, BX, BY, AW, AZ, CS and CT are tridians in a triangle ABC. Tridians generate all kinds of interesting relationships at various points of intersection. They also generate interesting relationships of various areas they cut. With the help of the Geometer's Sketchpad, make several conjectures about tridians in a triangle. Formulate and prove several of your conjectures. In particular, prove that Area (KLM) : Area (ABC)* = 1:7

Again, I asked students to develop their proofs relying on Affine coordinates. However, a frequent inappropriate solution was to calculate the lengths of the sides of the two triangles using the above mentioned "distance formula", and then use Heron's formula[1] to calculate the areas of triangles. Because of repeated approximations in the use of square roots, this solution did not result in a "clean" ratio, but offered a reasonable approximation.

Unhappy with these approximations, Isaac, one of the best students in class, suggested the following:

To avoid calculations with fractions, and without loss of generality, the following coordinates were assigned: $A(0,0)$ $B\,(0, 21)$ and $C\,(21,0)$. The equations of the lines, based on the pairs of points, were calculated and points of intersection were found: $K(3,12)$ $L(6,3)$ $M(12,6)$.

Until this point the solution proceeded as expected, correctly involving the use of Affine coordinates. But then Isaac used the following formula, that he recalled from his Linear Algebra class, to determine the areas of the triangles:

If vertex A is located at the origin (0, 0), and the coordinates of the other two vertices are given by $B = (x_b\,,\,y_b)$ and $C = (x_c, y_c)$, then the area S can be computed as ½ times the absolute value of the determinant

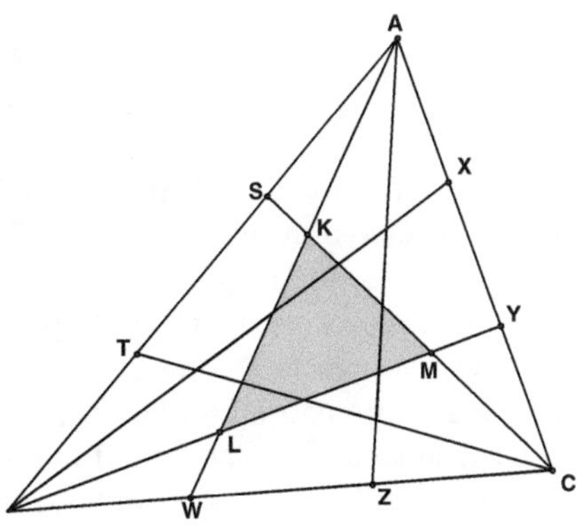

Fig. 8 Trideans in a triangle *ABC*

[1] For a triangle with sides a, b, c, the area A is determined by $A = \sqrt{s(s-a)(s-b)(s-c)}$, where s is a semiperimeter of the triangle, $s = \frac{a+b+c}{2}$.

$$\text{Area}\,(ABC) = \frac{1}{2}\left|\det\begin{pmatrix} x_b & x \\ y_b & y_c \end{pmatrix}\right| = \frac{1}{2}\left|x_b y_c - x_c y_b\right|.$$

Further, for three general vertices, the equation is as given below:

$$\text{Area}\,(ABC) = \frac{1}{2}\left|\det\begin{pmatrix} x_a & x_b & x_c \\ y_a & y_b & y_c \\ 1 & 1 & 1 \end{pmatrix}\right| = \frac{1}{2}\left|(x_c - x_a)(y_b - y_a) - (x_b - x_a)(y_c - y_a)\right|.$$

Applying these formulas to the triangles *ABC* and *KLM* (as labeled in Fig. 8), Isaac drew the following conclusion:

$\text{Area}\,(ABC) = \frac{1}{2}|21^2| = 220.5,$
$\text{Area}\,(KLM) = \frac{1}{2}|(12-3)(3-12) - (6-3)(6-12)| = \frac{1}{2}|-81+18|$
$=\frac{1}{2}\times 63 = 31.5,$
$31.5 \times 7 = 220.5$, therefore, $KLM{:}ABC = 1{:}7$
QED.

Though the use of tools from Linear Algebra was commendable, the method, though leading towards a correct ratio, was obviously wrong. The formulas that Isaac applied are based on the Cartesian coordinates and are not applicable in case of general Affine coordinates. In fact, similarly to the case of medians, Isaac proved one specific case of isosceles right angle triangles (derived by interpreting the coordinates of *ABC* in the Cartesian plane), but not the general case.

As mentioned above, I wondered why solutions for the specific cases, inappropriately provided by students, determined correct relationships for the general case. I turned to books and found help from no one other than Archimedes himself! Indeed, he did a lot of mathematics besides cry Eureka! (Stein, 1999) I learned that the general affine mapping is described as

$$x' = ax + by + e$$
$$y' = ax + dy + f.$$

In this sense, the transformation of Cartesian coordinates to Affine coordinates can be considered as an Affine mapping. Though lengths are not preserved in an Affine mapping, the mapping magnifies parallel line segments by the same factor and as such preserves the ratios of parallel line segments. Further, an Affine mapping does not preserve areas, but it preserves ratios of areas. This property is best illustrated by transforming the grid of squares to the grid of parallelograms (See Fig. 9). The idea that each one of the squares, that can be made as small as necessary, corresponds to similar parallelograms is the intuitive basis for the preservation of the ratio of areas.

More formally, adopting the proof from Stein (1999) the mapping given by

$T(x,y) = (ax + by + e, cx + dy + f)$ magnifies areas by $|ad{-}bc|$

Fig. 9 Intuitive view of an Affine transformation

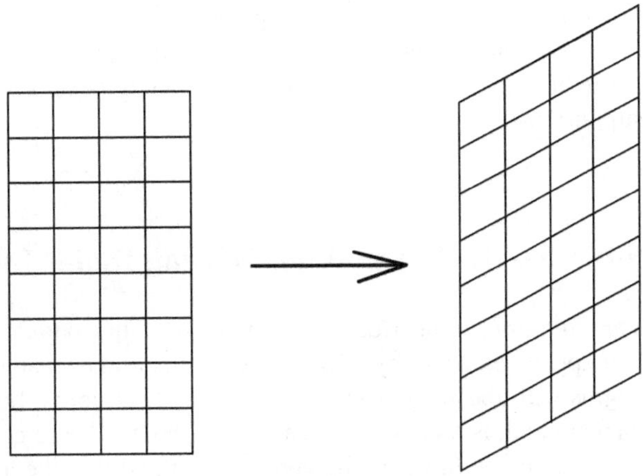

This is shown by considering a triangle *PQR*, with Cartesian vertices *P*(0,0) *Q*(1,0) *R* (0,1) that is transformed by *T* to triangle *P'Q'R'* with vertices *P'*(*e,f*) *Q'*(*a+e, c+f*) *R'*(*b+e, d+f*).

The area of the triangle *P'Q'R'* is given by considering the difference between the area of rectangle with sides *a* and *d* and the 3 triangles that complete *P'Q'R'* to this rectangle, as shown in Fig. 10.

$$\text{As such, Area } (P'Q'R') = ad - [{}^1/_2\, ac + {}^1/_2 bd + {}^1/_2 (a-b)(b-c)] = \\ = ad - [{}^1/_2 \text{ ad} + {}^1/_2 \text{bc}] = (ad-bc)/2.$$

The area of *PQR* is ½, therefore, comparing the areas of *PQR* and *P'Q'R'*, the magnifying factor of *T* for areas is (*ad-bc*). This explains why the ratios proved by students hold in a general case, despite inappropriately invoking tools that assume Cartesian coordinates.

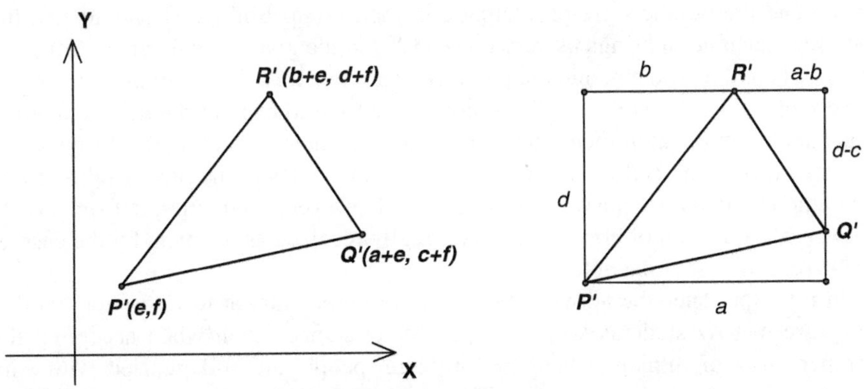

Fig. 10 Image of an Affine transformation of a triangle

What have I learned in this experience? To make sense of students' unanticipated errors I learned a new (to me) chapter in mathematics. The question that puzzled me I turn now to my students, and use my newly acquired mathematical knowledge to support theirs (mM).

Story 3 – Bellboy and the Missing Dollar (mP)

The "missing dollar riddle" or "missing dollar paradox" is a famous brain teaser that appears in a variety of published collection of mathematical puzzles. The riddle begins with the story of three men who check into a hotel. The cost of their room, they are told, is $30. So, they each contribute $10 and go upstairs. Later the manager realizes that he has overcharged the men and that the actual cost should have been only $25. The manager promptly sends the bellboy upstairs to return the extra $5 to the men. The bellboy, however, decides to cheat the men and pockets $2 for himself and returns $1 to each of the men. As a result, each man has now paid $9 to stay in the room (3 × $9 = $27) and the bellboy has pocketed $2 ($27 + $2 = $29). The men initially paid $30, so the question is *where is the missing dollar?*

Another version of this riddle changes the scene and the players – three ladies go to a restaurant for a meal. They receive a bill for $30. They each put $10 on the table, which the waiter collects and takes to the till. The cashier informs the waiter that the bill should only have been for $25 and returns $5 to the waiter in $1 coins. On the way back to the table the waiter realizes that he cannot divide the coins equally between the ladies. As they did not know the total of the revised bill, he decides to put $2 in his own pocket and to give each of the ladies $1. Now that each lady has been given a dollar back, each of the ladies has paid $9. Three times 9 is 27. The waiter has $2 in his pocket. Two plus 27 is 29. The ladies originally handed over $30. Where is the missing dollar?

The setting, the characters, and the currency involved is being altered depending on the setting in which the story is presented. What remains invariant are the numbers – and the numbers are problematic in their compatibility. That is to say, the incorrect calculation brings us very close ($29) to the given initial value ($30), and that is where the problem, and the perceived paradox, lies. A variety of experts on a variety of websites and forum discussions have tried to explain the miscalculation. The paradox in the aforementioned situations is created by adding the $2 pocketed by the waiter or the bellboy to the $27 paid by the ladies or the men. Adding these two amounts does not answer any question. However, subtracting 2 from the 27 answers the question of how much was actually received as payment by the cashier or the receptionist at the hotel desk.

In my experience the above explanation, or others similar to it, do not "work." They are met by students with a degree of resistance. Even when accepting the "proper" way of thinking about the problem, people are still puzzled with why "the other" way creates a paradox. The essence of the puzzle lies with the difference between the $29 that the story mentions and the desired initial $30, and so the search for the missing dollar continues. This is why the puzzle has survived for so many generations and, most likely, will continue to intrigue curious minds for many generations to come.

Pedagogical Approach Via Mathematical Variation

My students, prospective teachers, wondered how the puzzle can be best explained to students, identifying in their questions personal unease with the situation. I sought a way to offer an explanation that would be more accessible for students and teachers alike. Unlike other explanations, which stay with the story, I decided to alter the story not by introducing an alternative setting, but by implementing a numerical change. That is, I offered a different story – that keeps the same story plot, but uses different numbers.

Let us say the room cost only $20 and the bellboy was sent to return $10 to the men. For simplicity of division, he pocketed $1 and returned $3 to each of the men. In this situation the men paid $7 each, for the total of $21. The bellboy has $1. Adding the actual payment to the one pocketed dollar gives us $22. Would it make sense to suggest, starting with the initial collection of $30, that $8 is missing?

And if this is not convincing enough, it is possible to change the numbers in the story once again, giving the men a "Stay with us for 1/3 the price" coupon and sending the bellboy to return to them $20. By now, knowing the bellboy's desire for a simple and fair division, we have him pocket $2 and return $18 to the men, $6 each. In this situation the men paid $4 each, for a total of $12. The bellboy has $2. Adding the actual payment to the 2 pocketed dollars gives us $14. Would it make sense to suggest, starting with the initial collection of $30, that $16 are missing?

Varying numbers, whether large or small, helps in making sense of the situation. Numerical variation of the story could be more convincing than any attempt to explain the original one. The absurdity of the missing dollar in the original situation is brought to surface when we establish the general structure of adding the paid amount to the pocketed amount. If the general structure of "missing money" makes no sense, then neither does its specific example of the "missing dollar."

I suggest that dealing with the arithmetic of the specific puzzle introduces teachers to a valuable instructional strategy – that of numerical variation. As such, I present this as an example of mP – of using mathematics to enhance pedagogy. Once the strategy is recognized and adopted, it can be implemented by teachers in a variety of situations.

Example 1 – Division with Decimals

A pound of black sand cost $1.72. How much sand can you buy for $0.40?

When this or a similar problem is presented to either middle school students or preservice elementary school teachers, there is a significant number of individuals who make errors in setting up the division statement, that is, dividing 1.72 by 0.40

rather than 0.40 by 1.72. What is the best way to help them? Of course, pointing to their error is not helpful beyond the given problem.

The general multiplicative structure that a learner needs to recognize in order to solve this problem is the following: A pound of black sand costs *A*, how much sand can you buy for *B*? This is an example of a more general form of quotative (measurement) division, that is, division structure that determines how many times can *A* fit into *B* or how *B* can be measured with *A*.

Once the structure is recognized, the solution is given by $B \div A$. The question, however, is what is it, that can guide learners towards seeing the generality in this particular case (Mason & Pimm, 1984)? What I learned from teaching is that the best explanation is in changing the numbers. This can be seen in Polya's tradition as thinking of a simpler but similar problem. Here the similarity is obvious, and the simplicity is achieved by introducing compatible numbers.

> A pound of black sand cost \$2, how much sand can you buy for \$6?
> A pound of black sand cost \$2, how much sand can you buy for \$20?

The numbers in these examples are compatible, that is, easily manipulated and work well together. Learners seldom have problems with these kinds of questions, so using them as a starting point is beneficial. Once the general structure is established, it is possible to move to "more problematic" numbers involving fractions.

> A pound of black sand cost \$2, how much sand can you buy for \$0.50?

And then gradually return to the original problem.

This strategy can be seen as a modification of the "structured variation grids" (Mason, 2001; Mason, 2007) in that it is a gradual numerical variation for the purpose of prompting recognition of structure. So, why is the structure more readily recognized when numbers are compatible than when they are not? In Mason's terms, the source of the obstacle appears to be in the perceived range of permissible change. That is, the numbers in the initial problem are "too far" from the students' example space of problems that are associated, implicitly, with measurement division. Numerical variation assists in recognizing similarities and extending the general structure, a necessary step for the solution.

Example 2: "Big" Percentage

When someone wants to declare absolute confidence, he may claim being "120% sure." When someone wants to acknowledge an effort needed for a certain task, she may claim devoting to it 200% of her energy or 150% of her time. We accept these claims with a smile, as a tendency to emphasize a certainty or an effort rather than an accurate measure.

When a whole is 100%, what is indicated by a percentage higher than 100? Experience shows that when a high percentage appears in a mathematical problem situation, it often leads the learners away from recognizing the general structure. Consider for example the following problem:

> The price of a bottle of wine was \$10. However, as the wine got older, its price increased by 400%. What is the new price?

In a class of preservice elementary school teachers, about half of the students claimed that the new price was \$40, explaining that 400% meant "quadrupling." Once again, what we found helpful towards recognizing the general strategy is numerical variation:

> The price of a bottle of wine was \$10. It increased by 20%, what is the new price?
> The price of a bottle of wine was \$10. It increased by 35%, what is the new price?
> The price of a bottle of wine was \$10. It increased by 100%, what is the price now?

From considering students' approaches to the problem, it appears that their main difficulty is with the perceived range of permissible change (Mason, 2001, 2007). While 20%, 35%, or even 100% fits within what is expected – both in a familiar "real world" context and in a mathematics classroom context – the increase of 400% appears beyond a "reasonable" permissible change. Yet again, the purpose of the numerical variation is not only to help with a specific task at hand, but also to develop appreciation of this general pedagogical strategy.

What have I learned from this experience? In an effort to explain a puzzling situation, I learned to appreciate the explanatory power of compatible numbers. I was further able to extend this power to more common and curriculum based situation, and in such to equip teachers with a tool of numerical variation that can serve as a helpful pedagogy (mP).

Conclusion

As seen from the stories I shared, my learning through teaching was triggered by interaction with students' questions and students' work on problems. As such, it also became learning *for* teaching. That is to say, what I learned has become an integral part of my "teaching repertoire." I thank my students for the opportunity to learn. It would not have happened without them.

References

Baker, B., Hemenway, C., & Trigueros, M (2000). On transformations of basic functions. In H. Chick, K. Stacey, J. Vincent, & J. Vincent (Eds.), *Proceedings of the 12th ICMI study conference on the future of the teaching and learning of Algebra* (Vol. 1, pp. 41–47). Melbourne: University of Melbourne.

Borba, M. C., & Confrey, J. (1996). A students' construction of transformations of functions in a multirepresentational environment. *Educational Studies in Mathematics*, *31*(3), 319–337.

Bransford, J. D., Brown, A. L., & Cocking, R. R. (2000). *How people learn: Brain, mind, experience and school.* Washington, DC: National Academy Press.

Eisenberg, T., & Dreyfus, T. (1994). On understanding how students learn to visualize function transformations. In E. Dubinsky. A. Schoenfeld, & J. Kaput (Eds.), *Research in collegiate mathematics education* (Vol. 1, pp. 45–68). Providence, RI: American Mathematical Society.

Fehr, H., Fey, J., & Hill, T. (1973). *Unified mathematics.* Menlo Park, CA: Addison Wesley.

National Council of Teachers of Mathematics [NCTM]. (2002). *Principles and standards for school mathematics.* Reston, VA: NCTM.

Liljedahl, P., Chernoff. E., & Zazkis, R. (2007). Interweaving mathematics and pedagogy in task design: A tale of one task. *Journal of Mathematics Teacher Education*, *10*(4–6), 239–249.

Mason, J. (2001). Tunja Sequences as example of employing students; power to generalize. *Mathematics Teacher*, *94*(3), 164–169.

Mason, J. (2007). Structured variation grids. Available: http://msc.open.ac.uk/jhm3/

Mason, J., & Pimm, D. (1984). Generic examples: Seeing the general in the particular. *Educational Studies in Mathematics*, *15*, 277–289.

Stein, S. (1999). *Archimedes: What did he do besides cry Eureka.* Mathematical Association of America Washi.

Zazkis, R. (1999). Divisibility: A problem solving approach through generalizing and specializing. *Humanistic Mathematics Network Journal*, (21), 34–38.

Zazkis, R., Liljedahl, P., & Gadowsky, K. (2003). Students' conceptions of function translation: Obstacles, intuitions and rerouting. *Journal of Mathematical Behavior*, *22*, 437–450.34.

Dialogical Education and Learning Mathematics Online from Teachers

Marcelo C. Borba and Rúbia B.A. Zulatto

Introduction

More than 50 years ago, Paulo Freire (2005) proposed the notion that education is a dialogical process. The notion of dialogue for him is paramount and implies being open to listening deeply and responding with intentionality. Traditionally, teaching has emphasized ways in which students learn from teachers, which for Freire, is only one side of the coin.

This book is dedicated to analysis of how teachers learn from teaching. In this chapter, we will examine this phenomenon in the context of online education courses, discussing how we learn from teachers when they are in the role of students enrolled in continuing education courses. In this chapter, we will show how listening deeply to teachers has made us think about mathematical problems we have never thought about before and, consequently, learn mathematics in the process. Although we also learned much regarding pedagogical content during the course, it will not be the focus of this chapter.

We will also show how university professors who, like ourselves, are engaged in teaching teachers online how to use geometry software in face-to-face classrooms, need to be open to taking risks and being pushed beyond their comfort zone. In particular, we will emphasize that the risk is greater once the decision has been made to adopt an interactive-dialogical approach for an online course. However we will also argue that, once the virtual community has become dialogical, the risks diminish. We even suggest that one can grow accustomed to the risk and feel more comfortable with it.

Before we present the example about conics, we will present our theoretical perspective regarding the use of information and communication technology and of dialogical teaching education, as well as the context of the online course. We will then show how the problem-solving dynamic we set up for the course led one of the teachers to pose a problem that initially none of the participants knew how to solve.

M.C. Borba (✉)
GPIMEM and UNESP, Rio Claro-SP, Brazil
e-mail: mborba@ms.rc.unesp.br

R. Leikin, R. Zazkis (eds.), *Learning Through Teaching Mathematics*, Mathematics Teacher Education 5, DOI 10.1007/978-90-481-3990-3_6,

The research project that yielded the results presented here was born when a nation-wide network of schools owned by a private foundation approached us to help their teachers learn how to use geometry software. The schools had good computer laboratories, but they were surprised to find that the teachers were not using them. Studies have shown that the mere availability of laboratories is not sufficient to guarantee the implementation of the use of computers in regular elementary education. Since we had experience in implementing the use of software at the middle and high school levels as well as university levels, we accepted their invitation. We also proposed that, following an initial face-to-face course, we seek to use online courses as an arena for developing dialogical relationships with teachers. Our research group, GPIMEM,[1] based at UNESP (São Paulo State University), Rio Claro, Brazil, has developed a culture of software use in mathematics education for more than 15 years and, since 1999, has been engaged in investigating the possibilities of the online world in mathematics education.

We accepted the invitation to teach the online course and, at the same time, initiated a research project to study the possibilities of the online world as a means for continuing education. Since the first experience of the online course received a positive evaluation by the network of schools, we ended up teaching a total of eight online courses. This project also composes part of a broader research agenda aimed at documenting the role of different technologies in transforming mathematical practices developed by collectives of teachers, students, and technology, or as we have labeled it, collectives of humans-with-media (Borba & Villarreal, 2005). Although the goal of this project did not initially include the focus of this chapter, the reader will see how the example and the discussion presented follow from previous research we have developed regarding how risky it can be to implement technology in the classroom.

From Inter-shaping Relationship to the Notion of Humans-with-Media

Within the debate about the role of technology in cognition in the mathematics education community, Borba (1993, 1995) documented a series of examples in which students used software in ways that were completely foreign to the concerns of the *Function Probe*^{TM} (Confrey, 1991) design team, which he belonged to led by Jere Confrey at Cornell University, USA. The participation in the *Function Probe* design team and the development of experimental research with the software provided a singular opportunity for observing how students were shaping the software, and not only being shaped by it. While it was widely acknowledged at that time that software could influence students' cognition, there was little focus on the way students

[1]Technology, other Media and Mathematics Education Research Group. www.rc.unesp.br/igce/pgem.gpimem.html

might be transforming the software (such as Function Probe or another function or geometry software).

Inspired by Freire (2005) and Schutz (see Wagner, 1979), Borba developed the idea of a dialogical relationship between software designers and users, even though such a relationship is not synchronous like the human dialogue that Freire and Schutz believed to be so paramount. Based on these ideas and analysis of the role of different media (paper and pencil, graphing calculators, software, etc.) in the development of different kinds of mathematics (Borba & Villarreal, 2005), Borba and his colleagues in the GPIMEM research group came to view mathematical knowledge as being the product of collectives of humans-with-media. In this view, material culture such as inscriptions, (oral language), paper-and-pencil, books, and computer technology play an important role. Computer technology reorganizes human thinking in a way that is qualitatively different from language (Tikhomirov, 1981), as it allows for different ways of extending human memory (Levy, 1993). Thus, the notion of humans-with-media emphasizes an epistemological perspective that views knowledge as being constructed by a collective composed of humans and material artifacts. In this view, the artifacts – material culture – play an important role in knowledge production.

Knowledge is produced by humans, but also by different media such as, writing, or the new modalities of language that emerge from computer technology.

> We believe that humans-with-media, humans-media or humans-with-technologies are metaphors that can lead to insights regarding how the production of knowledge itself takes place (. . .). This metaphor synthesizes a view of cognition and of the history of technology that makes it possible to analyze the participation of new information technology 'actors' in these thinking collectives[2] (Borba & Villarreal, 2005, p. 23).

In the example that will be discussed in this chapter, we will argue that the geometry software and the platform used were also co-actors in an online collaboration in which the leaders of the course learned some mathematics with the teachers who were learning how to put geometry software to use in their regular face-to-face middle and high school classrooms throughout Brazil.

Making the Risk Zone More Comfortable: Online Courses to Teach How to Use Software in the Mathematics Classroom

Use of Information and Communication Technologies (ICTs) in the mathematics classroom has increased in recent years, contributing to growing debate regarding the possibilities and difficulties of its use by teachers. Authors including Borba and Penteado (2001), Zulatto (2002), and Kaput (1989) have highlighted the importance of facing the challenge of thinking about mathematics with ICT resources.

---

[2]Thinking Collective is a term used by Levy to emphasize that knowledge is produced by collectives composed of humans and non-human actors.

Penteado (2001) emphasizes that in conventional/traditional classrooms teachers often prefer to remain within a *comfort zone* in which they are able to predict and control almost everything; and although they may not be satisfied with the situation, they prefer to remain within this comfort zone rather than facing the challenge of entering a risk zone. This occurs because, in the latter, teachers experience situations that are not common in their practice, which makes them feel threatened because they are unable to foresee a path to follow.

Borba and Penteado (2001) present two examples of how teachers can enter the risk zone when they decide to use technology. The teacher's control in relation to the knowledge and the way the lesson develops, for example, becomes less constant, usually due to technical problems and the variety of paths the students may follow. In addition, questions often arise that teachers may not immediately be able to respond to, obliging them to say "I don't know," which was generally uncommon in their day-to-day practice.

Power relations are also affected by the presence of ICTs in the classroom, as information that used to be largely provided by the teacher and the textbook has become more readily accessible to students through software programs or the Internet. At times, students know more about the computer than teachers do, and it is necessary to recognize, as Penteado (2000, p. 31) puts it, that with the use of ICTs in the classroom, "the power legitimated by the domination of information is not only in the hands of the teacher: students gain increasingly larger spaces in the process of negotiation in the classroom." Students often know how to find original paths in the world of informatics to obtain certain information and generate doubts that even the teacher is unable to respond to.

Out of fear and insecurity, among other aspects, many teachers opt not to face the challenges of adopting ICTs in the classroom, or merely modify their lessons to include closed-ended activities with no room for exploration. The questions that emerge tend to be much like those that emerge in their conventional classes and are, therefore, easier for them to respond to. In other words, teachers may attempt to "domesticate" ICT, adapting their use as closely as possible to the way they are accustomed to using paper and pencil. As a result, they fail to take advantage of the great potential of ICTs. While this discussion is somewhat outdated, it still merits attention, since working in the risk zone continues to be a challenge for some teachers. Here we propose considering it in the context of online distance education. In the online courses we teach, we seek to avoid domesticating web technology and to help teachers avoid domesticating software when they use it in their face-to-face computer laboratories.

In Borba, Malheiros, and Zulatto (2007), we discussed issues related to the teaching and learning of mathematics based on the experiences of our research group over the past 10 years. We believe that teaching in online environments situates the teacher within a new model of risk zone with respect to the use of ICTs in the teaching of mathematics. New challenges arise: How to follow the progress of my student who is physically distant? How to discuss mathematics online? How to express my reasoning? What resource is most appropriate for each situation?

We have also investigated how different interfaces used in online courses transform the way mathematics is produced by participants. More recently, Borba & Zulatto (2006) elaborated on the conjecture that mathematics is also transformed as we change from a face-to-face context to online distance courses for teachers, pointing out the important role of the different tools available to students online. For instance, there are online environments in which writing in a chat or in a forum is the only way for participants to communicate among themselves. In our online course proffered to mathematics teachers, our analysis showed how writing shapes the mathematics discussed. The analysis was carried out based on the idea that when technology changes, the possibilities for mathematics are also altered. This is the main idea behind the notion that a collective of humans-with-media (Borba & Villarreal, 2005) is the basic unit that constructs knowledge, as discussed earlier in this chapter. Knowledge is always produced by humans, but in conjunction different media such as writing, or the new languages that emerge from computer technology.

In this way, we have shown how online courses that rely heavily on chats emphasize the role of everyday spoken language to express mathematical ideas (Borba & Santos, 2005; Santos & Borba, 2008). Our research has also shown that videoconferences can be conducted in ways that resemble traditional classroom lectures, or in ways that facilitate a dialogical relationship in which special tools transform collaboration among participants and provide collective problem solving activities (Borba & Zulatto, 2006). Next, we will present a detailed description of the context of the course in which we learned from the dialogical interactions with the teacher/students, followed by examples in which we, the professors of the course, learned mathematics from a problem posed by one of the participating teacher/students.

The Context

We developed a course, entitled "Geometry with Geometricks," that took place in response to a demand from mathematics teachers employed in a nation-wide network of schools sponsored by the Bradesco Foundation.[3] Teachers employed in their 40 schools, located throughout Brazil and including some in the Amazon rainforest, have access to different kinds of activities, including courses administrated by an educational center based in the greater São Paulo area. Following the improvement of Internet connections in Brazil, they realized that online courses could be a promising option, since transporting teachers from all corners of Brazil (which is larger in area than the continental U.S.A.) to a single location to participate in courses was

[3]The Bradesco Foundation is supported by the Bradesco Bank and has social objectives, as their schools are generally located in poor neighborhoods. Although they are a private foundation, their schools are free and they develop intense continuing education activities with their teachers. There is at least one school in each one of the 26 Brazilian states and in the nation's Capital, Brasília. We would like to thank them for their support for the research project we conducted together with our teaching.

ineffective cost-wise as well as pedagogically. Cost effectiveness is related to the size of the country, and the pedagogical consideration concerns the fact that teachers would usually participate in the courses with little or no chance of implementing new ideas while taking the course.

The first type of online mathematics courses offered to teachers by the Foundation was based on a model involving little interaction between the leader of the course and participants. This is one reason we had to overcome some initial resistance when we began, as the teachers were accustomed to having a much more passive role in online courses like this. Gradually the teachers came to accept and respect our model, based on online interaction combined with applications in their face-to-face classes in middle and high school followed by online discussion of their experiences.

The educational center of this network of schools approached us, requesting a course about how to teach geometry using Geometricks, a dynamic geometry software originally published in Danish and translated into Portuguese (Sadolin, 2000). Geometricks has most of the basic commands of other software such as Cabri II and Geometer Sketchpad and was designed for plane geometry. As we know from extended research on the interaction of information technology and mathematics education, the mere availability of software and a well-equipped laboratory with 25 microcomputers, as in the case of these schools, is not enough to guarantee their effective use.

We designed a course using an exploratory problem-solving approach divided into four themes within geometry (basic activities, similarity, symmetry, and analytic geometry). There was usually more than one way to solve the problems, and they could be incorporated at different grade levels of the curriculum according to the degree of requirements for a solution, and according to the teacher's preference. Both intuitive and formal solutions were recognized as being important, and the articulation of trial-and-error and geometrical arguments was encouraged. We "met" online for two hours for eight Saturday mornings over a period of approximately three months. Prior to this synchronous activity, a fair amount of e-mail was exchanged during the week to solve technical issues regarding the software as well as clarify the problems proposed or pedagogical issues regarding the use of computer software in the classroom (e.g. should we introduce a concept in the regular classroom and then take the students to the laboratory, or the other way around?). Pedagogical themes were also discussed during the online meetings, in particular in one session in which the students, rather than working on problems during the week, were assigned to read a short book about the use of computers in mathematics education (Borba & Penteado, 2001). Teachers from the same school as well as from different schools were encouraged to work together to solve problems face-to-face or online.

The headquarters of the Bradesco Foundation had already purchased an online platform that provided participants with access to chat, forum, e-mail, and videoconference and allowed them to download activities posted by us. In our course, participants could download problems and they could also post their solution if they wanted to, or send it privately to one of us (the professors of the course). The

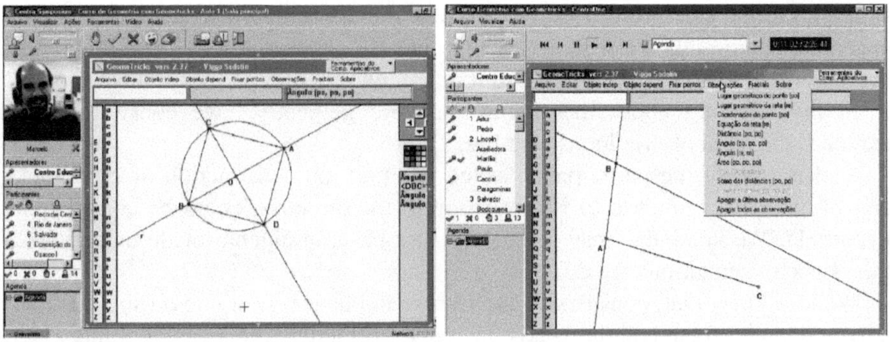

Fig. 1 Computer screen during the videoconference (displaying Geometricks software to every participant)

platform allowed the screen of any of the participants to be shared with everyone else.

For instance, we could start showing a screen of Geometricks on our computer and everyone else could see the dragging that we were performing on a given geometrical construction (Fig. 1). A special feature made it possible to "pass the pen" to another participant who could then add to what we had done on a Geometricks file (Borba & Zulatto, 2006).

Thus, all participants were able to share their geometrical constructions, ideas, and difficulties with others. During the videoconference meetings, Borba led the discussions and facilitated the dynamic interaction, while Zulatto managed the technical aspects and created the geometric constructions with the mouse. In this way, we acted collectively. Zulatto was the leader during the week, providing support by e-mail and following and commenting on activities sent by the students.[4]

The Example

With the proposal of developing a course based on interaction and dialogue (Freire, 2005; Alrø & Skovsmose, 2002), the activities the students engaged in during the week were discussed in online meetings on Saturdays. Everyone had a voice, with solutions being presented by the participants, who shared their reasoning. Thus, the course was not centered on us, the professors, but rather on the contributions of all the participants, in a process of collaborative education.

In this context, questions were raised not only by the teachers, but by the students as well, for everyone to reflect on. Thus, we were prepared for the risk that unknown

[4]To avoid confusion, we will refer to ourselves as the "professors" and to the teachers enrolled in the course as the "students."

Some figures are from:
http://www.algosobre.com.br/matematica/geometria-analitica-parabola.html
http://www.algosobre.com.br/matematica/geometria-analitica-hiperbole.html

situations might emerge, as occurred in one of the meetings when one of the students posed a question that we were not readily able to answer, despite having explored the activity beforehand and discussed it in a previous edition of the course.

In this meeting about the theme "analytic geometry," we constructed and explored the parabola via locus (see Fig. 2).

Gleice, one of the participants, asked whether the construction of a parabola, and of another symmetric to it in relation to the directrix, could be considered a hyperbola. We knew the answer was no. Our basis in algebra made us certain of this, since the equations are different.

We also knew that, considering the Cartesian plane *x*O*y*, a line *d* (directrix) and a fixed point *F* (focus) on the *x*-axis, the curve formed by the locus of points *P*(*x*,*y*) of the Cartesian plane is denominated a parabola (Fig. 3), such that

$PF = Pd$ where
$PF =$ distance between points P and F
$PP' =$ distance between point P and line d (directrix).

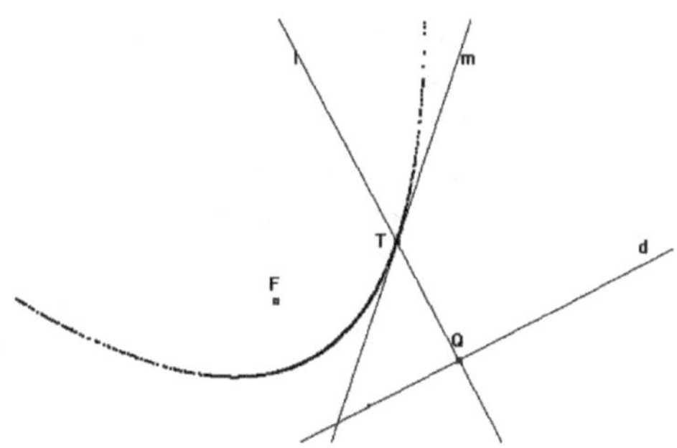

Fig. 2 Steps for construction of a parabola via locus

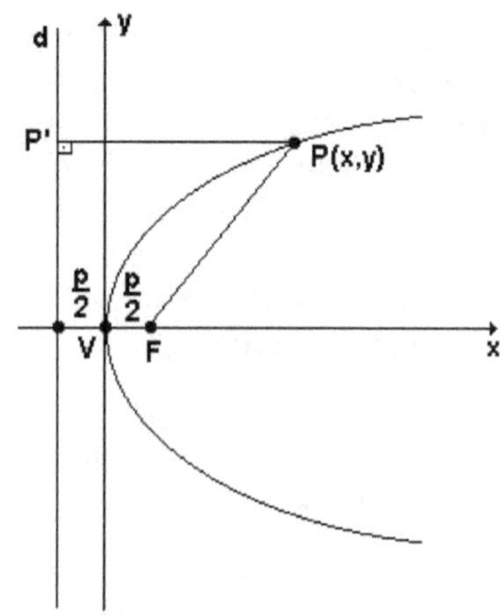

Fig. 3 Geometric definition of the parabola

The reduced equation of a parabola with 1 x-axis symmetry and vertex (V) at the origin is given by $y^2 = 2px$ where p is the measure of the parameter of the parabola, such that $VF = p/2$. The x-axis is the symmetry axis of the parabola and p is the distance between the focus F and the directrix d.

We also know, algebraically, that the reduced equation of the hyperbola is $\frac{x^2}{a^2} - \frac{y^2}{b^2} = 1$. This is because, given the fixed points F_1 and F_2, (foci) such that the distance between these points is equal to $2c > 0$, hyperbola is defined as the curve whose module of the difference between the distances of each of its points P and these fixed points F_1 and F_2 is equal to a constant value $2a$, where $a<c$ (Fig. 4):

Although we knew the algebraic differences between parabola and hyperbola, we did not know how to answer this question based on geometric justifications. The class was drawing to a close and everyone was given the task of researching and bringing geometric arguments to the next class.

We were definitely experiencing a risk zone. We had to say "we do not know," and this question made everyone study. We were fortunate that time was running out so that we could end the class. We left the question open for everyone to study. It was necessary to find a geometric justification.

Experiencing this risk can be natural in the teaching and learning process, but we knew that we would have to present a solution in the next meeting. We looked for books, specialized sites, etc. Some of the students also sent us solutions. After studying the concepts from the geometric point of view, we elaborated a justification that we presented to the students. What follows is a description of how we dealt openly with the problem in the next class: We started from two free points A and O. A circumference with center O and radius AO was drawn. A point Q was fixed in this circumference and a free point P was marked outside of it. Next, a perpendicular bisector of P and Q was constructed and a line drawn passing through O and Q. Point L was the intersection of these latter two lines. When point Q was moved, it was possible to visualize the locus of point L, as shown in Fig. 5.

We then sought to justify this construction. We went back to the definition of hyperbola, translating it to the computer screen in the virtual environment shared by all the students, which resulted in a construction like the one shown in Fig. 2. Next, we presented in the form of a problem the question of why a construction like the

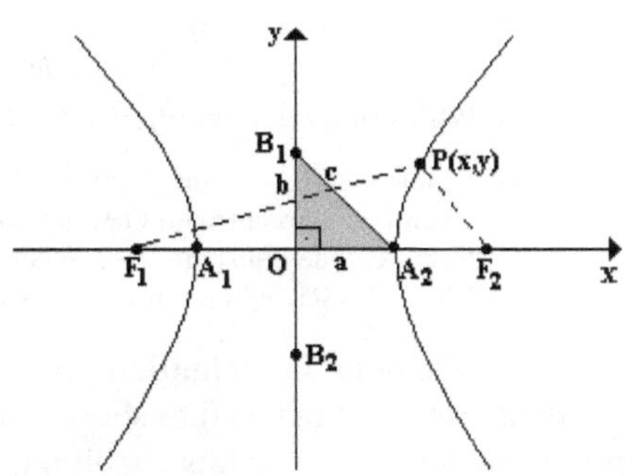

Fig. 4 Geometric definition of hyperbola

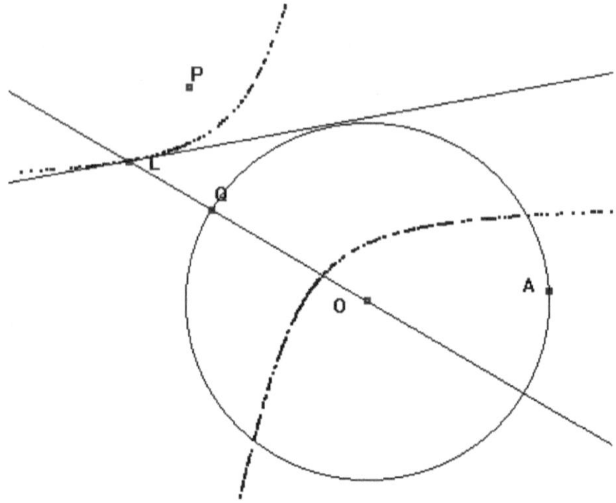

Fig. 5 Hyperbola construction via locus

one above, in fact, results in a hyperbola. Observe how we approached the question, and how we used the addition of the segment *LP* to aid in our argument:

> **Marcelo**: So when I take the distance [...] LO – LP has to be constant, the difference between them. It could also be LP – OL because I am only interested in the absolute value, OK!? This distance between them will be called 2a, in the same way that the distance between the foci is 2c [...]. I'll give you another minute to come up with some type of argument about why LO – LP is constant when I make this construction, when I drag point Q and generate this locus. Another hint to help you visualize it: let's add a line that is not necessary for the construction but that will make it clearer, so everyone sees that the distance LP is OL, OK?! (Fig. 6) [...] Now I'd like some of you to pronounce, or about an argument, try, no problem if it is not correct [...].

Gleice presented a justification:

> I think the distance remains constant because point Q and point O, they are constant, fixed; the distance between them is always the same. Then you made the construction of the perpendicular bisector from point Q to point P, obtaining point L, which is always the same distance from Q and P; that's why when you move, it's always the same distance, the difference is always constant.

Going back to Gleice's question, we felt it was pertinent to construct a parabola with concavity facing up and another with focus symmetrical in relation to the directrix, with concavity facing down, as illustrated on Fig. 7:

> **Marcelo**: Note that, in this construction, what is remaining constant, with the same distance, is RK, which is congruent with QR; now it's no longer like the other, that kept the radius of the circumference constant, see? When I drag, for example, point Q, QR with RK is constant, but PR – RK will not necessarily be constant. Did you see the difference?

To be a hyperbola, by definition, the difference between *PR* and *RK* should be a constant. Since we planned to show that the construction of these two symmetrical parabolas did not satisfy this condition, we needed only to find a counter-example.

Fig. 6 Geometric arguments about the hyperbola

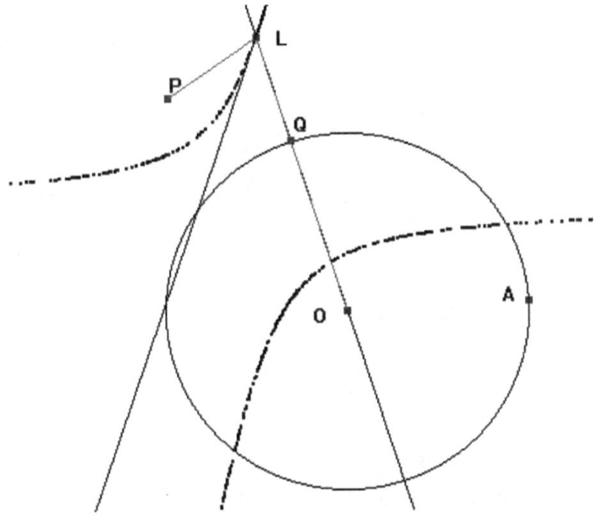

Fig. 7 Two parabolas under construction

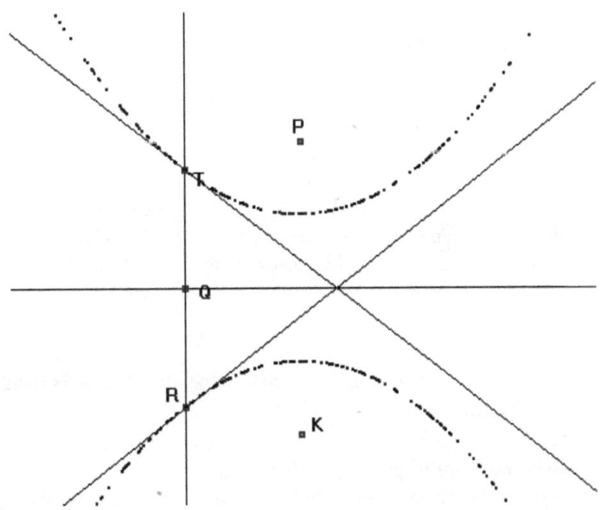

Using the software, it was possible to verify it by measuring the segments *PR* and *RK*. The first result was $PR - RK = 13.69 - 8.49 = 5.2$. Dragging point *Q*, these measures changed, and the difference should have remained constant. However, the result was $5.17 - 1.74 = 3.43$. We suggested that this type of counter-example could be explored in the first year of high school. We affirmed that, after showing the above solution, which uses a measure command of the software, discussion should be encouraged regarding the need for a more geometrical argument about why the above construction reflecting a parabola is a

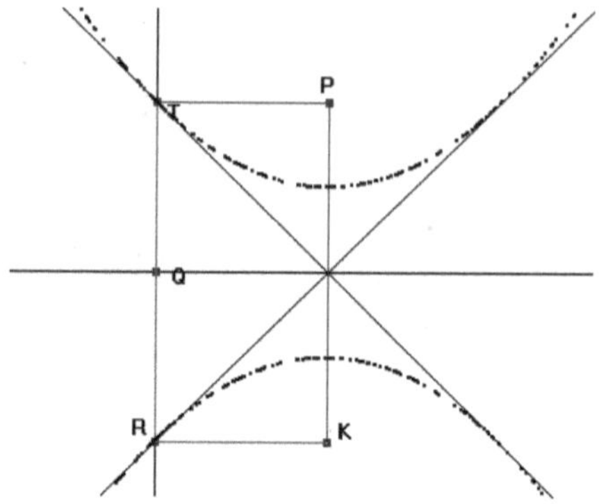

Fig. 8 Building a counter-example

hyperbola. We then presented a counter-example using only geometrical construction. To find this counter-example, we selected a position for point *Q* such that *KRQ* would be a right angle and we applied the Pythagoras theorem to the right triangle *PRK*. To facilitate visualization, we constructed the segments *PT*, *PK*, *RT*, and *RK* (Fig. 8).

Marcelo's argument below shows how he explained the counter-example:

> Since we constructed a parabola, [. . .] KR=QR, since this line is the perpendicular bisector between points P and Q, in both cases, OK!? [. . .] so now we have the rectangle KRTP and we can say that we have a measure of 2 by 1 (in reference to the segments TR and PK measuring double that of segments TP and RK, in a 2 to 1 proportion). [. . .] Note that this 2 by 1 is not using the same metric that the software uses in the observation menu, OK?! OK, so [. . .] we can see that RK=1 and this KP=2, so this diagonal here [PR], according to the Pythagoras theorem, will measure. Is that clear to everyone? And therefore, PR – KR will give $\sqrt{5} - 1$, which is a little more than 1, correct? It will give zero point something, it will give one plus something.

And proceeding with this reasoning, point *Q* was dragged to obtain another figure for *PKRT* because

> since we want to prove that this is not a hyperbola, we will find another point not on the 'hyperbola' (pseudo-hyperbola), of this that would be the hyperbola, and show that the difference between PR and KR is not one-point-something (Marcelo).

The other "special point" for *Q* is when *P*, *T*, *R*, and *K* are aligned (Fig. 9). Under these conditions, when the points are aligned, the segment *PR* will measure 1.5 (since the measure of *KR* = *RQ*). Thus, the difference between *PR* and *KR* is exactly 1.0 (since it is 1.5 – 0.5). Previously we had arrived at a value slightly greater than 1.0. However, we noticed that the difference did not remain constant, which contradicted the definition of a hyperbola.

All the participants felt this counter-example was sufficient to prove that a parabola reflected in relation to an axis is not a hyperbola despite appearing to be so.

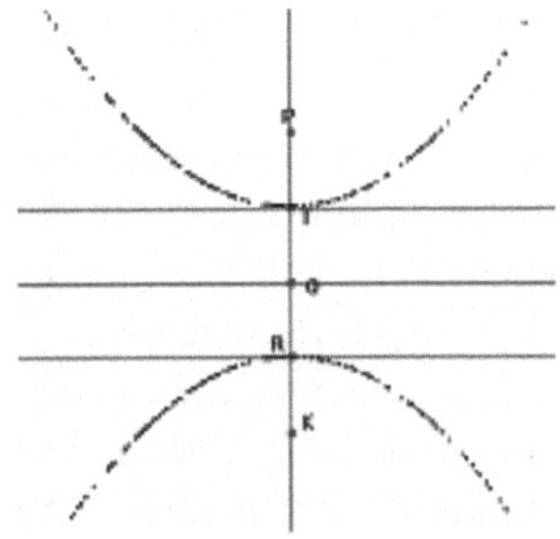

Fig. 9 Second step of the construction of a counter-example

Online Collaboration to Foster Our Learning of Mathematics

This problem posed by one of the teacher/students participating in a course designed to teach how to use geometry software placed us, the professors of the course, in a risk zone because we were unable to respond to the student's question, and "didactical contracts" in courses such as these presuppose that university professors know what they are teaching. We had prepared the content of the course, we had prepared ourselves to teach other contents, but as we have argued, the likelihood of entering a risk zone, and having to say "I do not know," increases when using software to teach. The original focus of our research was not to see how we learn mathematics from teachers, but to document the way online platforms transform collaboration among teachers (Borba & Zulatto, 2006; Borba & Gadanidis, 2008). In this perspective, we have shown how different interfaces – software, online platforms, etc. – are linked to changes in the mathematics produced by collectives of humans-with-media and how users shape technology according to their own uses.

In this chapter, we believe we have made the case that listening to teachers in online courses based on a dialogical approach is a way for all participants, and in particular for us, to learn mathematics. The fact that the teacher/student could easily, using a software command, reflect a parabola about an axis and "visually puzzle" all the participants and convince them that this problem was worth studying, shows the role played by Geometricks (or any geometry software chosen by the network of schools) in creating new problems together with the teacher.

Borba and Villarreal (2005) present many examples of the role of software or paper-and-pencil in co-constructing conjectures, reasoning, visualization, and proofs in mathematics education at different educational levels and in different contents. Extending this argument to tools such as online platforms is part of the agenda of our research group GPIMEM. One good example was presented at PME-30 (Borba & Zulatto, 2006). In the case presented here, the platform can be seen as merely a setting where professors learned in the process of teaching teachers, and we do not claim that the production of mathematical knowledge differs significantly when an online platform is used compared to traditional face-to-face environments. However, the fact that the Geometricks software could be manipulated and seen by all with this particular platform (unlike other platforms we have taught with), and that the voice of one of the participants could be shared, was important for the emergence of the case reported. Although this does not give the platform a strong role in our opinion, another important factor worth noting is that the platform enabled teachers in different locations throughout Brazil to work in groups, conjecturing and experimenting with Geometricks on their computers as we taught, without the physical presence of a university professor. Unlike face-to-face classes, social norms in an online context do not dictate that it is impolite to "talk" while the professor is "talking" and manipulating the software through the online platform. This role of the platform and the above conjecture regarding social norms will require further investigation.

We, as leaders, and the teacher/students were also very important participants and key actors, as we were able to break traditional didactical contracts and establish collaboration in a way that enabled us to enjoy being in the risk zone. This breaking of the didactical contract was what made it possible for teachers/students to pose new problems to university professors. Thinking about and developing a solution for this problem generated learning for the professors, as we had to articulate parabola and hyperbola construction in a new way in order to build an argument for a problem we were unfamiliar with.

The risk zone has become the comfort zone for all of us. We, as professors, are exhilarated by teaching in new environments, by fostering collaboration in online environments with mathematics teachers we have never (or not yet) met, and we were both already familiar with the exhilaration of working with software, since it often leads us into unanticipated situations in the classroom. The example presented illustrates how we can learn mathematics in our own teaching. It seems that teaching with technology and teaching with software in online environments require teachers who are more comfortable working in the risk zone while learning together with their students/peers! Like engaging in "radical sports," with practice, the risk zone can become comfortable.

Acknowledgments Although they are not responsible for the content of this chapter, we would like to thank Antonio Olimpio and Ricardo Scucuglia, members of our research group GPIMEM, for comments on earlier versions of this chapter. We would also like to thank Anne Kepple for her careful and insightful review of the English. Finally, we would like to thank, in memoriam, Geraldo Duarte, a colleague of the mathematics Department at UNESP, for chatting with us about the possible ways of solving the problem that was at center stage in this chapter.

References

Alrø, H., & Skovsmose, O. (2002). *Dialogue and learning in mathematics education: Intention, reflection, critique*. Dordrecht: Kluwer Academic Publishers.

Borba, M. C. (1993). *Students Understanding of Transformations of Functions Using Multi-Representational Software*. Dissertation in Mathematics Education. Ithaca: Cornell University.

Borba, M. C. (1995). Overcoming limits of software tools: A student's solution for a problem involving transformation of functions. In L. Meira & D. Carraher (Eds.), *Proceedings of the PME, 19*(2), 248–255.

Borba, M., & Gadanidis, G. (2008). Virtual communities and networks of practising mathematics teachers: The role of technology in collaboration. In T. Wood & K. Krainer (Eds.), *International handbook of mathematics teacher education: (Vol. 3). Participants in mathematics teacher education: individuals, teams, communities, and networks*. Rotterdam: Sense Publishers.

Borba, M. C., Malheiros, A. P. S., & Zulatto, R. B. A. (2007). *Educação a distância online*. Belo Horizonte: Editora Autêntica.

Borba, M. C., & Penteado, M. G. (2001). *Informática e Educação Matemática*. Belo Horizonte: Editora Autêntica.

Borba, M. C., & Santos S. C. (2005). *Educação Matemática: Propostas e Desafios, Pesquisa Educacional e Cotidiano Escolar*, *7*(2), 291–312.

Borba, M. C., & Villarreal, M. E. (2005). *Humans-with-media and the reorganization of mathematical thinking: Information and communication technologies, modeling, visualization and experimentation*. Dordrecht: Kluwer Academic Publishers.

Borba, M. C., & Zulatto, R. B. A. (2006). Different media, different types of collective work in online continuing teacher education: Would you pass the pen, please? In *Proceedings of the PME, 30*(2): 201–208.

Confrey, J. (1991). Function Probe©. Apple Macintosh® Computer. Designed by J. Confrey, F. Carroll, S. Cato, P. Davis, and E. Smith.

Freire, P. (2005). *Pedagogia do oprimido*, 45. Ed., Editora Paz e Terra Rio de Janeiro.

Kaput, J. J. (1989). Linking representations in the symbol systems of algebra. In S. Wagner & C. Kieran (Eds.), *Research issues in the learning and teaching of algebra* (pp. 167–194). Hillsdale, NJ: Erlbaum.

Levy, P. (1993). *As Tecnologias da Inteligência: o futuro do pensamento na era da informática*. Rio de Janeiro: Editora 34.

Penteado, M. G. (2000). Possibilidades para a formação de professores de matemática. In M. G. Penteado & M. C. Borba (Eds.), *A informática em ação:Formação de professores, pesquisa e extensão*. Editora Olho d'Água.

Penteado, M. G. (2001). Computer-based learning environments: Risks and uncertainties for teacher. *Ways of Knowing Journal*, *1*(2), 23–35.

Sadolin, V. (2000). *Geometricks*. Software de Geometria Dinâmica com fractais.

Santos, S. C. & Borba, M. C. (2008). Internet e softwares de Geometria Dinâmica como atores na Produção Matemática On-Line. *Zetetiké*, *16*(29), 93–116.

Tikhomirov, O. K. (1981). The psychological consequences of computerization. In J. V. Wertsch (Ed.), *The concepto f activity in soviet psychology* (pp. 256–278). New York: M.E. Sharpe Inc.

Wagner, H. R. (1979). *Fenomenologia e relações sociais. Textos Escolhidos de Alfred Schutz*. Rio de Janeiro: Zahar Editores.

Zulatto, R. B. A. (2002). *Professores de Matemática que utilizam softwares de geometria dinâmica: suas características e perspectivas*. Máster thesis. Rio Claro. Brasil: Universidade Estadual Paulista.

Role of Task and Technology in Provoking Teacher Change: A Case of Proofs and Proving in High School Algebra

Carolyn Kieran and José Guzmán

In their review of the emerging field of research in mathematics teacher education, Adler, Ball, Krainer, Lin, and Novotna (2005) have argued that we need to better understand how teachers learn, from what opportunities, and under what conditions. The research findings that we recount in this article provide a compelling case for the particular opportunities and conditions under which the knowledge and teaching practice of a mathematics teacher evolved.

The Context of the Present Study

When our research group[1] developed the program of research that included the present study, it was decided that the use of new technologies (i.e., Computer Algebra Systems – CAS) for the teaching of algebra would be one of its principal components. Another was the design of novel tasks that would both take advantage of the technology to further the growth of algebraic reasoning and focus on the interplay between algebraic theory and technique. The theoretical framework that underpins the research, one that we refer to as the *Task-Technique-Theory* frame (see Kieran & Drijvers, 2006, for details), draws upon Artigue's (2002) and Lagrange's (2002) adaptation of Chevallard's (1999) anthropological theory of didactics. From their research observations, Artigue and her colleagues came to see techniques as a link between tasks and theoretical reflection, in other words, that the learning of techniques was vital to related theoretical thinking. Based on this notion, our research group developed a research program that conceptualized algebra learning at the high school level in terms of a dynamic among task, technique, and theory within technological environments.

C. Kieran (✉)
Université du Québec à Montréal, Québec, Canada
e-mail: kieran.carolyn@uqam.ca

[1] Our appreciation to the other members of the research group, which included at the time that the program of research was elaborated, André Boileau, Denis Tanguay, Fernando Hitt, and Luis Saldanha, as well as the consultant, Michèle Artigue.

R. Leikin, R. Zazkis (eds.), *Learning Through Teaching Mathematics*, Mathematics Teacher Education 5, DOI 10.1007/978-90-481-3990-3_7,

At the same time that we began to create a series of tasks that would invite both technical and theoretical development in 10th grade algebra students, we also made contact with several practicing mathematics teachers to see if they might be interested in collaborating with us. The form of collaboration that we arranged was on several levels. First, the teachers were our practitioner-experts who, within a workshop setting, provided us with feedback regarding the nature of the tasks that we were conceptualizing. Second, after modifying the tasks in the light of the teachers' feedback, we requested that, at the beginning of the following semester, they integrate the entire set of tasks into their regular mathematics teaching and that they be willing to have us act as observers in their classrooms. Third, throughout the course of our classroom observations, which occurred over a five-month period in each class, we also offered a form of ongoing support to the participating teachers by being available to discuss with them whatever concerns they might have. In addition, we conducted interviews with some of them immediately after certain lessons that we had perceived to be worthy of further conversation, lessons that we had thought might even be considered pivotal moments in their practice. The following narrative concerns one such pivotal two-lesson sequence, taught by the teacher Michael.

Michael's Story

Some Background

Michael was one of the teachers involved in the project. Up to the time of the present study, we had already observed 15 of his classes, that is to say, each of the lessons in which he had thus far integrated a CAS-supported task from the set that had been created for the research project. Michael, whose undergraduate degree and teacher training had been done in the U.K., had been teaching mathematics for five years, but he had not had a great deal of prior experience with technology use in mathematics teaching, except for the graphing calculator. He was a teacher who, along with encouraging his pupils to talk about their mathematics in class, thought that it was important for them to struggle a little with mathematical tasks. He liked to take the time needed to elicit students' thinking, rather than quickly give them the answers.

We began to observe Michael's class from the very beginning of the Grade 10 school year. The students in this class had learned a few basic techniques of factoring polynomials (for the difference of squares and for factorable trinomials) and the solving of linear and quadratic equations during their 9th grade mathematics course. They had used graphing calculators on a regular basis; however, they had not had any experience with symbol-manipulating calculators prior to the onset of our project, which made use of the *TI-92 Plus* hand-held, CAS calculator. These students were already quite skilled in algebraic manipulation, as was borne out by the results of a pretest we administered at the beginning of the study; but we were informed that they had never engaged in any activity related to proving, either in geometry or in algebra.

This article concerns the two lessons that had involved the $x^n - 1$ task set (hereinafter referred to simply as the $x^n - 1$ task), the last component of which was a proof problem. We observed, and videotaped, both of these class lessons. The day after the close of the two lessons, the first author interviewed Michael. The next few paragraphs describe first the task and then our classroom observations of the proving segment of the task, followed by an analysis of this activity. Then we present extracts from the interview with Michael and a second analysis that draws on both his interview reflections and our earlier classroom observations.

The $x^n - 1$ Task

The design for the two-lesson sequence was an elaboration of earlier work carried out by Mounier and Aldon (1996) with their 16- to 18-year-old students on a task that involved conjecturing and proving general factorizations of $x^n - 1$. Our task activity had three parts. The first part, which involved CAS as well as paper and pencil, aimed at promoting an awareness of the presence of the factor $(x - 1)$ in the given factored forms of the expressions $x^2 - 1$, $x^3 - 1$, and $x^4 - 1$ (see Fig. 1), as well as leading to the *generalized* form $x^n - 1 = (x - 1)(x^{n-1} + x^{n-2} + \ldots + x + 1)$.

The next part of the activity involved students' *confronting* the paper-and-pencil factorizations that they had produced for $x^n - 1$, with integer values of n from 2 to 6 (and then from 7 to 13), with the completely factored forms produced by the CAS, and in *reconciling* these two factorizations (see Fig. 2).

1. Perform the indicated operations: $(x - 1)(x + 1)$; $(x - 1)(x^2 + x + 1)$.
2. Without doing any algebraic manipulation, anticipate the result of the following product $(x-1)\left(x^3 + x^2 + x + 1\right) =$
3. Verify the above result using paper-and-pencil, and then using the calculator.
4. What do the following three expressions have in common? And, also, how do they differ?
   $(x - 1)(x + 1)$, $(x-1)(x^2 + x + 1)$, and $(x-1)(x^3 + x^2 + x + 1)$
5. How do you explain the fact that when you multiply: i) the two binomials above, ii) the binomial with the trinomial above, and iii) the binomial with the quadrinomial above, you always obtain a binomial as the product?
6. On the basis of the expressions we have found so far, predict a factorization of the expression $x^5 - 1$.

Fig. 1 Some of the initial tasks of the activity

In this activity each line of the table below must be filled in completely (all three cells), one row at a time. Start from the top row (the cells of the three columns) and work your way down. If, for a given row, the results in the left and middle columns differ, reconcile the two by using algebraic manipulations in the right hand column.

Factorization using paper-and-pencil	Result produced by the FACTOR command	Calculation to reconcile the two, if necessary
$x^2 - 1 =$		
$x^3 - 1 =$		
$x^4 - 1 =$		
$x^5 - 1 =$		
$x^6 - 1 =$		

Fig. 2 Task in which students confront the completely factored forms produced by the CAS

Conjecture, in general, for what numbers n will the factorization of $x^n - 1$:

i) contain exactly two factors?
ii) contain more than two factors?
iii) include $(x+1)$ as a factor?

Please explain.

Fig. 3 Task in which students examine more closely the nature of the factors produced by the CAS

An important aspect of this part of the activity involved reflecting and *forming conjectures* (see Fig. 3) on the relations between particular expressions of the $x^n - 1$ family and their completely factored forms.

The final part of the activity (see Fig. 4) focused on students' *proving* one of the conjectures that they had generated during the previous part of the task. This

Prove that $(x+1)$ is always a factor of $x^n - 1$ for even values of n.

Fig. 4 The proving task

proving activity is the central component of the analysis of teacher practice and teacher change that we present in this article.

Our Classroom Observations

After students had completed the first two parts of the $x^n - 1$ activity, they were faced with the proving segment of the task: Prove that $(x + 1)$ is always a factor of $x^n - 1$ for even values of n. Mathematically experienced students might possibly have been able to generate a proof along the following lines:

$$\begin{aligned} x^n - 1 &= x^{2k} - 1 (\text{for } n \text{ even}) \\ &= (x^2)^k - 1 \\ &= (x^2 - 1)(x^{2^{k-1}} + x^{2^{k-2}} \dots + 1) \\ &= (x + 1)(x - 1)(\dots). \end{aligned}$$

However, our research team did not consider for an instant that such a symbolic form of proof might be forthcoming from the 15- and 16-year-olds in our study. Nevertheless, we did entertain the idea that some generic form of proof might be produced. For example, students might propose that the expression x^{18} (where the 18 represents any even integer) could be viewed as $(x^2)^9$, and thus that $(x^{18} - 1)$, which is equivalent to $((x^2)^9 - 1)$, could be factored according to the general rule for $(x^9 - 1)$, but with the x being replaced by x^2. As mentioned earlier, the students of Michael's class had not had any prior experience with proving in algebra. Such lack of experience with proving is not unusual for students of this age. This is reflected in the general absence of algebraic proving activity among high school students in the research literature. Nevertheless, some attention has been given to number-theoretic proofs (e.g., Healy & Hoyles, 2000; see also Mariotti, 2006), as well as to proofs involving geometric figures (Balacheff, 1988). However, we could find nothing that was closely related to algebraic proofs of the kind being proposed within our $x^n - 1$ task. Mounier and Aldon's (1996) report, which had stated that students generated four proofs for various factorizations of $x^n - 1$ where n is a positive integer, did not describe the actual activity of proving nor provide the steps of the students' proofs.

To return now to our observations of the unfolding of the proving activity in Michael's class, the students worked on this part of the task, mostly within small groups, for about 15 minutes. Some were using their CAS calculators, but most were just talking about how they might approach the task and occasionally jotting things down on paper. During that time, the teacher circulated and was heard to offer the following remarks to various groups (T = Teacher):

T: See if you can prove this and not just state it, as some people have done so far (picking up one student's worksheet and reading it to the class): "When n is greater than or equal to 2, $(x+1)$ is a factor because." Let's see if we can go a little bit beyond that. Can you write down what you come up with. ... Yeah, but you need more than just examples. . . . You need to get something written down. . . . Look, you need to think in order to answer this. This is the only hint I'm giving you, you need to think about where the $(x+1)$ comes from.

Getting students into this proving task was not straightforward, as they had never before engaged in such mathematical activity. However, with the teacher's encouragement, they did make progress. When he sensed that the majority of them had arrived at some form of a proof, he opened up a whole-class discussion, oriented around various students' sharing their work:

T: Ok, guys. Quite a lot of you got quite close in doing this. What I want you to do, and I've asked a couple of people who've done it in completely different ways, to see if they can put forward their explanation. I want you to be quiet, listen to their explanation, then we'll discuss it once they've got it done, once they've completed their little spiel, ok.

He invited selected students to come to the board, one at a time. As will be seen, the principal contributions of the students can be grouped into three distinct approaches. The first proof, which is presented immediately below, revolves around the idea of "difference of squares." Despite a follow-up counterexample involving the "sum of cubes," and a return to the validity of the notion of "difference of squares," the proof-giver never quite fills in the gaps to arrive at a full proof.

Proof 1: A General Approach Based on the Difference of Squares

Paul was invited to come to the front and to present his "proof":

Paul: Ok. So, my theory is that *whenever* x^n *-1 has an even value for n*, if it's greater or equal to 2, that, *one of the factors of that would be* x^2 *-1*, and since $x^2 - 1$ is always a factor of one of those, *a factor of* $x^2 - 1$ *is* $(x+1)$, *so then* $(x+1)$ *is always a factor*.
S2: Could you say it again? [other students react all at once, making many comments]
S3: Why don't you write it on the board?
T: Guys! Give him a chance.
Paul: You want me to write? [addressing the teacher]
T: Write down what you want to write down.
S4: Can you talk at the same time?

Paul then proceeded to write down at the board that which he had just stated orally. The teacher then asked: "Is everyone willing to accept his explanation?" While many seemed to agree with what Paul had proposed, a few voiced disagreement – to which the teacher responded: "Ok, guys, one at a time. Ok, start with Dan."

A Proposed Counterexample Involving the Sum of Cubes

Dan then came forward with what he considered a counterexample, $x^{12} - 1$, to Paul's proof. Dan proceeded by factoring $x^{12} - 1$ as $(x^6 + 1)(x^6 - 1)$, the latter of which he refactored as $(x^3 + 1)(x^3 - 1)$. His subsequent factoring of $(x^3 + 1)$ – a sum of cubes – yielded the sought-for $(x+1)$ factor (see Fig. 5). Thus, he maintained that the presence of $x^2 - 1$ was not necessary for a proof because he (Dan) had shown that, for even values of n, the factoring of $x^n - 1$ does not have to end up with a difference of squares. A sum of cubes could result, and it too would yield a factor of $(x+1)$. This led immediately to many students' voicing disagreement, to which the teacher remarked:

T: Ok, so, so this is good [he points to the third line on the board, which contains $(x^3-1)(x^3+1)(x^6+1)$]. This is good because, Paul, the problem I had with yours, is how do you get from here to here [he points to $x^n - 1$ and then to the $x^2 - 1$ of Paul's board work; he then draws a red arrow to highlight the gap between those two lines of the proof], does that follow? He's just given you a counterexample where it does not follow.

Some Students: It does though [with many students speaking at once].

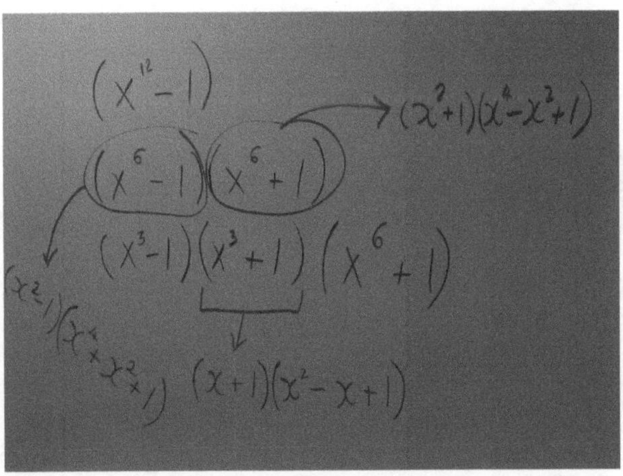

Fig. 5 Dan's counterexample in the central section, with the counterproposal by Paul that $x^6 - 1$ does indeed yield $x^2 - 1$ (at the *leftmost arrow*)

The Counterproposal Containing Seeds of a Generic Proof

Many of the other students, including Paul, contended that Dan's was not a counterexample, after all. They argued that the expression $x^{12} - 1$ could, in fact, produce $x^2 - 1$ if it were factored differently:

Paul: Isn't x^6+1 a sum of cubes? ... *So couldn't you also do the* $x^6 - 1$ *as the difference of cubes* [one student says "yeah"] *and that's* $x^2 - 1$.

T: [he circles (x^6–1) in red and draws an arrow on the left to show the alternate factorization being proposed. See the leftmost arrow and its accompanying factorization in Fig. 5]

Paul: [continuing what he was saying] So $x^2 - 1$ times whatever [the teacher writes $(x^2-1)(x^4+x^2+1)$ on the board]. So there's your x^2–1.

S5 (a student other than Paul): Even though it's not fully factored [referring to x^{12}–1], $x^2 - 1$ is still a factor of that.

Paul: Sir, it can be factored down

T: Yeah I know it can be factored down, and I am not saying you're wrong, what I'm saying is that your reasoning to get from $x^n - 1$ down to this [he points to the $x^2 - 1$ line of Paul's proof] is not complete. Do you agree (to Paul)?

Analysis of Proof 1

While Paul had seen that $x^6 - 1$ could be viewed as a difference of cubes, and thus that $x^2 - 1$ was a factor, he did not seem able to link this particular example with his general affirmation that for all even *n*s in $x^n - 1$, one would always arrive at $x^2 - 1$ as a factor. Yet, he was unbelievably close. Could he see that $x^6 - 1$ was equivalent to $((x^2)^3 - 1)$, even if he had never expressed it in quite this way? Or was his realization based solely on his experience with factoring the "difference of cubes" and merely with perceiving 6 as a multiple of 2 and of 3? If the former, why not see also that $x^8 - 1$ was equivalent to $((x^2)^4 - 1)$, ..., and more generally that $x^n - 1$ for even *n*s could be expressed as $((x^2)^p - 1)$ where $n = 2p$? And so if $x^n - 1$ has $(x - 1)$ as its first factor, why not then see that, similarly, $((x^2)^p - 1)$ would have $x^2 - 1$ as its first factor, and thus $(x + 1)$ as a factor? While Paul had certainly intuited some of this in offering his initial proof, the connections were likely still quite tentative and not yet able to be formulated in an explicit way. However, the teacher, Michael, had insisted that, for Paul's proof to be complete, there needed to be a theoretical link connecting the two main lines of the proof (the $x^n - 1$ line and the $x^2 - 1$ line): "Yes, we know we will get there eventually, but how do we know that we will eventually get there without doing all the actual factoring?" Paul's proof had a "gap" in it (see Weber & Alcock, 2005, for more on "gaps").

Proof 2: A Proof Involving Factoring by Grouping

The second approach to the proving problem was put forward by Janet. Janet's proof, which she and her partner Alexandra had together generated, was based on their earlier work on reconciling CAS factors with their paper-and-pencil factoring (for the tasks shown in Fig. 2). They had noticed that for even *n*s, the number of terms in the second factor was always even. Janet argued, as she presented the proof at the board using $x^8 - 1$ as an example, that it would work for any even *n*:

Janet: When n is an even number

T: Write it on the board, show it on the board.

Janet: [she writes "$x^8 - 1$" and below it: $(x\text{-}1)(x^7+x^6+x^5+x^4+x^3+x^2+x+1)$]

T: Ok, listen 'cause this is interesting [addressed to the rest of the class], it's a completely different way of looking at it, to what most of you guys did. Ok, so explain it, Janet.

Janet: *When n is an even number* [she points to the 8 in the $x^8 - 1$ that she has written], *the number of terms in this bracket is even, which means they can be grouped and a factor is always (x+1).*

T: Can you show that?

Janet: [she groups the second factor as: $x^6(x+1)+x^4(x+1)+x^2(x+1)+1(x+1)$]

T: Thanks Janet. Do we understand what she put out there?

A Student's Query Related to the Factor $(x + 1)$

As soon as Janet had finished the writing of her proof at the board, another student posed a rather insightful question: "But how do you know that the group is going to be $(x+1)$?" As no student could offer any response to this, the teacher Michael interjected with a general notation for Janet's proof, in the hope that this might perhaps help the questioner to see the logical necessity of the $(x + 1)$ factor. (It is noted that Michael remarked to us during the classroom observations that Janet's proof was one that he had not thought of before; yet, he was able to react quickly with a general formulation in response to the "$(x+1)$" question.):

T: You know it's going to be x^{n-1} plus x^{n-2} plus dot, dot, dot, plus x plus 1 [he writes on the board as he speaks] and you know there's an even number here yeah? [he points to the series of dots in the polynomial]. Yes? So you know that, in there, if we take this [he points to the x^{n-2}] as the term outside. You know that these two [he points to x^{n-1} and x^{n-2}] can be factored and it's just $(x+1)$ as the other factor of these two [he wrote $x^{n-2}(x+1)$], yeah? And that would be the case for any and all in between [he points to the series of dots], and including this [he points to the "$+ x + 1$" at the end of the sequence and writes $1(x+1)$ on the far right of the $x^{n-2}(x + 1)$ that he had already written; see Fig. 6].

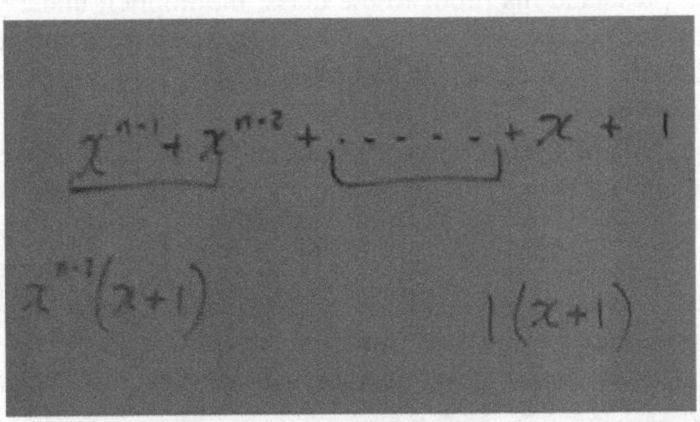

Fig. 6 A general notation illustrating that $(x+1)$ will be a factor upon grouping

Analysis of Proof 2

Janet's proof, which was generic in that it embodied the structure of a more general argument and was a representative of all similar objects (Balacheff, 1988; Bergqvist, 2005), was one that seemed to be understood and appreciated by most of the students in the class (see Weber, 2008, for related discussion). It also provided insight as to why the proposition holds true not only for that single instance but for all related cases (Rowland, 2002). Janet had been able to explain how the terms of the second factor (the factor beginning with the x^7 term) could be grouped pair-wise, yielding a common factor of $(x + 1)$, even if she did not complete the factoring process:

$$\begin{aligned} x^8 - 1 &= (x-1)(x^7 + x^6 + x^5 + x^4 + x^3 + x^2 + x + 1) \\ &= (x-1)[x^6(x+1) + x^4(x+1) + x^2(x+1) + 1(x+1)] \\ &= (x-1)(x+1)(x^6 + x^4 + x^2 + 1) \\ &= (x-1)(x+1)(x^2+1)(x^4+1) \end{aligned}$$

Janet's proof had appealed to her classmates' common experience in factoring by grouping. But, as has been discussed by Balacheff (1987), generic proofs such as these may often use rather imprecise tools and be defective in certain respects – as was, for example, pointed out by her classmate's question as to how Janet knew that $(x + 1)$ would appear in the grouping, or the unposed question as to how she knew that there would always be an even number of terms in the second factor of the first line of her proof. Nevertheless, it was a clever proof with a degree of elegance that indicated to the teacher, Michael, that his students could go much farther in the activity of proving than he had initially thought possible (compare with Bergqvist, 2005, where teachers were reported to believe that only a small number of students can use higher level reasoning).

Proof 3: A New Conjecture Involving $x^n + 1$ Where n Is an Odd Integer

When Paul had presented his proof to the class, the implicit underlying argument was that when one begins with $x^n - 1$ where n is an even integer, and if one continually takes the even exponent and treats it as a difference of squares, then one eventually arrives at $x^2 - 1$. Shortly after Janet had finished explaining her proof, the issue of Paul's proof came up once more. To provoke the students, the teacher offered the following counterexample: "Just out of interest, what would happen if this was $x^{14} - 1$?" [he wrote $(x^{14} - 1)$ under the $(x^n - 1)$], to which a student easily responded: "(x^7-1) times (x^7+1)." The teacher wrote at the board $(x^{14} - 1) = (x^7 - 1)(x^7 + 1)$ and then wondered aloud: "Where does that leave your proof, Paul?" However, rather than leaving the class stymied, this question provided an opening for another student who had been conjecturing something new:

Andrew: See, when it's a prime number, then the first part here is x+1 as a factor. . . . From, like x^5+1 you get, $x^4-x^3+x^2-x+1$, like when you factor it on the calculator, that's what you get.

T: Ok.

Andrew: x+1 times $x^4-x^3+x^2-x+1$.

T: Say it again Andrew [he is ready to write down Andrew's verbalizings at the board]

Andrew: When you factor $x^{10}-1$ on the calculator, you get ($x-1$) times (x+1) times ($x^4+x^3+x^2+x+1$) times ($x^4-x^3+x^2-x+1$).

T: Yeah [while completing the writing of Andrew's factorization at the board]. So, just go back a bit. That was these two together [tracing an arc joining ($x-1$) and ($x^4+x^3+x^2+x+1$)] to give you the x^5-1.

Andrew: Yeah, and the next two would be (x+1) and ($x^4-x^3+x^2-x+1$) (See Fig. 7).

T: So you're going into something that we haven't looked at in this class. You're setting up another hypothesis. What is your hypothesis?

Andrew: Well, that's what I was trying to get at. . . . If the division by 2 gives an odd number, then it goes (x+1).

T: So you're saying that, for the second hypothesis, something like this [he writes down $(x^5+1)=(x+1)(x^4-x^3+x^2-x+1)$, just as the bell rang]. And you're saying that's true for all odd numbers?

Andrew: That's what I think.

T: So if we could prove this, then we've got it. But we've run out of time.

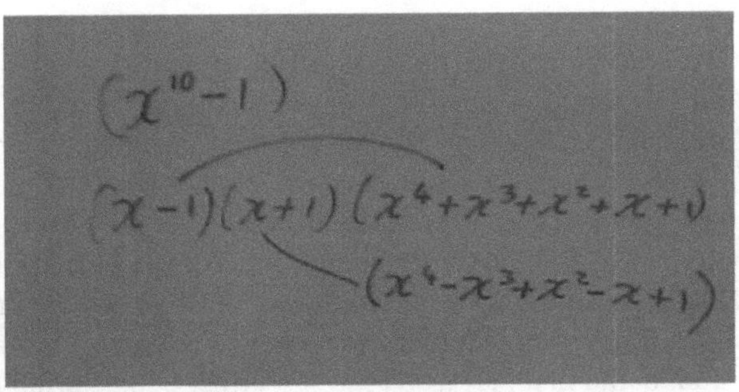

Fig. 7 Moving toward a conjecture involving x^n+1 for odd ns

Analysis of Proof 3

When Andrew had been working earlier on the second part of the x^n-1 task, which had involved the reconciling of his paper-and-pencil factoring with the CAS factoring, the $x^{10}-1$ example had presented a surprise. He had first factored it with pencil and paper as $(x^5+1)(x^5-1)$, and then refactored the (x^5-1) according to the newly-learned general rule, but had left the (x^5+1) factor as is. But the CAS produced as its factored form for $(x^{10}-1)$: $(x-1)(x+1)(x^4+x^3+x^2+x+1)$ $(x^4-x^3+x^2-x+1)$. Andrew noticed this additional factoring by the CAS, that is, that $(x^5 + 1) = (x + 1)(x^4 - x^3 + x^2 - x + 1)$. So, he began to conjecture and test the more general rule:

$$x^n + 1 = (x + 1)(x^{n-1} - x^{n-2} + ... - x + 1), \text{ when } n \text{ is odd.}$$

It is interesting that, as Andrew was explaining his conjecture to the teacher, it was clearly a new idea for the teacher too. While Andrew's conjectured new rule did not address the gap in Paul's proof, it did provide a worthy response to the teacher's $(x^{14} - 1)$ counterexample: that is, that it did not matter if the "difference of squares" approach led to exponents that were odd integers; by taking the "plus" factor (e.g., x^7+1), one would still end up with a factor of $(x + 1)$!

A Few Remarks Regarding the Proving Part of the Activity

Keeping in mind that the proving attempts that we have just witnessed were generated by 15- and 16-year-olds with no prior experience in algebraic-type proofs, their work is indeed remarkable. Hanna (2005) points out that, "While in mathematical practice the main function of proof is justification and verification, its main function in mathematics education is surely that of explanation" (p. 47). She adds that, "A good proof, however, must not only be correct and explanatory, it must also take into account, especially in its level of detail, the classroom context and the experience of the students" (p. 48). While the explanatory power of Janet's and Andrew's proofs was in a sense stronger than that of Paul's, even his had the seeds of a powerful explanation.

How might we account for the richness of the students' work with respect to proving? Mariotti (2002) has argued that there is no proof without theory. In the same vein, Mariotti and Balacheff (2008) have emphasized that the proving process as a complete whole necessarily starts with the production of conjectures before moving on to proof. In this regard, it is noted that the entire two-lesson sequence that was devoted to the $x^n - 1$ task involved an interplay between theory and technique. Such is the backbone of all the task activities developed within the present project. For the given task, the development of student conjectures was requested right from the second question in the first part (see Fig. 1). Further conjecturing that was more explicitly related to the proving task was also called for in the second part of the activity (see Fig. 3). Thus, the ideas that the students generated during the proving task were those that they had been conjecturing about and playing with throughout the entire activity.

The findings of this study contrast with some of the prior work on proof and justification that has been reported in the literature. For example, Healy and Hoyles (2000) found from their study of 14- and 15-year-olds' conceptions of proving within the number-theoretic domain that students were more likely to prefer empirical to algebraic arguments: "Regarding explanatory power, arguments that incorporated algebra were most likely to be viewed neither as showing why the given statement was true nor as representing an easy way to explain to someone who was unsure" (p. 414), "… students were put off from using algebra because it offered them little in the way of explanation; they were uncomfortable with algebraic arguments and found them hard to follow" (p. 415). These findings of Healy and Hoyles are consistent with the results of Lee and Wheeler (1987) who found that high school students preferred numerical examples to algebraic proofs and did not view algebra as a tool for justification and proof.

So why were the students of our study so impressed with the explanatory power of the algebraic proofs generated by themselves and their peers? There are a couple of major differences between the kind of proving tasks used in the two studies referred to above and the task used in this study, differences that can explain the divergence in the findings. First, as already pointed out above, the $x^n - 1$ task had built into it a great deal of prior conjecturing activity that was related to the ideas that were integral to the proving part of the task. This is in contrast to the proving tasks used in the Healy and Hoyles survey that were not preceded by prior student activity in developing related conceptual ideas. Students of the Healy and Hoyles study were confronted with the survey instrument "out of the blue," so to speak. Second, and this is a critical difference, the Healy and Hoyles problems presented students with statements, such as, "When you add 2 even numbers, your answer is always even," followed by choices of proof responses that included numerical, algebraic, and pictorial approaches. That students tended to choose numerical justifications as being more convincing is not surprising, given the numerical aspect of the initial problem statement. However, the $x^n - 1$ task is not a task that suggests numerical exemplification as does the above number-theoretic task. Furthermore, most of the tasks in the Healy and Hoyles study were presented to the students in a verbal rather than an algebraic form, the latter of which was the case in our study and which may have induced students to embrace an algebraic form of proof. In fact, the proving activity of the $x^n - 1$ task did not evoke the usual dialectic between the numerical and the algebraic – as is often the case in algebraic activity – but rather a higher-level dialectic between specific algebraic examples (e.g., $x^2 - 1$) and more general algebraic formulations (e.g., $x^n - 1$). The students in the current study remained at an algebraic level throughout – one that had been supported by a great deal of prior work involving related conjecturing.

Hanna and Barbeau (2008) have advanced the notion that "proofs yield new mathematical insights, new contextual links and new methods for solving problems, giving them a value far beyond establishing the truth of propositions" (p. 346). We would add, in closing this section and in introducing the next, that these new mathematical insights were found in our study to flow both in the direction of the proof-giver and in the direction of the proof-receiver – the proof-receivers being both students and teacher.

The Subsequent Interview with Michael

The 35-minute interview with Michael took place at the close of the proving activity. It inquired into a range of issues related to his views on the research project, as well as his impressions of the most recent activity involving the $x^n - 1$ task. The main thrust of his reflections, which are captured in the following verbatim extracts, focus first on his initial expectations, then on his changed views after having experienced a few months of classroom activity with the project tasks and CAS technology, and finally on related future plans.

His Initial Expectations – Extract 1

Interviewer: Did you have any expectations or apprehensions about the proposed use of symbolic calculators in your math class before this project started?

Michael: Hmh, I guess I wasn't sure how it was going to go; I was apprehensive to some degree. I was a little bit concerned about how the students would take to it and whether they would see it as being dragged away from what they needed to do. I was a little bit worried about how the parents would take it, but that's been no issue at all. In parent-teacher interviews, a lot of them said they were quite pleased that we're doing some of these things and pushing the kids a little bit further. My feeling about the project itself was that we had enough time to do it, so it couldn't be bad. I'd figured you guys had put some thought into what you were doing and there was a good chance that it was going to be successful and to help them a little. I don't know if my expectations were that huge, but I was hoping there would be something there.

His Changed Views – Extract 2

Interviewer: Do you now see this technology as playing a different role in your class from the time before the project started?

Michael: Yes, for sure, because before the project started, like I said, I hoped it would be good, but my expectations were not that high about it. I certainly have been very pleasantly surprised with what's happened and I don't think I would have considered when we did this in June last year – when we went for the training days – I don't think I would have considered that I would be at this stage. I didn't think I would have been in a situation where I'd be saying to you: "I want to use this again next year." I don't think that those were my expectations, I thought it would be ok and kind of fun, and a nice diversion, but I didn't think we would be quite at the level that we are. I guess my expectations were a lot lower than what we've achieved.

Brief Commentary on Extracts 1 and 2

Michael had not had high expectations at the outset of the project. This makes the results all that much more interesting and persuasive. Some mathematics educators have been heard to express some reservation regarding the role that technology can play in the learning of mathematics, a few even suggesting that it is the already-converted with their "rose-tinted glasses" who are technology's greatest proponents. Yet, here we have a teacher who was not already "converted" at the start of the project and who, as the project progressed, became highly impressed with the way in which the technology was serving to enrich the mathematical learning of his students.

The Tasks and the Technology – Extract 3

Interviewer: If I were to ask you to describe what you think has been the impact on the students of this project both mathematically speaking and technologically speaking, how would you describe this impact?

Michael: I think the biggest impact, and the thing I've been most happy with is the way you guys have designed the activities. It's the way that we've challenged their [the students'] thinking and actually made them think about a process that maybe they knew how to do, but made them think about why they're doing it that way. And I think that's what the calculator has helped them to do and helped them to really, really look at whether they understand the material – basically the meta-cognition kind of idea of thinking about the process you're going through yourself, the thinking you're going through. That's something we don't do enough of in mathematics, I think we should do and I really like to do it. So in a mathematical sense I would say that's been the biggest thing. . . . The learning through the technology was amazing. But it's the amount of work that you put into these activities, and that's why they were so successful. The technology is nothing by itself. That's why it's been such a pleasure to do this and why I have really enjoyed it – it's because you people clearly know what you're doing and have spent a lot of time organizing these. Like you said, we [the teachers] were involved, but only to a small extent. And it's been really good to see how the kids have developed with these [the tasks] and worked with them.

Change in His Teaching – Extract 4

Interviewer: Has this project affected your style of teaching in any way?

Michael: [shy-laughing] I don't know. I think it's made me think more, or made me realize that what I like is making them think a little bit more. And I think I did that anyway, I remember when you came into class last year that there were some things similar happening, but it just made me, just consider a little bit more: Can I let them come through this themselves, let them try this out themselves a little bit more, which I think I always did – but just seeing these activities work, it's made me realize there's more scope to it than I have done in previous years. There is much more scope to let them really go and really know the material properly. So, [to answer your question] I think so, a little bit.

Pushing Students to Go Farther Mathematically – Extract 5

Interviewer: Has the project altered your view of the nature of the mathematics content that can be taught at this level?

Michael: Yes. Because some of the things that you had in those activities I wouldn't have touched. Such as, especially the last activity [the $x^n - 1$ task set], you know there's no way I would have gone anywhere with that. It was way beyond anything that they need to know, but just doing that activity was such a fulfilling experience for, not just for me, I spoke with some of the kids afterwards, and they really enjoyed it. They really did! Just going way beyond what they needed to do [in the math program for that grade level] and they were all able to do it. The really nice thing about that activity is that, at the end of it, everyone had something. Even if they didn't all have as nice a little proof as Janet and Alexandra, all of them had worked someway along the lines to get to something. So, so yeah, it certainly opens up things and they couldn't have done that without the technology. So, so for sure is the answer to your question.

Brief Commentary on Extracts 3, 4, and 5

While the CAS technology was deemed by Michael to be essential to the changed nature of his students' mathematical learning, he was quick to point to the role played by the task activities. He emphasized that the technology by itself would not have produced that which he and his students experienced in this new learning environment. It was the tasks and the way in which they pushed students beyond what is normally asked of them in their mathematics program that seemed crucial. The task sheets included questions that were not only different, and also rich and challenging extensions of that with which they were already somewhat familiar, but which proved to be quite feasible because of the presence of the CAS technology. In addition, the content of the tasks provided Michael with the grist needed to pose additional questions and to encourage his students to think hard about difficult ideas. The intertwining of novel and substantive mathematical tasks, and technological tools appropriate for these tasks, led to mathematical activity that the students quite enjoyed and from which they learned a great deal. This, in turn, promoted the development of new awarenesses on the part of the teacher, awarenesses that will be discussed shortly.

Using Technology to Increase Student Involvement and Promote Learning – Extract 6

Michael: With this technology, the learning is not the same [as with my teaching at the board]. Learning goes much further, it is much more involved. That's why I have really, really enjoyed it. Normally, I'd be involving about two or three of them, but not the entire class. With this tool, it gives them the extra level of ability, and it involves more students. It gets them into it a lot more.

Interviewer: You mentioned that the technology in combination with the activities made them think a lot more about their mathematics, and some of the different steps in the process. What if they hadn't had the technology, could you have seen or imagined that the same sort of progress would take place with similar activities but not incorporating technology?

Michael: I guess with some of the activities it would have been possible, but I think with some of them it would have been either impossible or very, very fake 'cause you'd have to give them answers anyway, you'd have to give them results from the calculator. If you take the last activity, the activity 6 [the $x^n - 1$ task] that we did. The only way you could have done that without the technology would have been to give them what the calculator gave them itself. So then it becomes less hands-on, they don't get into it as much. The fact that they derived these things and went through the process of "there it is [gesturing to the right to suggest a paper-and-pencil result], that's what the calculator gave [gesturing to the left – emulating a comparison], how do I reconcile the difference, how do I factor this, how do I do it?" I think that without the technology it would be so artificial that it would lose them. And basically, you would have had to use the technology at some level anyway to give you the answers, so the fact that they could discover things themselves is a valuable effect.

Future Plans – Extract 7

Interviewer: Do you see yourself using this technology and these activities perhaps next year in the same class?

Michael: I was actually saying to David [his colleague and department chair] that I fully intend using them. You know that our school is in the process of moving toward laptops and it's interesting that when you do that, there is a lot of what I would call – to use a derogatory term – "fluff" to it: "Ok, that's a nice powerpoint, that's a nice little internet site to look at this, or you can use a smart board and you can highlight this and that looks really pretty." So, when I asked you for the DVD of the class lessons on the $x^n - 1$ task, the reason why I asked for it

> is that I want to show people [my colleagues] what can actually be done. Even when we were not actually using the technology all the time [in class], the learning through the technology was huge.

Brief Commentary on Extracts 6 and 7

In orchestrating his classroom teaching, Michael did not always use the technological artifact of the classroom view-screen hooked up to a TI-92 CAS calculator. Yet, he was firmly convinced that the "learning through the technology was huge." Not only did it allow the students to go farther mathematically, it encouraged them to be more active and more involved participants in the process of learning. Michael's impressions of the project activity were so favorable, in fact, that he wanted to continue using the tasks and the technology the following year; he also wanted to share the video of the lessons of the last two classes on the $x^n - 1$ task with his colleagues, just so that they could see what is possible with this technology and with the kinds of tasks that were developed within the project.

The above interview-highlights are now analyzed in the light of what we observed during the two classes, with a view to focusing more particularly on the teacher and his learning.

Analysis and Discussion

As stated by Michael, it was his participation in the research project, a project involving tasks and technologies that were new to him, which led to new awarenesses regarding his practice of teaching algebra. From an analysis of the two observed lessons in conjunction with the follow-up interview, we too became aware of the changes that had occurred in Michael. This section focuses first on what changed in Michael and, second, on what enabled those changes.

What Changed in Michael

Zaslavsky and Leikin (2004) have pointed out that, by observing their students' work and by reflecting on this work, teachers learn through their teaching. We have found this to be the case for Michael, as well. In particular, Michael's learning was in three areas: his knowledge of mathematics, his knowledge of mathematics teaching and learning, and his practice of teaching.

His Knowledge of Mathematics

Before his participation in the project, Michael had never really had the opportunity to think about a general rule for factoring the family of polynomials of the form

$x^n - 1$. The prior workshop sessions between the members of the research team and the participating teachers had included discussions on this task and on one of the ways that its proving component might be thought about. However, the mathematics that Michael learned from engaging in the two class lessons on this task went beyond that which he had learned during the workshop. It involved specifically certain ways to approach the proof problem; moreover, this new learning on Michael's part was provoked by the students themselves.

The proof produced by Janet and Alexandra, which involved factoring by grouping with the generic example for $n = 8$, was one that had not occurred to Michael before his students actually generated it. He found it a "nice little proof," to use his own words. A second contribution to the mathematical learning of Michael was occasioned by the proving attempt of Andrew. While Andrew was describing what he had observed for the CAS factoring of $x^{10} - 1$, it became clear that not only was this an interesting finding for his classmates, but also for the teacher, Michael. The new pattern that Andrew had noticed regarding the factoring of x^5+1 as $(x+1)(x^4-x^3+x^2-x+1)$ paralleled the pattern that he had noticed earlier regarding x^3+1 as $(x+1)(x^2-x+1)$. Although time did not allow for the proving of his conjectured new rule nor for its integration into the previous proofs that had been put forward during that lesson, there was no doubt that this was a new piece of mathematics for Michael.

His Knowledge of Mathematics Teaching and Learning

As reported by Zaslavsky and Leikin (2004), teachers' engaging in learning activities designed for student mathematical learning can be an effective vehicle for their professional growth. An additional factor that has been emphasized by Mason (1998) is that it is one's developing awareness in actual teaching practice that constitutes change in one's "knowledge" of mathematics teaching and learning. Although Michael did participate in our professional development workshop prior to his integrating the novel tasks and technology into his teaching, it was his actual practice with these materials that had the greater impact regarding his "developing awarenesses" in the area of mathematics teaching and learning. We note five of these new awarenesses.

- Michael developed a new awareness of what students at this grade level can accomplish mathematically – given appropriate tasks – as well as the realization that they can go further mathematically than expected (Extracts 4 and 5).
- Michael developed a new awareness of the role that technology can play in the mathematical learning of students (Extracts 5, 6, and 7).
- Michael developed a new awareness that students' mathematical knowledge changes with the combined duo of "task-technology" (Extract 3).
- Michael developed a new awareness of how he might further provoke mathematical reflection in students; that is, that he could go even further than he usually went in his questioning, given appropriate tasks (Extract 4).

- Michael developed a new awareness regarding the culture of the class: It changes when technology is present and is used in the classroom. Students become more involved; they are more autonomous (Extract 6).

His Practice in Subsequent Mathematics Classes

It was not only Michael's awarenesses, which developed during this project. These new awarenesses were translated into practice. As he had said during his interview, he fully intended to take advantage of the wider scope offered by the project tasks and technology and to use them in subsequent years to push his students into thinking more deeply about their mathematics (Extracts 2 and 4). We continued to observe several of Michael's classes during the two years following this study. We witnessed, just as he had hinted he would do, a continuing development of his approach to encouraging students to go a little further in their thinking. This reflected his realization, which he stated during the interview, that he could push his students to think a little bit more about their mathematics. In addition, we also observed that he never stopped using the tasks and CAS technology that he mentioned he had so much enjoyed using during the present study. Thus, we were able to note a further evolution in his teaching practice – a practice characterized by the newly-acquired awareness of the role that technology, when accompanied by appropriate tasks, can play in the development of students' mathematical learning. Regular conversations with him during the ensuing years, including one quite recently, have highlighted his and his students' successes with, in particular, the $x^n - 1$ task with its proving component.

What Enabled These Changes

We can point to several factors that enabled the changes that we observed, as well as those that were disclosed to us by Michael himself. These enabling factors were found to include the following: (a) access to the resources and support offered by the research group; (b) use of CAS-supported tasks whose mathematical content differed from that usually touched upon in class; (c) Michael's disposition toward student reflection and student learning of mathematics; (d) the quality of the reflections of his own students on these tasks; (e) Michael's attitude with respect to his own learning. The first two factors relate principally to the role played by resources "from without," while the remaining three could be said to be "from within" in that they concern the given teacher and his students. However, as will be argued, it is the interaction of the two dimensions that promoted teacher change.

Access to the Resources and Support Offered by the Research Group

As was noted above, the change in Michael's knowledge of mathematics was occasioned by two different, but related, experiences. The first of these involved his prior discussions with members of the research group. These discussions had focused on new tasks and thus new mathematical awarenesses, which thereby constituted a first round of change with respect to Michael's existing mathematical knowledge. While the ideas for, and initial design of, the tasks came from the research group, these were shared with the participating teachers in a workshop setting that involved their working on the tasks themselves. They were then invited to offer feedback and to suggest changes to the tasks. These changes were subsequently integrated into a modified design for the tasks. Thus, the first exposure to the mathematical ideas inherent in the tasks occurred during the workshops that preceded the integration of the tasks into the teachers' classroom lessons. It was at these workshop sessions that Michael initially encountered the mathematics of the $x^n - 1$ task. This was also his first introduction to the use of CAS technology.

As the project unfolded and the researchers became a regular presence in Michael's class, there was ample opportunity to provide ongoing support to Michael. The researchers were able to offer assistance of a pedagogical, mathematical, and technical nature, whenever Michael so desired. In actual fact, such requests for support were quite rare, as each of the tasks was accompanied by a teacher version that included suggestions regarding discussion ideas, as well as additional information of both a mathematical and didactical sort. The normal interaction between Michael and the researchers after each lesson tended to be informal and conversational, much like that between collaborators.

Use of CAS-Supported Tasks Whose Mathematical Content Differed from That Usually Touched upon in Class

Michael had expressed the fact that the $x^n - 1$ task with its exploration of the factors of this family of polynomials for integer values of n, along with its proving component, was a type of task that went far beyond what the students "needed to know." (It was also a new type of task for him.) While he might never have presented such a task to his students in the past, the experience of doing so convinced him that such tasks are indeed not only feasible, but also enjoyable to the students and lead to deeper mathematical reflection on their part (Extract 5). Based on our observations, we contend that novel tasks such as this one can change in a positive manner the usual teaching-learning dialectic of the mathematics classroom and are at the heart of both student and teacher learning.

Watson and Mason (2007) have argued that, "factors which influence the effectiveness of a task in promoting the intended kind of activity include ... established practices and ways of working; students' expectations of themselves and of each other as influenced by the system and their pasts ..." (p. 207). While "novel tasks" may in fact be part of some teachers' "established practices and ways of working," we think rather that novel tasks – especially those involving proofs – in which algebra students have never before engaged, are likely the exception and not usually the norm. The very absence of "established practices" or student "expectations" may, in fact, lead to the success of novel tasks, and thus to some nuancing of Watson and Mason's statement.

Michael's Disposition Toward Student Reflection and Student Learning of Mathematics

Michael worked very hard at encouraging his students to reflect, at giving them time to reflect, at listening closely to their reflections, and at having them share their reflections with the rest of the class. Even if he expressed the realization that, with the help of the activities designed by the research group, he could do even more in this regard, he was already predisposed to such practice. This predisposition was of course related to the importance Michael ascribed to students' learning to think for themselves. One of the signs of this didactical stance on mathematical learning was the way in which he often presented counterexamples to challenge students' thinking rather than immediately correcting them or giving the right answer. He aimed at having students develop their mathematical reasoning and critical thinking.

In her study of one teacher's practice of listening and responding to students' solution strategies, Doerr (2006) found that, "as the teacher asked for students to describe and explain their thinking, this not only contributed to the teacher's understanding of their thinking, but it created a situation where the students could refine their thinking and shift to a new way of thinking about the problem" (p. 20). As the case of Michael suggests, not only does listening to students support the development of students' thinking, it also leads to new awarenesses and professional growth in the teacher. Had Michael been a teacher who did not encourage the voicing of his students' mathematical ideas, he would hardly have come to know that "their learning through the technology was huge" (Extract 7), nor would he have realized the pedagogical role that technological tools can play in enhancing mathematical learning. Thus, as Leikin and Levav-Waynberg (2007) have pointed out, a teacher's pedagogical principles can provide support for the growth of his/her knowledge regarding not only student learning but also teacher learning. In fact, Michael's pedagogical disposition with regard to mathematical activity (i.e., his view on encouraging the voicing of student reflection) served to beget not only new pedagogical knowledge regarding mathematics teaching and learning but also new mathematical knowledge for him.

The Quality of the Reflections of His Own Students on These Tasks

During the interview, Michael mentioned on several occasions how struck he was by the quality of the mathematical contributions of his students, contributions such as those by Janet and Andrew, which had evoked new mathematical insights within Michael, as well as within the students of his class. He was clearly a teacher who could learn from his students, just as did the teachers in the Leikin and Levav-Waynberg (2007) study; these researchers reported that "teachers who are sensitive to their students and flexible in their interactions with them, and who grant students autonomy in learning, end up learning mathematics from their students' replies" (p. 366).

In addition to Michael's development of mathematical knowledge from his interactions with his students, so too was the development of his knowledge of mathematics teaching and learning enhanced by the quality of his students' reflections. New awarenesses, such as, "that which students at this grade level can accomplish mathematically" and "the role that technology can play in the mathematical learning of students," were occasioned by the students themselves. This is in contrast to Monaghan's (2004) findings that some teachers in his study noticed that "tasks in technology-based lessons led their students to focus on the technology and at least three of the teachers felt an 'is this maths?' tension when their students attended to technological details at, in their opinion, the expense of the mathematics" (p. 336).

Michael's Attitude with Respect to His Own Learning

Each time that Michael said during the interview, "it's made me realize . . .," we interpreted this to indicate Michael's openness to learning from his project participation. He experienced a great deal of joy – mentioned many times throughout the interview – at seeing how the students were positively responding to the tasks and, thus, in his learning not only about the mathematical levels they were reaching from their experience but also about the ways in which the technology and the tasks themselves were encouraging this response. He was in fact reflecting on his students' reflections. It is also noted that Michael was open to participating in the project right from the start, to learning something from it – even if he was not sure initially whether it would lead to new learning for his students. Watson and Mason (2007) have emphasized that "to become an effective and professional mathematics teacher requires development of sensitivities to learners through becoming aware of one's own awarenesses" (p. 208). There is little doubt that the professional awarenesses Michael developed throughout the project, and which he shared with us, constituted a heightened sensitivity regarding his learners.

Reflections on What Changed in Michael and on What Enabled These Changes

We stated above that the changes in Michael's knowledge of mathematics, of mathematics teaching and learning, and in his practice of teaching were enabled by two kinds of factors, those from without and those from within. However, as our study progressed, it became clear that it was in the interaction of these two types of factors that change was promoted. Had it not been for the "from-without" factors, that is, the access to the resources and support offered by the research group and, consequently, the use of CAS-based tasks whose mathematical content differed from that usually touched upon in class, then the "from-within" factors, such as, the quality of the reflections of his own students on these tasks, would not have been put into play. Similarly, had it not been for "from-within" factors, such as Michael's disposition toward student reflection and student learning of mathematics, as well as his attitude with respect to his own learning, then the "from-without" factors related to the research team's contributions would not have taken root and flowered. Both types of factors supported each other in a mutually intertwining manner.

This is of interest from a theoretical perspective. It suggests firstly that the integration of novel materials and resources that have been designed to spur mathematical learning is more likely to be successful when the teachers who are doing the integrating are able to see that these resources are having a positive effect on their students' learning. Secondly, the novel materials and resources have a greater likelihood of producing this positive effect on student learning when the teacher doing the integrating engages in teaching practices that encourage student reflection and mathematical reasoning. The synergy between the two types of factors was found to be a positive force in the development of Michael's professional awareness, and one that constituted change not only in his knowledge of mathematics and mathematics teaching/learning, but also in his practice.

One final remark in this section concerns the role of the CAS technology on Michael's learning. As mentioned earlier, Michael's prior experience with classroom technology had included mainly graphing calculators, but not the CAS. Before the unfolding of the project in his own classroom, he never imagined the impact of this technology on his students' mathematical learning, and thus on his own learning of what his students could accomplish. At the heart of his coming to see that the student learning through the technology was huge was his realization that the presence of the technology changes the nature of the questions that can be asked of students, and thus the kind of mathematical reflection they engage in. While the tasks themselves were a crucial component of Michael's learning within his own practice, the actual design of the tasks was set up in such a way as to work hand-in-hand with the affordances of the technology. In fact, the first two parts of the $x^n - 1$ task set, which were foundational to the proving part of the activity, could not have been managed without the CAS. In this respect, the CAS technology was central to Michael's learning.

Concluding Remarks

In conclusion, we wish to emphasize briefly only two issues. One is that, while this study fits into the broad research domain of teachers learning from their own practice (e.g., Jaworski, 2006; Zaslavsky & Leikin, 2004), a significant feature has been that the practice was nourished by input coming to a large extent from outside. The second issue concerns the mathematical activity that was stake in the study, that of proving.

With respect to the first issue, much of the research related to teachers' learning from their own practice emphasizes teachers' planning of their interactions with students, followed by their subsequent reflective analysis of these interactions. Considerably fewer studies (exceptions include, e.g., Leikin, 2006) follow the path that we did where the majority of the planning of the instructional interaction with respect to the mathematical content and the task questions to be posed to the students had already been elaborated in advance by the research team, even if in partial collaboration with the participating teachers. This, we feel, added a dimension to the study that does not often come into play in research on teaching practice. As a consequence, the teacher's reflective analysis of his interactions with the students had to take into consideration – in a somewhat different manner than would otherwise be the case – the worthiness, or not, of the particular mathematical content at stake, the way in which it was elaborated, and the technological tools that were used to support its approach. The positive nature of the reflections shared by Michael during the post-lesson interview with one of the researchers suggests that the integration of resources coming from without can be a powerful stimulus to teachers' learning from their own practice.

With respect to the second issue, only rarely does the teaching of algebra in high school include activity with proving. The teacher featured in this study, Michael, could be said to have been very courageous in agreeing to integrate into his teaching of algebra the $x^n - 1$ task with its proving component. He had never before included proofs within his algebra teaching; nor had his students ever engaged in this form of algebraic activity. Nevertheless, the success that he and his students experienced with it went way beyond his (and likely their) expectations. Hanna and Barbeau (2008) have raised the following query: "Approaching proof as more than a formal way of certifying a result is bound to make increased demands on the teacher and involve more engagement by the students; the long-term value would seem to be clear, though not quantified, but can the increased demands be managed?" (p. 352). Michael's and his class's experience with the proving segment of the $x^n - 1$ task provides a strong existence proof of the notion that the increased demands can indeed be managed.

Acknowledgments The research presented in this chapter was made possible by grants from the Social Sciences and Humanities Research Council of Canada (Grant # 410-2007-1485), from the Québec Ministère des Relations Internationales, and from CONACYT of Mexico (Grant # 80454). We express our appreciation to our research colleagues, as well to the class of students and their mathematics teacher who participated in the research.

References

Adler, J., Ball, D., Krainer, K., Lin, F.-L., & Novotna, J. (2005). Reflections on an emerging field: Researching mathematics teacher education. *Educational Studies in Mathematics*, *60*, 359–381.

Artigue, M. (2002). Learning mathematics in a CAS environment: The genesis of a reflection about instrumentation and the dialectics between technical and conceptual work. *International Journal of Computers for Mathematical Learning*, *7*, 245–274.

Balacheff, N. (1987). Processus de preuve et situations de validation. *Educational Studies in Mathematics*, *18*, 147–176.

Balacheff, N. (1988). Aspects of proof in pupils' practice of school mathematics. In D. Pimm (Ed.), *Mathematics, teachers and children* (pp. 216–235). London: Hodder & Stoughton.

Bergqvist, T. (2005). How students verify conjectures: Teachers' expectations. *Journal of Mathematics Teacher Education*, *8*, 171–191.

Chevallard, Y. (1999). L'analyse des pratiques enseignantes en théorie anthropologique du didactique. *Recherches en Didactique des Mathématiques, 19*, 221–266.
Doerr, H. (2006). Examining the tasks of teaching when using students' mathematical thinking. *Educational Studies in Mathematics, 62*, 3–24.
Hanna, G. (2005). Challenges to the importance of proof. *For the Learning of Mathematics, 15*(3), 42–49.
Hanna, G., & Barbeau, E. (2008). Proofs as bearers of mathematical knowledge. *ZDM Mathematics Education, 40*, 345–353.
Healy, L., & Hoyles, C. (2000). A study of proof conceptions in algebra. *Journal for Research in Mathematics Education, 31*, 396–428.
Jaworski, B. (2006). Theory and practice in mathematics teaching development: Critical inquiry as a mode of learning in teaching. *Journal of Mathematics Teacher Education, 9*, 187–211.
Kieran, C., & Drijvers, P., with Boileau, A., Hitt, F., Tanguay, D., Saldanha, L., & Guzmán, J. (2006). The co-emergence of machine techniques, paper-and-pencil techniques, and theoretical reflection: A study of CAS use in secondary school algebra. *International Journal of Computers for Mathematical Learning, 11*, 205–263.
Lagrange, J.-b. (2002). Étudier les mathématiques avec les calculatrices symboliques. Quelle place pour les techniques? In D. Guin & L. Trouche (Eds), *Calculatrices symboliques. Transformer un outil en un instrument du travail mathématique: Un problème didactique* (pp. 151–185). Grenoble, France: La Pensée Sauvage.
Lee, L., & Wheeler, D. (1987). *Algebraic thinking in high school students: Their conceptions of generalisation and justification* (Research report). Montreal, Quebec: Concordia University, Mathematics Department.
Leikin, R. (2006). Learning by teaching: The case of Sieve of Eratosthenes and one elementary school teacher. In R. Zazkis & S.R. Campbell (Eds.), *Number theory in mathematics education: Perspectives and prospects* (pp. 115–140). Mahwah, NJ: Lawrence Erlbaum Associates.
Leikin, R., & Levav-Waynberg, A. (2007). Exploring mathematics teacher knowledge to explain the gap between theory-based recommendations and school practice in the use of connecting tasks. *Educational Studies in Mathematics, 66*, 349–371.
Mariotti, M. A, (2002). La preuve en mathématique. *ZDM Mathematics Education, 34*, 132–145.
Mariotti, M. A. (2006). Proof and proving in mathematics education. In A. Gutiérrez & P. Boero (Eds.), *Handbook of research on the psychology of mathematics education* (pp. 173–204). Rotterdam, The Netherlands: Sense Publishers.
Mariotti, M. A., & Balacheff, N. (2008). Introduction to the special issue on didactical and epistemological perspectives on mathematical proof. *ZDM Mathematics Education, 40*, 341–344.
Mason, J. (1998). Enabling teachers to be real teachers: Necessary levels of awareness and structure of attention. *Journal of Mathematics Teacher Education, 1*, 243–267.
Monaghan, J. (2004). Teachers' activities in technology-based mathematics lessons. *International Journal of Computers for Mathematical Learning, 9*, 327–357.
Mounier, G., & Aldon, G. (1996). A problem story: factorizations of $x^n - 1$. *International DERIVE Journal, 3*, 51–61.
Rowland, T. (2002). Generic proofs in number theory. In S. R. Campbell & R. Zazkis (Eds.), *Learning and teaching number theory* (pp. 157–183). Westport, CT: Ablex Publishing.
Watson, A., & Mason, J. (2007). Taken-as-shared: A review of common assumptions about mathematical tasks in teacher education. *Journal of Mathematics Teacher Education, 10*, 205–215.
Weber, K. (2008). How mathematicians determine if an argument is a valid proof. *Journal for Research in Mathematics Education, 39*, 431–459.
Weber, K., & Alcock, L. (2005). Using warranted implications to understand and validate proofs. *For the Learning of Mathematics, 25*(1), 34–38.
Zaslavsky, O., & Leikin, R. (2004). Professional development of mathematics teacher educators: Growth through practice. *Journal of Mathematics Teacher Education, 7*, 5–32.

Learning Through Teaching, When Teaching Machines: Discursive Interaction Design in Sketchpad

Nicholas Jackiw and Nathalie Sinclair

Introduction

Seymour Papert famously asserted that in technology-based classrooms, the distinction between teachers and learners would be blurred. In *Mindstorms*, he suggested several specific ways in which this might happen. For example, he pointed to the way in which children and their teachers would work on problems together, blurring the novice/expert distinction that pervades mathematics classrooms. He also discussed the way in which the curriculum—or at least the development and flow of content—might be controlled, at least in part, by the learner. Since the computer provides a generative set of materials to think with, learners can engage generatively with these ideas, thereby producing new ones. In his work with Idit Harel, he showed how students engaged in learning-through-teaching, by using Logo to teach the computer how to "do" fractions (see Papert & Harel, 1991). Indeed, learning mathematics was done by programming Turtle Logo—an activity Papert metaphorically described as teaching the turtle new words.

In this chapter, we extend Papert's work on blurring this distinction, between students and teacher, by focusing on the discursive interaction between different players in the technology-enhanced mathematics classroom. Instead of the physical and motivational roles of teachers and learners, we pursue the discursive and narrative roles that they adopt, first in a generic computer-enhanced classroom, then more specifically in interactions with *The Geometer's Sketchpad* (Jackiw, 1988) dynamic geometry environment. There we find evidence not only of Papert's *blurring* of roles, but, moreover, of a role *inversion*, in which computers become "students" and learners become "teachers" through their discursive interaction. We explore this claim in detail by analyzing specific linguistic mechanisms through which Sketchpad's design constructs and shapes its users' experience. Where typical DG

N. Sinclair (✉)
Simon Fraser University, Burnaby, BC, Canada
e-mail: nathsinc@sfu.ca

This material is based in part upon work supported by the National Science Foundation under Grant No. 0521981.

R. Leikin, R. Zazkis (eds.), *Learning Through Teaching Mathematics*, Mathematics Teacher Education 5, DOI 10.1007/978-90-481-3990-3_8,

research focuses on visual technology representations of mathematics, our emphasis here on verbal representations better allows us to pursue social, rather than cognitive, implications of technology design. Language, the participationists claim, is not only the critical tool for analyzing social roles in a classroom, but the medium in which those roles are defined and perpetually renegotiated. A stronger claim, from Wittgenstein through Vygotsky and Sfard (2008), is that language forms both the center and horizon of all mathematical knowing. In bringing Sfard's discourse approach to DG, we seek not to displace claims about the importance (to that technology) of visual cognition but rather to extend those approaches with critical tools better suited to the analysis of social roles and their meanings.

Discursive Roles of the Computer

What is the role of the computer in the classroom? Many models and metaphors have been proposed (Pimm, 1983, offers a useful catalogue of such metaphor). Perhaps the least subtle way to conceive of the computer's structural role in the mathematics classroom is simply as a novel medium in which mathematics is represented and manipulated—in the tradition of paper, blackboard, or—reaching back—papyrus and clay tablet. This is the conception that underlies descriptions of electronic blackboards, digital notebooks, or indeed the broad term of art multi-media. As a metaphor, this construction resituates to the computer *where* one "does the math," but leaves the more active classroom roles of the teacher and the student—as instructor and apprentice, in the mathematics' "doing"—powerfully untouched by the computer. A more content-oriented technological incarnation of this idea is the paradigm of *computer as textbook*—again, essentially a medium combined with authoritative curricular inscriptions in that medium. And it is easy to read Sketchpad's presentation of a blank sketch with drawing tools in this metaphor—basically as a digitally-enhanced medium for creating and depicting geometric drawings. And yet, when the student drags a vertex in Sketchpad, or more generally, executes a command in a computer environment, the computer *responds*. This takes us almost immediately beyond the realm of all prior media, which "respond" only to the degree they obey certain physical laws. In other words, the computer must be seen as taking on some discriminatory agency[1] in the classroom, and so—though powerful in its emphasis on representation—the metaphors of "computational media" do not help understand this fundamental interactivity.

[1] Pickering (1995) argues for the central role of agency in scientific work, and even accords material agency to the tools, machines, and artifacts of the scientific laboratory. While he also discusses individual human and disciplinary agency in the specific context of mathematics, he discounts any possibility of material agency there. A re-interpretation of the possible role of material agency in the development of mathematics might accord the various media traditionally used in mathematics a much more important role than merely unresponsive devices for inscription and representation.

An early metaphoric attempt to invest the computer with greater classroom agency explored the possibility of *computer as teacher*. This construction is explicit in the no longer fashionable program of "computer-aided instruction" (CAI), where the computer substituted for a human teacher in the presentation of test material to a human learner. Over the past twenty years, CAI has been replaced as a term by "computer-aided learning" (CAL), but the metaphor's agenda remains the same: The learner in question is the human student user; and software "aides" the learner by instructing her. We can see this language-slippage in the titles and abstracts of many present-day journals articles.[2]

Much of this disfavor into the term of CAI has fallen reflecting the degree to which early attempts at positioning *computer as teacher* failed to respect the sensitive and complex role of the human teacher in the mathematics classroom. And yet, the paradigm survives, even though the language changes to hide its tracks: Twenty years after "end" of CAI, the broadest conception of educational software still operative in schools is software that presents subject matter, drills technique, and assesses quantitative response (Becker, Ravitz, & Wong, 1999).

Whether these computational activities are hallmarks of "the teacher" does not interest us. Instead, we claim that in order for the computer's fundamental interactivity to be explained through metaphors of computational teaching, that it must not simply *act* as a teacher (that is, wield authority, command knowledge) but *interact* as a teacher (in its communication with and response to students). Thus to test the legitimacy of this metaphoric paradigm,[3] we seek in the following sections to identify discursive markers characterizing teacher interactions with students, and then to evaluate the computer's interactions with students in those same terms.

Organization of Discourse Around Evaluation

According to Stubbs (1979), the most conspicuous characteristic of a teacher's discourse—in terms of speech events in the classroom—is its constant organization around organizational and evaluative commenting. Pimm (1994) notes that these "meta-comments" arise from the teacher's stance that "the utterances made by pupils are seen as appropriate items for comment themselves" (p. 139). Such meta-commenting might include echoing or re-voicing what a student has said; or showing how it fits, or does not fit, with the rules and norms of the classroom. (Stubbs highlights the way in which this type of discourse is exclusively characteristic of teaching; in any other context, "it is generally regarded as boring, incongruous, inappropriate, pedantic, devious!").

---

[2] Consider this article focusing on CAL-based arithmetic instruction: "This study develops and implements a computer-assisted learning (CAL) program [featuring] both multiplicative facts *and the instruction of meanings behind those facts* (Chang, Sung, Chen, & Huang, 2008, p. 2904, our emphasis)."

[3] Pimm (1983) also offers several metaphors for the computer such as super-calculator and image-maker, but also computer as human being.

In writing about specific types of meta-commenting that occur in classrooms around the use of precise mathematical language or rules, Hewitt (2001) describes the example of students working on plotting the point (2, 3) on the Cartesian coordinate system. While student utterances such as "you go two up and three over" are perfectly well-understood by the teacher and could be executed successfully by any reasonable interpreter, we would not be surprised if the teacher were to re-voice this as "yes, you go three *over* and two *up*" or even "yes, you go three in the positive direction horizontally and two in the positive direction vertically," repeating and normalizing sequence, structure, and vocabulary not explicit in the equivalent student experience.

Breaking Conversational Maxims

Given Stubbs' point that this core aspect of teacher meta-commentary would seem inappropriate in most other conversational settings, we might not be surprised to find other ways in which discursive patterns in teaching break conversational norms. Grice's (1975) conversational maxims, which flesh out the cooperative principles by which normal speakers extract meaning from normal conversations, offer a means for articulating additional characteristic features of teacher-talk.[4] The first of Grice's four maxims is that of quality, in which one speaker assumes the other is representing himself honestly. (The other maxims involve quantity, relevance, and manner.) This first maxim is violated when speakers lie or intentionally contribute something factually erroneous, but also, perhaps more subtly, when they do something such as feign incomprehension or incompetence.

Teachers routinely violate Grice's maxim of quality when they ask a question to which they already know the answer. Since in normal conversation, speakers assume the maxim of quality, the fact that the teacher regularly asks such questions—and moreover, that students know the teacher is inauthentic in her feigned ignorance of the answer—marks the classroom conversational contract as highly unusual. Just how unusual is seen when we substitute students not already acculturated to the teacher's duplicity, as illustrated in Crawford's (1996) study of Australian aboriginal schoolchildren first encountering anglo-saxon school mathematics. "Aboriginal communities find the educational practice, used frequently by teachers of mathematics, of asking students questions when the answer is already known to the teacher, extremely puzzling and distasteful (p. 135)." Thus we place "didactic duplicity" along with "meta-commenting" as hallmarks of teacher interaction with learners.

---

[4] See Gerofsky (1999) for a discussion of the application of Grice's maxims to the use of word problems in mathematics.

Authority and Imperatives

Pimm (1994) asks the question of how different teachers might manifest the control they have through their discourse. Clearly, one way of doing this is through the meta-commenting, since the constant evaluation of student utterances allows the teacher to retain a position of power. More forcefully, teachers' discourse often relies on the imperative voice: do this, stop doing that, *factor* here and *solve* there—than any peer-to-peer discourse. As evident from these examples, the commands apply not only to classroom management but also to mathematical actions. Regardless of the form of instruction (constructivist, traditional, or other), the issuing of commands by the teacher places the student in the position of being commanded and as responsible for carrying out the instruction of the teacher.

That the imperative voice is a standard predicate construction in mathematics writing makes it an even more notable discursive choice for the mathematics teachers. Students also encounter the imperative voice in their textbooks, which like the teacher, become sources of authority. Herbel-Eisermann (2009) points to the way in which the dynamic of power and authority between the teacher and the textbook plays out. And while sometimes authority is given to the textbook, the teacher remains the director, and the student the directed.

Neither Medium Nor Teacher

If the computer is to be seen as playing the role of the teacher, then we would expect it to engage in patterns of discourse similar to the three described above: To take student utterances as opportunities for evaluative commenting; to engage in forms of conversation that are marked by didactic dishonesty (such as asking questions to which they already know the answers); and to retain and deploy authority through the use of the imperative voice. However, in considering the interaction between the student and Sketchpad, we find that Sketchpad engaging in markedly different—and even, perhaps, opposite—discursive patterns.

Consider first the discursive characteristic of meta-commenting in relation to Hewitt's example of plotting points. If the student were to enter (2, 3) on the computer, expecting to go 2 up and 3 over—an expectation that teachers tolerate and correct, in student conversation—the computational result would not match the student's inarticulate intent. In its response pattern—producing the wrong result—the computer does not re-voice; it doggedly executes. This is indeed feedback, as is Hewitt's teacher's revoicing, but it is feedback of a fundamentally different type. One sense of "feedback" is rich in judgment; it is the criticism of an able or worthy expert, and the sort implied by sentences such as "I really value your feedback." Separate from this is "feedback" as used, say, in biomechanics, where the term describes a raw phenomenon, a Newtonian response. Meta-commentary is the first form of feedback; computer execution, the second. That the latter feedback is fundamentally non-evaluative, in a psychological sense, is critical. The student does not wonder whether she was actually right or wrong in her interaction with the computer, nor whether her solution is actually valued. Instead, the student must decide whether what Sketchpad did was what the student wanted: The student evaluates Sketchpad.

Again referring to the example of plotting points, we note that the computer does not violate the maxim of quality. It does not pretend to misunderstand; it does not ask, "Now, how do we plot points?" Instead, the computer responds exactly as it has been instructed to do, not calling into question the intention of the student. But the computer has indeed responded, and the next utterance will be the student's. While some may feel that the computer's response lacks flexibility, Hewitt argues that students find it psychologically easier to follow the arbitrary and concise rules of the computer than those of the teacher, who does indeed have much greater communicational flexibility.

While Sketchpad does not ask question to which it already knows the answer; it does ask questions. For example, when a user selects a shape, and then selects the rotation command, a dialogue box appears, which asks the user for the angle of rotation to be used. Sketchpad does not already know the answer to this question and relies on the user to provide it. In fact, it is the user that frequently asks Sketchpad questions to which they already "know" the answer: measuring an angle that has already been constructed to be 90°, calculating the sum of a triangle's three angles, then dragging its vertices around, when it is already "known" that the sum is invariant. By repurposing the software's tools (or features) supporting inquiry and conjecture instead to the more vainglorious purposes of didactic demonstration and even self-indulgence, students engage in exactly the duplicity of the teacher feigning ignorance of methods, or surprise at conclusions, with which she is—in fact—already entirely familiar.

Finally, regarding the third characteristic of teacher talk, we observe that the computer issues no instructions or commands. In fact, an important role reversal occurs in students' interactions with Sketchpad, in which it is the student who must take on the responsibility of issuing commands: Nothing will happen on the blank sketch without the student first issuing a command, through a literally imperative voice. As we develop in the next section, student utters statements like *Construct Midpoint*, *Plot New Function*, and *Measure Circumference*.

In all of this analysis, we of course do not deny that there is, in fact. often—almost always—a real and human teacher who has instigated the students' work with the computer. Equally often, this work has been initiated with that teacher's command ("Students, construct an equilateral triangle."), or with didactic duplicity ("Students, who can help me remember how to construct an equilateral triangle?"), or with other teacherly language. We are not concerned here with that teacher and the student's relation to her; what we are drawing attention to is the way in which, once the student engages with the software, he himself becomes invested with the power to decree, direct, and demand. It is no longer possible to see the role of the computer here as that of a teacher. And indeed, we reach a surprising insight: Not only is the machine clearly *not* acting as "teacher" in this relationship, but the student clearly *is*. In terms of the evaluative, conversive, and imperative markers of the interaction, the student—the mathematics learner—learns from the relationship not by being *taught by* Sketchpad, but rather, by *teaching* Sketchpad.

The Sketchpad Trajectory Through Language

Is this premise credible? If *Sketchpad* cannot speak, in what sense can it participate in discourse? We are imputing rough parity between human and machine in terms of their capacity for discursive engagement—a move many science-fiction authors warn is ill-advised. And yet, we claim there is a coherent and well-defined linguistic trajectory to users' interactions with *Sketchpad*, an explicit interplay and evolution of language (about the student's desires, capabilities, comprehensions, and motivations) that, at the outset, enables our claim of "discourse," and in the denouement, legitimates it. In this second section, we turn to take a detailed look at the design of *Sketchpad*, to situate and explain this linguistic trajectory in specific reference to the program's user interface and modes of use. In this analysis, we adopt the perspectives of "interaction design" or "experience design" (Moggridge, 2007), methodologies and critiques that differ from conventional graphic design and industrial design by viewing the designed object as fundamentally temporal in nature, capable of responding to its user in ways that stimulate iterated responses to those (first-order) responses—in short, design philosophies that highlight the *interactivity* of, and in, designed objects.

In bringing such a perspective to *Sketchpad*, we must name our human subjects with care. "Students" and "teachers" are distinct actors defined by the drama of the classroom, just as "learner" and "expert" are roles potentially defined by a specialized observing assessor. But in interactions with software like *Sketchpad*, there is only one human role, that of "the user," and it is for this distinct role that the software constructs an experience and situates the conditions of its use. Thus as we track an individual's engagement with this constructed experience, we can expect to see them occupy multiple roles simultaneously (student *and* user; or teacher, user, *and* expert), and can expect those roles to occasionally coincide and occasionally conflict.

Pre-verbal Origins

Before we consider *Sketchpad's* "language," we acknowledge that readers only somewhat familiar with *Sketchpad* may well be puzzled by the idea that there is *any* language in the software. Most relevant research literature characterizes *Sketchpad* as a "Dynamic Geometry" environment—indeed, the term was invented to describe Sketchpad—and thus positions it as primally concerned with moving and interactive images, rather than with words, language, or discourse! Of course, dynamism and visualization are integral to the *Sketchpad* experience, and our focus here on language in no way attempts to deny their primary roles in shaping students' experience with the software. Indeed, the significant features of the Sketchpad screen on first starting the program are entirely visual: a large, blank drawing canvas (the user's "sketch") and a set of graphical icons representing drawing tools (compass, straightedge, and point plotter) for placing and constructing (visual representations of) geometric objects directly in the sketch. Thus, a student taking first steps in *Sketchpad* operates pre-linguistically, pre-symbolically, and almost pre-semiotically: The meanings of interface elements (such as the compass or a straight edge tool) are iconic in form[5] and the properties of geometric objects under consideration (such as length or area) are inherent physical dimensions of their on-screen representations. As the student begins to wield these tools first experimentally, and then with more confident dexterity—placing endpoints here and here, inscribing a triangle, ah, just there—there is a growing sense of interaction, but not yet of language-based communication.

The Emergence of Vocabulary

The first words a student is likely to encounter in *Sketchpad* are those of the program's menus. Reading in English's left-to-right order, the overall menu bar begins with several menus—File, Edit, etc.—so conventional in appearance (from other programs) that they operate almost iconically for any reasonably experienced computer user.

The first menu to promise geometric or mathematical distinctiveness is the Construct menu (Fig. 1), which in Sketchpad's mathematical design, consolidates all of the program's strictly Euclidean capabilities.[6] Beyond the curricular obviousness of that choice, for our purposes, there are several interesting characteristics of choices in this menu. Most importantly, they are composed of and described by words. This is perhaps obvious, but such words act not merely to distinguish among choices; they also introduce or support some normative mathematical vocabulary with which the student may or may not already be familiar. Second, many of these choices cause actions that can be equally accomplished using only the (visual) drawing tools. "Points on Objects" and "Points of Intersection," for example, can be

[5] In Pierce's semiotics (1977), the idea of iconic representation is considered the simplest or crudest form of semiotic register; in Saussure's (1916) more linguistically-oriented semiology, the iconic (which Saussure prefers to call the symbolic—that is, the sign whose meaning is fixed by its visual or morphological homology to an extra-lingual system of meanings, rather than only to intra-lingual structural variation) is deemed entirely pre-semiotic.

[6] Later menus—again from the left-to-right trajectory—introduce transformational primitives, and then analytic geometry, and finally some tools for graphing and algebra. All of these tools operate geometrically in a broad sense but are "not Euclidean" in that they represent mathematical capabilities (or, said differently, axiomatic foundations) beyond those which give rise to Euclidean geometry.

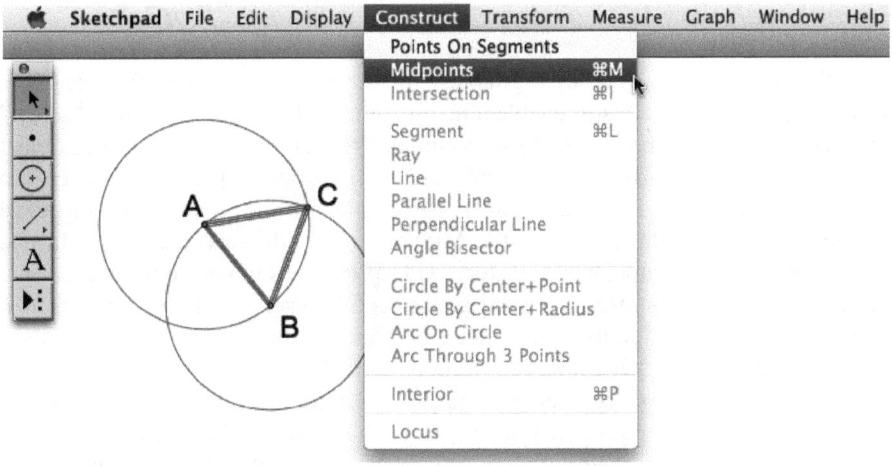

Fig. 1 The basic Sketchpad interface: a sketch (*center*), the toolbox (*left*), and the menu bar (*top*). Menus display both enabled and disabled choices, based on current sketch selections

performed by clicking the Point tool on the relevant objects or their crossings. Thus these and six other commands on the menu are straightforward analogs for known operations and offer simple new representations (though not, since verbalized, necessarily "equivalent" representations) of those known functions. More generally, all commands on this menu perform some construction that can be achieved with a multi-step construction of the drawing tools. (For example, the command "Point at Midpoint (of a segment)" command is itself equivalent to a construction involving two circles of which the given segment is a radius, followed by a line through those circles' intersections.) Menu choices are thus short-cuts for these more lengthy possible (or, depending on the student, actual) pre-linguistic work experiences. Third, read together with their menu title in the attentional flow of expressing a menu choice within Sketchpad's interface, the names become full imperative commands—"Construct point of intersection! Construct Parallel Lines"—with the computer as the implicit subject of these sentences. ("You construct a point of intersection!") The speaker of these imperatives is, of course, the user; and indeed, since every actual menu choice is preceded by the formulation of an intent to choose, the user's full *psychological* expression becomes the sentence "*I want* you to construct a point of intersection; I want *you* to construct parallel lines."

Taken collectively, these observations suggest that the prototypical student envisioned by this design encounters language in *Sketchpad* first as a naming vocabulary for familiar operations, but also perhaps as promising a more succinct or effective representation of operations than can be realized through other, more physically embodied, modes of effort. Finally, all of this language exists in the *service* of the user, for articulating commands that the computer then executes; and our introduction to the power of verbal symbolization comes through the act of *speaking*, rather than of being *spoken-to*.

From Vocabulary to Grammar; From Grammar to Plot

Not all commands, however, are available at the same time. A central principle of the software's operation is that to enable a menu command, the user must first "select" any mathematical prerequisites of that command. Selection—highlighting an object with the pointer tool—is the computer's *deixis*, it elevates a particular object into the special focus of the here and now. For example, the user selects one (or more) segments—and Sketchpad enables the Midpoint (or Midpoints) command. In one sense, this action extrapolates a gesture widespread in software interface design. (Thus, to italicize a phrase in a word processor, the user selects the phrase, then chooses "Italics" from some Style menu or button.) But in another sense, it's a stricter distillation of that trope, into a more formal and mathematical process of reasoning from input (a segment) to output (its midpoint). This is perhaps less obvious—indeed, other geometry packages such as *Cabri* (Baulac, Bellemain, & Laborde, 1988) and *GeoGebra* do it exactly backwards to this (for more discussion on the question of order, see Laborde & Laborde, 2008, p. 37). (There, the user first chooses the desired output—the "Midpoint" command—and then hunts around looking for relevant inputs.)

Selective (and selection-based) command enabling has several immediate consequences in the Sketchpad experience. For genuinely new users, it is not uncommon to first encounter the Construct menu when absolutely nothing in it is enabled (because nothing is selected). The user cannot speak, unless she has something to speak about! At the same time, students are not thrown into (or, more poetically, off of) a tower of Babel: One's options at all times reflect operations that make sense in one's local context, rather than the full and unordered catalog of the software's complete functionality. As one gains dexterity with this interaction paradigm, it develops a sense of mathematical contingency and dependency—segments *imply* midpoints—which in turn imposes discipline (at times the user must reflect a moment: "What uniquely defines the parallel line through this point here?"), but also at times opens an educational opportunity ("ah ha—and of course the same thing defines a perpendicular line!").

Ultimately, by determining which commands—which words—are appropriate and available in which contexts, the selection protocol imposes an operational *grammar* on the raw vocabulary of the menus. Selecting just a point does little to activate the Construct Menu. However, selecting a point and a segment not only fits into the grammar of constructing a parallel line—or a perpendicular one—but also affords the chance to construct a "Circle by Center+Radius"—a construction that uses the compass to copy given segment lengths. Grammar thus in turn implies—linguistically, at least—the possibility of speakers engaging in more powerful, subtle, or sensitive speech acts than the simple gruff commands of "construct this! construct that!"

We see this possibility unfold in the larger temporal trajectory of a *Sketchpad* construction sequence. Vocabulary and grammar combine, in time, to form a mathematical narrative or dialogue between the user and the tool. This dialogue—what Sedig and Sumner (2006) perspicaciously refer to as Sketchpad's "menu-based conversation" (p. 29)—emerges spontaneously through the effects of the selection protocol over time. While users must select their intended input before issuing a command, the software responds to that command by constructing *and selecting* its result. Thus, the output of one operation (the selected result) is already poised, by the software's conversational response, to act as the input to the user's next conversational salvo. If the student engages in any forward-reasoning trajectory—a chain of dependent constructions, a derivation, or a proof—the conversational ball pursuing this "line of thought" moves forward in the form of selection, and is handed back and forth, from student to machine, from machine to student. This narrative gives a sense of momentum—of mathematical "plot" or "story arc"—while at the same time accomplishes a *focusing away*, a suppression of all the things of which it makes no sense to speak at this moment in the story.

Plot Summary and the Moral of the Story

Finally, we consider the far end of the designed Sketchpad construction trajectory, in the experience of creating what is known in Sketchpad as a "custom tool." Custom tools—like procedures in Logo, or like scripts and macros in early versions of *Sketchpad* and *Cabri*—encapsulate some set of operations for reuse later. In *Sketchpad*, one creates a custom tool by selecting a constructed example of the object that one wishes to conveniently recreate and choosing to *Create New Tool* from a relevant menu. Sketchpad responds by prompting the user for a name for that encapsulated operation (and an optional descriptive comment). The user may inspect the "script" defining the tool—a written transcript of the operations used to define it, formalized in a conventional, written geometric notation—but more importantly, the user may now use the newly-created tool, which appears with her previous tools in the program's toolbox, to apply the construction quickly to arbitrary inputs. (Figure 2 displays the script of a tool created from straightforward construction of an equilateral triangle—in this case, named by the user "my first equilateral Δ"; Fig. 3 shows several ornamental constructions rapidly made using that new tool to generate equilateral triangles.)

Again, language plays a prominent role in the interface design of this summative construction experience. First, and most dramatically, the representation of a student's construction presented in script view is fundamentally textual in nature (Fig. 2). In its substantive revoicing of the student's geometry and in its distilled sequential logic, notational flourish, and somewhat ostentatious mathematical language, it has the flavor not just of some work but of an *opus*. This seems teacherly. And yet, while of course *Sketchpad* generates this script, the software is only acting as a sophisticated copyist. In an important sense, the script is the student's own opus: The steps are the steps she herself used to construct her own equilateral triangle—or her own construction, whatever it may be that she has constructed. This is the document of her entire effort; and announces to the world "the whole story" of how she accomplished the (remarkable!) result on the left. In this sense, the mere presence of the script's language is perhaps more important than what it has to say:

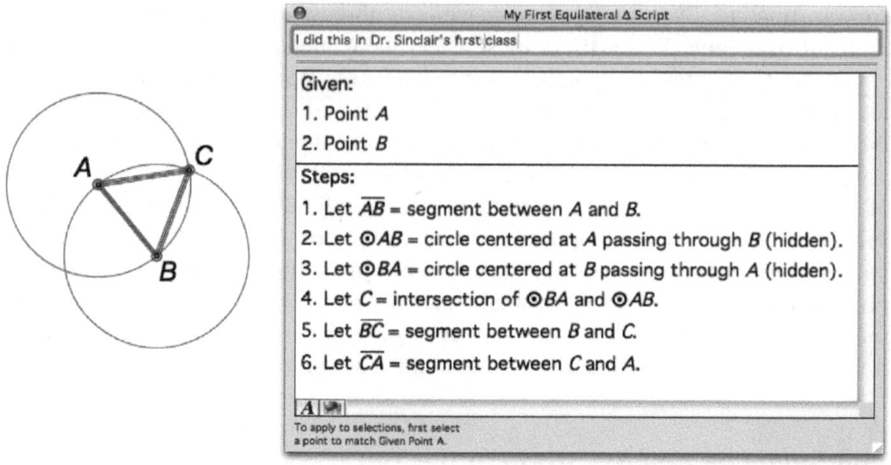

Fig. 2 A student construction and the script describing it as a custom tool. The title and comment, at top, are generated by the student; the description of the script's logic is generated by *Sketchpad*

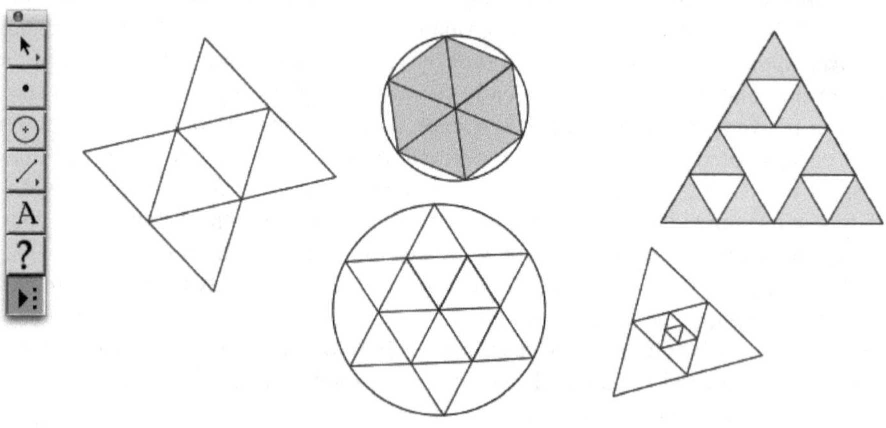

Fig. 3 Various motifs created with the new *My First Equilateral* Δ tool

After all, what it says is already so much more palpable and viscerally present in the construction selected at left. At this moment, all of the ornate mathematical language fades somewhat, revealing the most important piece of language as that which the student typed herself: the tool's name (and comment). This name becomes the name by which the tool is forever after accessed in the software's interface—the detailed script is quickly hidden away. Thus the name serves as the summary of the entire narrative the student has just authored, as the moral of the story—or its "takeaway"—in the sense of that distilled essence of a plot that we hope persists past any recollection of its detailed narrative unfolding, and which we can take into new contexts as a new and enhanced understanding, as we complete our move from the mathematical particular of our first constructed equilateral triangle to the mathematical generalization of all future and all possible equilateral triangles.

Thus, to summarize, we see a complete, evolutionary, and narratively-contoured linguistic trajectory in the design intent of a student's experience in (and "through") the software. This trajectory begins in a pre-verbal exploration of core and simple drawing tools and moves into language with the introduction of menu vocabulary and commands. A specific and systemic protocol enables commands, leading to a linguistic grammar of interaction, and this same protocol equally describes those commands' results, leading to a language-based sense of conversation or even plot. Successfully managed, the result is at once a formal piece of mathematical language, a demonstration of achievement, and an operational enhancement to the software's functionality. At the end of the story, a cycle completes: All of the narrative flowering of the construction trajectory folds back in on itself, and its verbalized structure collapses and distills itself back into a new, pre-verbal drawing tool (the *My First Equilateral* Δ tool, new in the toolbox) where it waits, poised in anticipation of the next construction episode.

This is clearly an idealized description. Readers familiar with Sketchpad will know that while our overall description of this trajectory may feel piecewise correct, that trajectory rarely manifests itself in the complete structure we describe here. Pitfalls abound. Students lose track of which tool is which. For some, reading the menu text is a tremendous barrier. Others cannot figure out how to construct what they want to construct; and others again—who construct it perfectly—go on to reconstruct it by hand five times in a row without ever defining a reusable new tool. We are, after all, describing an interaction design rather than a classroom observation. Even from that perspective, our comments are highly limited in their scope. In this hermeneutic analysis of Sketchpad's user interface, we have ignored completely the whole premise of Dynamic Geometry—which is surely the main point of this particular software—and even in terms of language within it, we have scarcely heeded the way in which the program's language acts to introduce and reify standard vocabulary for school geometry, or the way in which computer-based language provides a context in which formal and obdurate constraints on acceptable inputs are perceived as completely unobjectionable, by students (who are familiar with it from all other software interactions), rather than as some unnatural and unwelcome sinister game played on them by math teachers. Finally, from a "close reading" of this particular software's user interface design, we reach conclusions applicable only to *Sketchpad*, and not to other software at large, or even to other programs that seem *Sketchpad*-like. Nonetheless, we put these various objections aside to focus on what is core to our claims. Even if never fully realized by a single student in a single setting, the overall intended design trajectory of a technology is still relevant in the contour and meaning with which it endows (and sometimes combines) the fragments of actual experience that make up use of the software. Evaluating the potency of those meanings in turn helps inform our consideration of Sketchpad's impact on individual students. In Sketchpad's interaction design, we find the presence of a totalizing discursive structure co-fabricated by the user and the machine. In the linguistic form of this structure, Sketchpad determines both a vocabulary and a grammar to the language of communication, and in this sense acts as an authority. But it is the user who actually breathes sentences and intentions into the software's grammar, who enunciates its subjects, enacts its predicates, and arrives at its objects. In the narrative form of this discursive structure, the user is the first-person subject of their own mathematical autobiography, and—on the other side of the screen—Sketchpad acts as that narrator's occasional foil, able listener, and expert scribe.

Conclusion

In this chapter, we have considered the issue of learning through teaching from a discursive perspective on learners' teaching machines. Our analysis considered, first, the nature of language interactions at the classroom level, in terms of the roles of the different players in the classroom, and second, the individual language of a student in conversation with the computer. In terms of the former, we examined the discursive characteristics associated with mathematics classroom teachers and found that, from this point of view, the computer is not discursively analogous to the teacher; and that, moreover, the language and interaction of the user with the computer positions the student as the teacher. We might still ask, if the user is teaching, what is the *mathematical* learning involved in learning through teaching? After all, the goal of the mathematics classroom is not to turn students into teachers!

Our answer to this question is embedded in the discursive perspective adopted by our critique. Following Sfard's definition (2008) of learning as a change in one's pattern of communication, we explore here how teaching *Sketchpad* one's desires and intentions relates to the more prosaic task of learning geometry. In the previous section, we illustrated the user's interaction with the software along a discursive evolutionary trajectory, from pre-verbal origins, through vocabulary and grammar, to plot and narrative. Intertwined in this trajectory are layers of mathematical meanings, or of mathematics that *becomes meaningful* at the moment it becomes discursively recognizable and articulable. Thus in meeting new vocabulary, students meet new mathematical concepts (segments, circles) and definitions. The syntactic structures in which students learn to wield new terms correspond to the mathematical structures in which those concepts interact. In *Sketchpad* "menu conversations" about rotation, users mark angles and fix centers, then select pre-images and rotate them (by those angles, with respect to those centers) into transformed images. These structures indicate how specific mathematical pieces (points, angles, turns) fit together, and how new specific concepts and definitions (such as rotation) emerge by making the coming-together of mathematical pieces precise. Where specific software syntax helps frame specific mathematical structures, the more general grammar of software interactions introduces users to the much broader structure of mathematical reasoning. The software move from selected prerequisites to constructed results mirrors the causal structure of mathematical derivation, in which independent givens (quantities, axioms, postulates) are bound together through logically robust and general templates into new results (solutions, theorems, proofs).

At an even broader level, by situating users explicitly as authors and editors of mathematical narrative, the software encourages users to the practice of mathematical analysis (which we might constitute, discursively, as what has not yet been said, that should be), and to mathematical inquiry (what has not yet been said, that might be).

Thus, we see two distinct roles of language in a discourse-analytic close-reading of the *Sketchpad* experience. First, language reveals itself as a means both of oppression and of liberation within the learning environment, in constructing individual subject positions and in giving individuals the tools to reconstruct and reconfigure them. Second, rich strata of mathematical content, process, and culture are embedded in discursive practice and students' induction into, and eventual fluency in, mathematical language mirrors their introduction and eventual facility within these domains. In considering how these twin roles of language relate to students' learning mathematics, we find Papert's lofty aim—of students' "learning to speak mathematics" (p. 13) functions not only as a metaphor, but as a mechanism.

References

Baulac, Y., Bellemain, F., & Laborde, J. M. (1988). *Cabri-géomètre, un logiciel d'aide à l'enseignement de la géométrie, logiciel et manuel d'utilisation*. Paris: Cedic-Nathan.

Becker, H. J., Ravitz, J., & Wong, Y. T. (1999). *Teaching, learning and computing 1998 national survey*. University of California, Irvine: Center for Research on Information Technology and Organizations.

Chang, K., Sung, Y., Chen, Y., & Huang, L. (2008). Learning multiplication through computer-assisted learning activities. *Computers in Human Behaviour*, *24*(6), 2904–2916.

Crawford, K. (1996). Cultural processes and learning: Expectations, actions, and outcomes. In P. Nesher, L., Steffe, P. Cobb, G. A. Goldin, & B. Greer (Eds.), *Theories of mathematical learning* (pp. 131–147). Mahwah, NJ: Lawrence Erlbaum Associates, Inc.

Gerofsky, S. (1999). Genre analysis as a way of understanding pedagogy in mathematics education. *For The Learning of Mathematics*, *19*(3).

Grice, P. (1975). Logic and conversation. In Cole, P., & Morgan, J. (Eds.), *Syntax and semantics* (Vol. 3). New York: Academic Press.

Herbel-Eisenmann, B. (2009). Negotiation of the "presence *of* the text": How might teachers' language choices influence the positioning of the textbook? In J. Remillard, B. Herbel-Eisenmann, & G. Lloyd (Eds.), *Mathematics teachers at work: Connecting curriculum materials and classroom instruction*. New York, NY: Routledge.

Hewitt, D. (2001). Arbitrary and necessary: Part 2 assisting memory. *For the learning of mathematics*, *21*(1), 44–51.

Jackiw, N. (1988). *The geometer's sketchpad*. Emeryville, CA: Key Curriculum Press.

Laborde, C., & Laborde, J. (2008). The Development of a dynamical geometry environment: Cabri Géomètre. In G. Blume & K. Heid (Eds.), *Research on technology and the teaching and learning of mathematics*. (Vol. 2, pp. 31–51). Charlotte, NC: Information Age Publishing, Inc.

Moggridge, B. (2007) *Designing interactions*. Cambridge, UK: MIT Press.

Papert, S. (1980). *Mindstorms: Children, computers and powerful ideas*. London, UK: Harvester Press.

Papert, S., & Harel, I. (1991). *Situated constructionism*. Norwood, NJ: Ablex Publishing Corporation.

Pickering, A. (1995). *The mangle of practice: Time, agency and science*. Chicago, IL: The University of Chicago Press.

Pierce, C. S. (1977). In Charles H. (Ed.), *Semiotics and significs*. Bloomington, IN: Indiana University Press.

Pimm, D. (1983). Self images for a computer. *Mathematics Teaching*, *105*, 40–43.

Pimm, D. (1994). Spoken mathematical classroom culture. In S. Lerman (Ed.), *Cultural perspectives on the mathematics classroom* (pp. 133–147). The Netherlands: Kluwer Academic Publishers.

Saussure, F. de. (1916). Cours de linguistique générale. In C. Bally & A. Sechehaye, with the collaboration of A. Riedlinger (Eds.), Lausanne and Paris: Payot (trans. W. Baskin), *Course in general linguistics*, Glasgow: Fontana/Collins, 1977.

Sedig, K., & Sumner, M. (2006). Characterizing interaction with visual mathematical representations. *International Journal of Computers for Mathematical Learning*, *11*, 1–55.

Sfard, A. (2008). *Thinking as communicating: Human development, the growth of discourses, and mathematizing*. Cambridge, England: Cambridge University Press.

Stubbs, M. (1975). Organizing classroom talk. Occasional paper 19, Centre for Research in the Educational Sciences. University of Edinburgh.

What Experienced Teachers Have Learned from Helping Students Think About Solving Equations in the One-Variable-First Algebra Curriculum

Robin Marcus and Daniel Chazan

Introduction

Our chapter proceeds in five parts. We begin with a task that teachers face, deciding how to begin to introduce students to complex ideas. Arguably, this task of teaching may be responsible for the growth of the unique mathematical knowledge for teaching that teachers possess (Cohen, 1993, argues for the role of "unpacking" in the growth of teachers' knowledge; Ball and colleagues, e.g., Hill, Schilling, & Ball, 2004, provide evidence for mathematical knowledge that is uniquely the province of teachers.). We then turn to the particular cast this task of teaching takes in the context of introducing students to equations in school algebra. The next two sections then focus on the context in which the two teachers teach and their responses to a set of interview prompts and their actions in classroom observations. The final section reviews the interview and observation data and underscores the evidence we see for the teacher learning that informs the practice of these two experienced teachers.

The Difficult Task of Choosing Explanatory Starting Points

As teachers, and others responsible for the development and delivery of curriculum, work to introduce students to complex mathematical notions, starting to convey complex ideas to students is always difficult. Starting places for an explanation can be too abstract or formal, and as a result divorced from students' experience (as critics argued with New Math set theoretic definitions of number). Or starting places can be too specific and concrete. In such a case, students must abstract the essential aspects of a concept from examples that have many other aspects, and they may often include unintended aspects of the example in their definition (as suggested

R. Marcus (✉)
University of Maryland, College Park, MD, USA
e-mail: marcus_robin@hotmail.com

R. Leikin, R. Zazkis (eds.), *Learning Through Teaching Mathematics*, Mathematics Teacher Education 5, DOI 10.1007/978-90-481-3990-3_9,

by Vinner, 1983 in his introduction of the term concept image, in addition to concept definition). While a powerful and concrete example can serve as an important conceptual anchor, as researchers have found (e.g., Vinner and Dreyfus, 1989, with respect to students and functions and Even, 1993, with respect to preservice teachers and functions) such concept images can later become pedagogical obstacles (in the sense of Sierpinska, 1992). Students may have to disconnect a particular image from a concept in order to develop a broader sense of a concept, one that matches its definition and not a particular example (Lakatos' 1976 rational reconstruction of the history of the Descartes-Euler conjecture finds a similar dynamic in the work of mathematicians).

One goal of this chapter is, in the context of algebra instruction, to identify classroom actions by teachers that reflect the mathematical knowledge they have developed through teaching. We will use the pedagogical challenge of starting places, as well as the notions of concept image, concept definition, and pedagogical obstacle to identify such teacher actions—to identify teachers' mathematical knowledge in action. This exploration involved interviews of teachers in a US high school and observations of their teaching. In the paper, we illustrate the terms decompressing, trimming, and bridging outlined in the framework for knowledge of algebra for teaching developed by Ferrini-Mundy, Floden, McCrory, Burrill, and Sandow (2005). We begin with a curricular analysis and then provide background context on the teachers we interviewed and observed.

A One-Variable-First Perspective on Solving Equations and Systems of Equations

Equations appear throughout mathematics in different guises. They are sometimes thought of as mathematical objects worthy of exploration in their own right (e.g., the now-defunct mathematical field of study called Theory of Equations). At other times, they are representations of other mathematical objects (e.g., as signaled by the locution "the equation of a function" or "equation of a line"). From his study of school algebra curricula and his experience as a writer of curricula, Zalman Usiskin (1988) suggests that in the US school curriculum, equations are treated in five different ways: as formulae, equations to solve, identities, properties of numbers, or the equation of a function (p.9). A common school textbook definition avoids these subtleties by offering a criterion for recognizing an equation when one sees one: An equation is a string of symbols with an equal sign in it.

Mathematics educators have studied equations as they appear in a range of school mathematics contexts. Researchers interested in students' transition from arithmetic to algebra have expressed concern about ways in which the equal sign in arithmetic is a call for carrying out an action on the expression on the left in order to know what to write for the expression on the right, while in algebra students are asked to act on both sides of the equation. Pedagogical interventions based on this concern (e.g., Herscovics & Kieran, 1980) are premised on the notion that overcoming a particular view of the equal sign will help students be more successful in algebra. In this chapter, we will make a similar argument within the kinds of equations that students and teachers typically see in school algebra; in particular, we will focus on examples that all fall within a single type in Usiskin's framework, equations to solve, what Freudenthal (1983) calls interrogative algebraic sentences (p. 310) and for which he invents a new notation.

In particular, in this chapter, we will focus primarily on issues involving equations in one and two variables, and peripherally on systems of equations involving two variables. Though we are aware of critiques of school algebra (such as, Fey, 1989) and of reform perspectives (like Heid, Choate, Sheets, & Zbiek, 1995; or Yerushalmy & Schwartz, 1993), we will focus on issues that arise in traditional US curricular approaches to developing students' flexible understandings of equations and solving in the context of these equations and systems of equations.

The equations that students in the US meet in introductory algebra (Algebra I or its equivalent) are usually equations in one or two variables, first equations in one variable and then equations in two. These equations have an equal sign, may or may not represent a function, they may or may not admit to closed form solution, and for the most part, with the exception of units on quadratic equations and absolute value, they are linear.

In making this short list of characteristics of equations, we suggest that equations in one and two variables, as treated by the curriculum, are quite different one from another. As others have argued with respect to arithmetic and algebra equations (e.g., Nathan & Koedinger, 2000), we would like to suggest that these differences matter for teaching and learning.

Most US curricula introduce students to equations in algebra by starting with linear equations in one variable. Some texts organize these into one-step or two-step equations, others organize them by the types of operations used to solve them (e.g., McConnell et al., 1990). By way of contrast, functions-based approaches to algebra (like Fey & Heid, 1995), begin with functions, a subset of equations in two variables, before introducing equations in one variable to solve. In typical textbooks, students are asked to solve linear equations in one variable. Starting with this comparatively simple task can challenge students to distinguish three related goals (isolating the variable, finding values to make true sentences, and representing a solution set) that will only become distinct with more complex tasks (Fig. 1).

Students are taught procedures to solve equations by operating on both sides of the equation to isolate the variable. They are then taught to check their solution by

Meaning of solve	Equations like: $3x + 2 = 4x - 7$
Isolate the variable	Taught to students
Find member(s) of the solution set	Implicit part of checking
Represent the solution set	Implicit, until more than one solution

Fig. 1 The goals of solving linear equations in one variable

substituting the solution back into the original equation. This check procedure builds on the notion that solving the equation is meant to find value(s) of the variable for which the equation will be a true statement. Finally, work with linear equations in one variable can support a view of the variable as an as of yet unknown number. In "most" cases, there is only one such number. The notion of a solution set, and solving as representing the solution set to an equation, is implicit until attention changes from linear equations to quadratic equations or to inequalities, or when linear equations with no solution or infinitely many solutions are addressed.

Intriguingly, the typical school algebra curriculum takes a very narrow approach to equations in one variable. Students, and teachers, rarely see equations of one variable that cannot be solved by isolating the variable, equations like: $4(x-1)-x = 3x - 4$, $2^x = x^2$, or $|x-2| = |x+1| + 3$. While such examples drive a wedge between the "isolate the variable" and "find values that make the statement true" or the "describe the solution set" meanings of solve,[1] they do not regularly appear in classrooms.

When the curriculum moves to equations in two variables, there is a large transition for students. A linear equation in two variables defines a function, explicitly or implicitly. Equations in two variables challenge the as of yet unknown number view of literal symbols like x and y. And, solving changes (Fig. 2). Building on the "isolate a variable" meaning of solve, linear equations in two variables can be solved for a particular variable, but there is a disconnect between isolating the variable and finding member(s) of the solution set. With linear equations in one variable, isolating the variable had resulted in finding the (unique) solution, except in "special cases"; whereas, with equations in two variables, isolating the variable may facilitate finding members of the solution set, but it is just a first step toward that end. Isolating the variable is now signified by "solve for," and finding members of the solution set now requires a different set of actions—one must choose a value for one variable and then solve the resulting equation in one variable in order to be able to write down an ordered pair that is a solution to the equation.

The "represent the solution set" meaning of solve is more available, though it is easy to overlook. Any equation in two variables represents a solution set that can be graphed on the Cartesian plane. Isolating a variable may make the task of graphing the equation easier, but even without doing any manipulation, the equation defines the solution set (as it did as well earlier with equations in one variable). While all of these different meanings of solve are still at play with equations in two variables, the command "solve" is generally not used; instead it is replaced by *solve for*, *create a table of* (x, y) *values*, or *graph*. Similarly, checking a *solution* becomes checking that an ordered pair *satisfies* the equation. This shift in language potentially obscures the connections among these actions and with the concept of solution.

---

[1] In what follows, we are concentrating on the solving of equations. When we use the phrase the "meaning of solve," we intend in the context of solving equations.

Meaning of solve	Equations like: $3x + 2 = 4x - 7$	Equations like: $9x + 3y = 12$	Systems like: $\begin{cases} 2x+3y=7 \\ x-y=-4 \end{cases}$
Isolate the variable	Taught to students.	Becomes "solve for".	A step in possible solution procedures.
Find member(s) of the solution set	Implicit part of checking.	Not part of "solve for." Signified by "create a table".	Stated goal. Part of checking.
Represent the solution set	Implicit, until more than one solution.	More readily available. Signified by "graph".	Implicit, until more than one solution.

Fig. 2 Tracking the meanings of "solve" across the one-variable-first school algebra curriculum

Finally, when the curriculum turns to a short unit on systems of equations, the meaning of the word "solve" takes yet another turn. With systems of linear equations, the notion of isolating the variable becomes a step in one solution procedure, no longer a defining meaning for the operation of solving. While the notion of solving for a member of the solution set moves to the background with equations in two variables, this meaning returns to the forefront with the move to systems of two linear equations in two variables. Students find the coordinated values of x and y that satisfy both equations. With systems of two linear equations, the notion of representing the solution set can be lost again, as it was with the linear equation in one variable, though it can be resurrected by examining systems involving non-linear equations, or systems with no solutions, or infinitely many, or systems of inequalities.

Our point in cataloguing the meanings of "solve" as one moves through the curriculum is to emphasize the transitions, or discontinuities (in the sense of Tall, 2002), required of students as they move through a one-variable-first curricular approach to equations in school algebra (A functions-based approach to algebra would have a different, but related, set of transitions, see Yerushalmy & Chazan, 2008). In seeking to identify teacher learning through teaching, we focus on how teachers deal with the challenges of helping students with these curricular transitions.

Before turning to examples of teachers grappling with their own understandings of these transitions and how to communicate with their students, we give a little background on the teachers and our interaction with them.

Context

We would like to juxtapose our analysis of equations and solving in the school algebra, one-variable-equations-first way of introducing algebra with interviews and observations of teachers in a US high school. We precede examination of the thinking and actions of these teachers with some background information that situates them in their local context and indicates how we interacted with them.

The teachers we interviewed and observed are from one high school in a district that ranks among the 20 largest school districts in the US. Like many schools that are described as urban schools in the US, this high school has a predominantly minority population (96%), high turnover among teachers (only 39% of classes are taught by a "highly qualified" teacher), a high rate of poverty among the families of students (51% on free/reduced meals), a high rate of student mobility (24% in, 17% out), poor student achievement on exams (a passing rate of 16% on a state Algebra 1 test in 2004), and a low graduation rate (65%).

The teachers we interviewed and observed were teaching introductory school algebra (Algebra I) to high school students (mostly grade 9, age 14 and above). Students in the state must pass high school algebra and geometry courses in order to graduate. Students are required to pass an algebra end of course exam for graduation.

The Text

The adopted textbook in the school district is Prentice Hall's *Algebra: Tools for a Changing World* (Bellman et al., 1998). This text combines a fairly standard approach to solving equations, as we described earlier, with an earlier-than-standard introduction of functions of one variable and their defining equations.

In the first chapter of the Prentice Hall algebra text, equations are defined (p. 11) as indicating that two expressions are equal. The first equations that students meet are *function rules*, defined as equations (of two variables) that describe functions. For example, on page 11, the first example of an equation is $t = s - 2$, but the example of an equation in the glossary at the back of the book is $x + 5 = 3x - 7$. The remainder of the first chapter then deals with order of operations, arithmetic with integers, properties of real numbers, experimental probability, and introductions to matrices and spreadsheets. Chapter 2 returns to functions and their representations.

After the material on functions of one variable, chapter 3 focuses on equations of one variable. It begins with a balance scale analogy for equations and describes solving an equation containing a variable as finding "the value (or values) of the variable that make the equation true" (p. 108). *Solutions* are defined as these values. The chapter contains six sections that develop techniques for solving linear equations in one variable by "get[ting] the variable alone on one side of the equal sign" (p. 108) using inverse operations. Students first solve one-step equations, then two-step equations, then equations that contain like terms on one side, parentheses, fractions, and percents.

In chapter 4, students meet linear equations in one variable with a variable on both sides of the equal sign. In the main body of the text, the technique for solving linear equations in one variable with a variable on both sides involves applying the properties of equality "to get terms with variables on the same side of the equation" (p. 164), initially aided by the use of algebra tiles. "Special types of equations," those having no solution and those that are identities, are introduced and defined. In the homework exercises, students are asked to write *identity* or *no solution* when they encounter such equations. At the end of the exercises for this section, there is a Self-Assessment journal prompt: "Summarize what you know about solving equations with variables on both sides by writing a list of steps for solving this type of equation" (p. 168). Following, there is a Technology page that provides a list of steps and practice exercises for using a graphing calculator to solve an equation with the variable on both sides by graphing and finding the x-coordinate of the point of intersection (p. 169). This chapter also contains one section on solving equations in more than one variable for a particular variable.

In chapter 5, students take a closer look at linear equations in two variables, with an emphasis on graphing linear equations and writing equations for lines in both slope-intercept and standard form. Then, in chapter 6, they meet systems of linear equations in two variables.

Our Interview and Classroom Observation

Our interview was designed to explore how teachers think about what an equation is and how to teach students about equations. With these goals in mind, we focused on understanding:[2]

- how teachers conceptualize "x" in their work with students,
- whether they believe that it is reasonable in an algebra class to conceptualize equations as questions about or comparisons of functions,
- how they distinguish for themselves between solving in the context of equations of one and two variables,
- how they conceptualize the role in an algebra class of equations in one variable in which the variable cannot be isolated,
- how they talk with students about expressions and equations, and
- how these issues interconnect and interrelate in their thinking about the design of instruction in a year-long course.

After a few introductory background questions, we began by asking teachers how they explain to beginning algebra students what an equation is. Then we presented the first major item: a card sort task. Teachers were presented with nine equations (Fig. 3) on nine index cards and asked to discuss them in relation to their conception of equations, to compare and contrast them, to order the cards in the ideal order in which they would want students to meet these examples, and to explain the reasoning behind their choices.

[2] Our interview is based on earlier work by Chazan, Yerushalmy, and Leikin (2007; 2008).

Fig. 3 Equations presented to the teachers for a card sort task

- $y = 3x - 4$
- $9x - 3y = 12$
- $4(x - 1) - x = 3x - 4$
- $g(x, y) = 3x - y$
- $3x^2 + x - 2 = 0$
- $3x - 4 = 12$
- $f(x) = 3x - 4$
- $3(x - 2) + 2 = 4x - 4(x - 3)$
- $y = 3x^2 + x - 2$

Subsequent items included tasks that asked teachers the following:

- to compare and contrast the following three tasks:
  1. Solve $3x - 4 = 12$.
  2. If $y = 3x - 4$, what value of x makes $y = 12$?
  3. If $f(x) = 3x - 4$, what value of x makes $f(x) = 12$?
- then to respond to two student solutions: a numerical solution based on the use of tables and a solution that involved operations on both sides of an equation.
- how they would talk with students about how to solve an equation like: $4(x - 1) - x = 3x - 4$.
- how they would solve equations like $\sqrt{(3x - 4)^2} = 3x - 4$, $2^x = x^2$
- and whether they would think that equations like this belong in an algebra class.
- how they think about solving $9x - 3y = 12$
- and how they might respond to students who, when asked to simplify, set an expression equal to zero and solved it.

The responses to these tasks provide rich glimpses of how the teachers understand what an equation is and how they imagine one might gradually introduce students to the complexities involved in this mathematical construct.

The following fall we observed the teachers we had interviewed while they were teaching a unit on solving linear equations in one variable (chapter 3 in their text). This is when the textbook first introduces solving equations. For some of the students, it may have been their first experience with solving equations. At the time of our observations, both teachers had been teaching the solving of linear equations for about one week.

In the next section, we choose to focus on two teachers for whom we have both an interview and a classroom observation and who gave evidence for learning from their teaching. Ms. Alley was in her third year of teaching at the time of the observation, and she had taught Algebra 1 each year. She had also taught pre-algebra and pre-calculus. Ms. Alley has an undergraduate degree in mathematics and describes herself as a mathematician. Ms. Lewis was in her sixth year of teaching, and she primarily teaches Algebra 1; although some years she taught pre-algebra as well. Ms. Lewis changed careers to become a math teacher; her undergraduate degree was in finance.

Both teachers explained that a primary reason why they are assigned to teach Algebra 1 every year is because they both have a reputation as effective classroom managers. It became evident during our conversations that both teachers care deeply about their students and have given a lot of thought to how to best prepare their Algebra 1 students for future mathematics courses and success on high stakes exams.

Ms. Alley and Ms. Lewis

On Equations: What Are They? What Is Their Place in the School Algebra Curriculum?

Early in the interview, we asked the teachers how they explain to students what an equation is. An excerpt from Ms. Lewis' response follows:

> Now that's a good question. Have I ever explained what an equation is? Now, when we set up an equation from a word problem, I mean, I can tell them where everything goes, but – what is an equation? Um. Hmm. Let me think about that. Good question. Um. <pause> Well we solve an equation to get a value for the variable. So an equation would be... hmm. Now I understand the concept of it, but putting it into words, okay, let me think about that. Can you help me out?... I mean, I know what an equation is, I know what the ultimate goal is... of an equation: to find a value for the variable that would make the equation true.... But they understand that I solve for the particular variable that will make this equation true; but, to give them a definition, I mean, that would really be hard to think of what a definition would be.

Ms. Alley gave examples of equations: $x + 5 = 7$ and $x - 5 = 12$. She explained that she first gives students examples with blanks instead of x; then she tries to have students extend their informal solving to the formal procedures. When presented with the nine equations on index cards in the first task, Ms. Alley distinguished equations one would "solve as equations" (those containing one variable) and "equations that are functions" (those containing two variables).

Ms. Lewis made many more distinctions: equations with variables on both sides, two-step equations, equations in slope-intercept form, and two-variable equations. For Ms. Lewis, functions are distinct from equations; although she used *equation* to describe linear equations in two variables (i.e. Ms. Lewis distinguishes $f(x) = 3x - 4$ as a *function*, but describes $y = 3x - 4$ as an *equation*). In ordering the equations, she considered the amount and type of solving involved—the earlier the equations move in her order, the more solving is involved and the easier it is to isolate the variable. For example, an equation of the form $ax^2 + b = c$ is "like a two-step [linear] equation, and then you just take the square root at the end; so equations of that sort will occur fairly early." This may suggest that an equation of the form $ax + b = c$, a "two-step equation," is her concept image of equation because that becomes the basis for "moving up" other equations.

Ms. Alley's primary focus in describing and explaining equations was also *solving*; however, she indicated that she wants students "to understand how to write a function first—what it is to put something in and get something out—so they understand that there is a solution." Despite this statement early in the interview, there was little evidence later in the interview, and in the classroom observation, of an impact from the early introduction to functions on the treatment of linear equations in one variable.

Both teachers agreed that non-routine equations do not warrant much attention in an introductory algebra course. Ms. Lewis was not at all comfortable with the idea of including equations for which isolating the variable does not solve the equation in the school algebra curriculum; except perhaps for honors or gifted/talented students, after they have "mastered" techniques for solving. When probed, she was not at all concerned that a very tiny portion of all possible equations can be solved by known symbol manipulation techniques. Similarly, when asked about the importance of providing opportunities for students to grapple with making sense of linear equations that are identities or contradictions, as will be outlined below on page 182, Ms. Alley indicated that this issue is not central to her instruction.

Advanced users of mathematics readily interpret the various types of equations they encounter using mathematical context to inform their interpretations; however, for students in the process of developing their concept definition of equation, such interpretations are not immediately correct—sometimes differences between equations are given too much weight and on other occasions important differences are overlooked. In order for teachers to help their students develop increasingly sophisticated understandings of equations, they must first interrogate their sophisticated understandings and then examine equations from the information available to their students in order to understand the complexities their students face. This process of examining a mathematical object, concept, or skill from an advanced perspective and then from a novice perspective, revealing complexities that are obscured by more sophisticated mathematical understanding, is described as *decompressing* in the Knowledge for Algebra Teaching Framework (Ferrini-Mundy et al., 2005).

On Solving: What's the Name of the Game? Isolate the Variable!

Both teachers emphasized how important it is for algebra students to learn how to solve equations. When asked about her primary goals for her Algebra I classes, Ms. Lewis responded, "Definitely solving equations I would say." She further explained,

> ...what's more important for them: to understand it, or just to be able to punch it in their calculator and write something that they really don't understand? ... I know the teacher who teaches Trig and Algebra 2/Trig and the Trig Analysis; he's an old teacher. Ok, so you know how the older teachers are – solve, solve, solve. ... So for my kids to be successful when they go to him, they need to know how to solve.

During her interview, Ms. Alley commented, "You can only make solving equations so exciting; whereas, that's the basis of everything you learn after it, and you have to know it."

Both Ms. Alley and Ms. Lewis commented on the difficulty students have with solving (via symbol manipulation), even with "simple" equations. In their attempts to help their students manage this difficulty, they both place a heavy emphasis on the phrase "isolate the variable" as articulating the goal of solving when they first introduce students to solving in the context of linear equations in one variable.

When we observed Ms. Alley teaching a lesson on solving linear equations in one variable, she repeatedly asked the class, "What's the name of the game?", to which the students responded in chorus, "Isolate the variable!" During an observation of a lesson at the same point in the algebra curriculum, Ms. Lewis explained to her class, "The number that is on the same side as the variable is the one we want to get rid of. . .so we can isolate the variable. . .'cause remember: that's the goal."

All the teachers we observed used language about removing "zero pairs," a term for *additive inverses* that is intended to be student-friendly, as part of the procedure to isolate the variable. This language is introduced in the textbook when explaining how to use algebra tiles to model solving a linear equation in one variable.

One might wonder what drives the teachers' decisions to emphasize solving as isolating the variable, rather than as finding the value or values of the variable that make the statement true. The teachers' responses to two different hypothetical students' solutions to questions that could be answered by solving the equation $3x - 4 = 12$ provide some insight. Sam's solution involved operating on both sides of the equation to isolate x, while Karim's solution involved creating a table of sequential $(x, 3x - 4)$ values to approximate the value of x for which $3x - 4$ equals 12. Both teachers expressed a preference for Sam's solution.

Ms. Alley commented, "It isn't that I wouldn't accept it [Karim's solution], I'm saying I would be less inclined to accept it." She further explained that she viewed Karim's solution as "guesstimating"; she sees rounding, Sam's method for approximating the value of x, as better estimation. When asked how her view would change if Karim used proportional reasoning to interpolate a precise value for x, she said that that would make his solution completely acceptable, but she would never teach this method because "No student would ever be able to see it—like, you have to give the students ways that they'll be able to see, and after teaching for 2 years, they would never see it that way. They don't think critically." Karim's solution builds upon the early experience with functions Ms. Alley advocated at the start of her interview; however, her comments here suggest that not only does she not use thinking about functions in her teaching of solving equations in one variable, she does not believe that it is feasible to do so.

Similarly, Ms. Lewis could not understand why Karim would use a table to solve $3x - 4 = 12$. She explained that she would never teach students to solve this way; she would teach them by solving (i.e. using symbol manipulation to isolate the variable). She views Karim's solution as using higher order thinking as compared to Sam's, which she calls "basic, basic solving." Ms. Lewis further explained that when two variable equations are introduced, students would create tables and answer

Fig. 4 A problem presented by Ms. Lewis

Which student solved correctly?

Kendra:	Tony:
$x - 3 = 12$	$x - 3 = 12$
$+3 \quad +3$	$-12 \quad -12$
$x \quad = 15$	$x - 15 = 0$
	$+15 \quad +15$
	$x \quad = 15$

related questions about inputs and associated outputs of functions; however, this is not connected to solving one-variable equations.

The comments from both teachers about Sam's and Karim's solutions suggest that the goal of finding the values that make the equation a true statement may receive less emphasis when teaching students to solve linear equations in one variable because such an emphasis may be seen as encouraging guesstimating or other undesirable strategies. Even though they have observed students struggle to solve "simple" linear equations via symbol manipulation, their comments further suggest that they believe that the usual algorithm is the easiest way to help students learn to solve equations.

During our interview, Ms. Alley commented, "it's just like I tell my kids, there's not one way to do every problem. So, . . . I think every student finds a way that they like the best, and then uses that." However, though this rhetoric suggests valuing multiple solution strategies, in the context of solving linear equations in the classroom, perhaps as a result of students' difficulties with solving equations, there was a focus on the usual algorithm. In fact, during our observation, a student started two or three times to explain that she had solved an equation a different way and arrived at the same solution. Both times Ms. Alley interrupted the student, "This is the *only* way," referring to a very specific procedure for isolating the variable by operating on both sides of the equation.

Similarly, when we observed Ms. Lewis, she presented a problem that asked students to evaluate two hypothetical students' solutions of a one-step linear equation to her class (Fig. 4). In the discussion that followed, Ms. Lewis focused on the strategic importance of isolating the variable and indicated that Kendra had solved correctly. According to Ms. Lewis, Tony had solved incorrectly because his first step did not progress him toward the goal of isolating the variable; therefore, Tony required two steps to solve the equation, rather than just one step.

Both teachers emphasized a particular procedure for solving equations in one variable that involves operating on both sides of an equation to isolate the variable in the least number of moves possible. At this point in the curriculum, "solving" is synonymous with this procedure; other procedures that might find the value(s) of the variable that make an equation a true statement are not considered *solving*. And, though they will come in later, the notions of finding members of the solution set, or describing the solution set as a whole, are not emphasized. This emphasis on isolating the variable seems like an attempt to ease the transition to solving equations in two variables for a particular variable.

Thus, as they choose starting points and seek to build on these to develop sophisticated understandings in their students, both teachers attempt to reduce the complexity of solving linear equations in one variable in order to make the content accessible to students. Ferrini-Mundy et al. (2005) call this practice trimming, when the essence of the mathematics is preserved, but simplified, in order to help students gain access to complex ideas.

On Identities and Contradictions: Assigning a Label

Early in the interviews we asked teachers to consider the equation $4(x - 1) - x = 3x - 4$. We asked them what questions or difficulties they expected students to have and how they would help students to interpret a "solution" of $0 = 0$. Ms. Lewis explained,

> I try to make them think of *the same* as being identity, and *different* as being no solution. . . . so that's the way that I try to make them understand that . . . They know that if my variable cancels, that my variable's gone, that it's one of two things. I look at what's left. If I have, . . . numbers that are same, it's 'identity', or numbers that are different, it's 'no solution'.

When asked whether students understand that for an identity, x could be any number, she replied that they do not; "that's higher order thinking." Ms. Alley explained,

> They'll think because it doesn't have '$x =$ something' that it implies no solution. And that's why . . . I *show* them on their graphing calculator. I'm like, 'well graph it, and . . . see what it comes out to when you're talking about functions; . . . put this as one, and put this as another, and see it's the same line, and that means that it's always true.' So, if they say 'always true' or 'never true'—I try not to say 'no solution' and 'infinite', but that's just the mathematician in me—so, what I should say: 'always true' and 'not true'.

When we described using a graphing calculator to create a table and/or graph of both sides of the equation to help students understand that x could be any number, Ms. Lewis said,

> That could be taking it a step higher, to make them understand that it could be any number. Because I think that's the big picture. . . . I think I concentrate so much on the little things, that. . . the big picture just never comes into view. . . . You made a good point when you said that, I was like – oh, that would be wonderful to do – but, would they really understand that? Now while I'm looking at that on a calculator, yeah, I really do. Because, I mean, . . .when we get to. . .systems and the equations come out to be the same, they understand that it is one line, which both things are the same. I mean they understand that concept. But I never thought of doing it in that manner and putting it in the calculator. That would be a good idea. That would be a good idea. I never thought of that.

In a later discussion of a system of two linear equations in two variables, Ms. Alley said of the equation $4(x - 1) - x = 3x - 4$,

> Usually these equations are far and few between. So, you might get one or two on a test, or three or four on a homework assignment. So it's just for them to know: hey, if I get 'something = something', it means either 'no solution' or 'infinite solutions'. So graphing, I wouldn't really emphasize graphing for [the one-variable equation], where I would for [the system of equations].

Linear equations in one variable that are identities or contradictions challenge the isolate-the-variable meaning of solve and require thinking about solving as finding the value(s) of the variable that make the equation a true statement in order to make sense of the results obtained by the procedure for isolating the variable. Both teachers, following the approach taken in the textbook, avoid this complexity by teaching students to label these "special cases" as *identity* or *infinite solutions* or *no solution*. The teachers' strategies with identities and contradictions are another example of these teachers attempting to trim complexity in order to make mathematical content accessible to their students.

On Solving Equations in Two Variables: Changing the Name of the Game

When we asked Ms. Lewis how she thought about solving equations like $9x - 3y = 12$, she explained that equations in two variables are first introduced in Chapter 5 of the textbook, where students learn to solve for y and also to find values for x and y. Ms. Lewis was the only teacher we interviewed who explicitly mentioned both notions of solving equations in two variables. We followed up by asking her if she found it difficult to explain to students what it means to solve for y, since she had earlier defined solving as finding the values of the variable that make the equation true, even though she had emphasized that isolating the variable was the way to find such values. Even though she focuses on isolating the variable as the name of the game in solving, she responded that this is a difficulty for students and further explained,

> ...when you say solve, they're looking for *one* answer, that x is equal to something or y is equal to something. And then when you say 'Solve for y'... So *I* made it a point, I never say 'solve for y'...if I say 'solve', I think that they would understand that I'm looking for a value for x and a value for y. But...I try not to say 'solve for y'...I would just say...'write this in the slope-intercept form'. I won't say 'solve for y'. Because when you say 'solve for y' they're gonna say '$y = 4$' or...something like that.

Ms. Alley focuses on finding values of x and y when discussing solving linear equations in two variables, and she heavily emphasizes thinking about this process as evaluating functions, whether explicitly or implicitly defined. Referring to the equations $y = 3x - 4$ and $9x - 3y = 12$, Ms. Alley explained, "I would use function boxes where I would have an x and y, an input and output." She contrasted this way of thinking to how she thinks about solving equations like $3x - 4 = 12$, which she would "solve as a two-step equation." She went on to clarify, "Or if I had a system of equations I could solve for an x and a y."

When asked how she explains to students what "solve" means when they are asked to "solve for y," Ms. Alley responded, "...The solving for x and y's, and you're given an input, you have to find an output, ...that's more of like a function rule, a function box, an in-and-out; whereas solving in that case is you're solving for *a* variable." Several times during the interview she referred to solving an equation in two variables for a particular variable as "transforming" the equation into a function; and thus, Ms. Alley, like Ms. Lewis, is able to avoid using the language "solve for y."

Even though solving for a given variable in a two-variable equation is procedurally the same as isolating the variable in a one-variable equation, both teachers find that students encounter difficulty because their "answer" is not "an answer." Even though both teachers characterize solving equations in one variable as isolating a variable, students still are focused on solving as producing a numerical solution.

With the introduction of equations in two variables, Ms. Alley and Ms. Lewis try to keep two meanings of solve in play. They want their students to be aware both of the method, isolating the variable, and the notion of identifying members of the solution set. Ms. Alley and Ms. Lewis were the only two teachers whom we interviewed that acknowledged the notion of solving as finding members of the solution set in the context of equations in two variables, in addition to the isolate the variable notion of solving. As experienced teachers, both felt that the transition from solving equations in one variable to solving equations in two variables is big enough that it warrants changing the name of the game in a sense. Since their students had come to expect a numerical solution as the result of solving, both teachers referred to solving in the context of equations in two variables as finding numerical values for x and y. Recognizing the importance of the skill of isolating a variable, the teachers employed other language (rewriting the equation or transforming it into a function) to help their students navigate the transition to solving a multivariate equation for a given variable, which yields a solution that is an algebraic expression rather than a numerical value.

In this manner, Ms. Alley and Ms. Lewis illustrate another of the Knowledge of Algebra for Teaching actions (Ferrini-Mundy et al., 2005), bridging, establishing connections or links between students' and teachers' understandings and between different mathematical concepts or skills. With their terminology, they attempt to help their students navigate the transition from solving linear equations in one variable to solving linear equations in two variables.

Evidence for Learning Through Teaching

In this paper, we describe how two teachers grapple with a problem that all teachers confront, helping students build more sophisticated understandings from more limited ones. Specifically, in teaching algebra, every teacher or curriculum designer needs to introduce equations to students—do they deal with equations in one variable first, or functions in one variable (defined by equations in two variables)? When teaching equations in one variable, do they start with equations of the form $ax + b = cx + d$, or build up to that? When do they introduce the idea that not all equations can be solved? And, depending on these decisions, as complexity is added, or greater generality is sought, how does one help students make the transition from where they have come to where they will now go?

In line with the district curriculum guide and the textbook, the two teachers we have described, Ms. Alley and Ms. Lewis, use similar strategies, some of which are common strategies in the US in Algebra 1 classrooms. They start with equations in one variable first; they carefully build up from less complicated equations to more complicated equations. And, they trim the complexity of the mathematics to aid their students; for example, they intentionally do not explore equations that are not amenable to closed form solutions.

Ms. Alley and Ms. Lewis teach in a particularly challenging teaching context where students' achievement is low. Both have a number of strengths as teachers. Both teachers know the school algebra curriculum well; both teachers revealed sensitivity to student difficulties in learning to solve equations that suggests a capacity to learn from teaching; and both teachers respond to these observed difficulties through their teaching in the best ways they know how. In particular, both teachers give evidence of learning from their teaching by recognizing the difficulties their students have when facing the transition from solving equations in one variable to solving equations in two variables. This transition is difficult largely because students have observed that solving an equation in one variable, except in "special cases," yields a unique numerical solution. Challenging this perception, Ms. Alley utilizes a graphical representation to initially help students "see" that some equations in one variable have no solutions and others have infinitely many.

In the context of solving equations in two variables for a particular variable, both teachers attempt to alleviate students' discomfort with isolating a variable and not obtaining a number as the result by departing from the standard instructions that appear in their textbook and using more descriptive commands that more clearly indicate the desired form of the result. For example, rather than use the standard instruction "solve for" that appears in their text, and declare that when solving equations in two variables, one cannot solve without being told for what variable one is solving, Ms. Lewis instructs students to "write this [linear equation] in the slope-intercept form," and Ms. Alley describes solving in this context as "transform[ing] the equation into a function." And, more idiosyncratically, to maintain continuity with their students' expectations, they call identifying members of the solution set (now pairs of coordinates) "solving" the equation in two variables. These departures from standard practice are important windows into teachers' knowledge developed from practice.

While some might criticize the teachers for trimming inappropriately by overemphasizing isolate the variable as the meaning of "solve" in the case of linear equations in one variable, or for bridging inappropriately by using nonstandard language to talk about solving in the context of linear equations in two variables, for us, these classroom actions reveal sensitivity both to students and to mathematics. The sensitivity displayed by these classroom interactions is one that developed for these teachers in teaching; through teaching, they became aware of the pedagogical obstacle presented by the transition between solving in the context of equations in one variable and solving for a variable in the context of an equation in two variables; they identified a pattern in student difficulty.

This pattern in student difficulty is around a transition in the curriculum, a transition that perhaps seems unimportant from the viewpoint of more advanced understandings. From a more advanced standpoint, both sorts of solving involve isolating a variable and the difference in the result of isolating the variable is unimportant. But, for a teacher, this sort of distinction becomes important as they work to help students eventually come to see this difference in result as unimportant. These teachers are aware of the similarity between these two contexts of solving and indeed do their best to emphasize it to their students. But, they are aware that this is not enough. They seek other ways to deal with the transition that their students must make between having the result of isolating a variable be a number or an expression. They have decided that, for their students, differences that are not salient to experts are important differences.

While Ms. Alley and Ms. Lewis used nonstandard language to help their students bridge from solving equations in one variable to solving equations in two variables, they kept two meanings of solve active. The standard approach of insisting that one cannot solve an equation in more than one variable unless told for which variable to solve suppresses the meaning of solve as finding members of the solution set, which may in turn suppress connections between an equation and its graph. Other researchers have found that students often fail to make important connections between equations and their graphs, in particular that every point on the graph of an equation is a member of its solution set (Knuth, 2000 with respect to high school students in college-preparatory mathematics classes ranging from first year algebra to calculus and VanDyke & White, 2004 with respect to entering college calculus students).

By emphasizing solving an equation in two variables as finding coordinated pairs of values that make the equation true, Ms. Alley and Ms. Lewis are laying the groundwork to emphasize connections between the equation and a table of values and the graph of the line. We did not have the opportunity to observe Ms. Alley or Ms. Lewis during instruction on linear equations in two variables, but it seems to us that finding members of the solution set is a potentially useful bridge to the problem of finding *all* members of the solution set, which can be represented as a graph—a line when the equation is linear. For us, these teachers' actions serve not only to create a bridge between students' understanding of solving equations in one variable to solving equations in two variables, but also, potentially, to create a bridge between linear equations and their graphs. At the same time, Ms. Alley and Ms. Lewis bridge equations in two variables with students' prior knowledge of function rules through their use of alternative language for "solve for." Both teachers acknowledge alternate meanings of solve in the context of equations in two variables, and the language they use for these meanings points to important mathematical connections.

Without advocating that the particular teacher decisions outlined in this paper are the optimal decisions to make, methodologically, we suggest that classroom actions in response to tasks of teaching, like decompressing and trimming complex mathematical ideas for students and bridging those ideas, are places where one can find teachers' knowledge in action and where one find the results of learning from teaching. Moreover, in the context of instruction where much is mandated by textbook and curriculum guide, teachers' knowledge is easiest to identify when teachers make choices that depart from what is either mandated or what is the norm. Even in circumstances where much is mandated, teachers still make many choices about how and when to introduce concepts and skills and what to emphasize; teachers have to decide when to introduce complexity and when to remove it. Fully weighing the potential short- and long-term payoffs, as well as tradeoffs, of such instructional decisions requires not only knowledge of the procedures and definitions of algebra, but also knowledge of various representations and the connections among them, of alternative approaches to teaching and learning, and of more advanced mathematics and its requisite skills, concepts, and habits of mind, as well as knowledge of one's students and patterns in the difficulties that they experience in their learning. We suggest that in the context of such work, one can find the kinds of knowledge of mathematics and sensitivity to nuances of mathematical differences that perhaps are uniquely the province of teachers, rather than mathematicians or engineers. The challenge that faces teacher educators and professional developers is how to create contexts that facilitate the learning of such knowledge.

Acknowledgements The research reported here was supported by NSF REC 0106709. This chapter represents the views of the authors and not the National Science Foundation.

References

Bellman, A., Chavis Bragg, S., Chapin, S., Gardella, T., Hall, B., Handlin, W., Sr., et al. (1998). *Algebra: Tools for a changing world.* Englewood Cliffs, NJ: Pearson Prentice Hall.

Britton, T., Paine, L., Raizen, S., & Pimm, D. (Eds.), (2003). *Comprehensive teacher induction: Systems for early career learning*. Dordrecht: Kluwer.

Chazan, D., Yerushalmy, M., & Leikin, R. (2007). *Cognitive complexity in school algebra?! Evidence from Teachers' Pedagogical Content Knowledge and Equations in One and Two Variables*. Unpublished manuscript. University of Maryland.

Chazan, D., Yerushalmy, M., & R. Leikin. (2008). An analytic conception of equation and teachers' views of school Algebra. *The Journal of Mathematical Behavior*, *27*(2), 87–100.

Cohen, D. K. (1993). *Teaching practice and its predicaments*. Unpublished manuscript, Ann Arbor: University of Michigan.

Even, R. (1993). Subject-matter knowledge and pedagogical content knowledge: prospective secondary teachers and the function concept. *Journal for Research in Mathematics Education*, *24*(2), 94–116.

Ferrini-Mundy, J., Floden, R., McCrory, R., Burrill, G., & Sandow, D. (2005). Knowledge for teaching school algebra: Challenges in developing an analytic framework. Retrieved March 17, 2009 from, https:// www.msu.edu/~mccrory/publications.htm

Fey, J. (1989). School algebra for the year 2000. In S. Wagner & C. Kieran (Eds.), *Research issues in the learning and teaching of Algebra* (pp. 199–213). Hillsdale, NJ: Erlbaum.
Fey, J., & Heid, M. K. (1995). *Concepts in Algebra: A technological approach*. Chicago: Everyday Learning.
Freudenthal, H. (1983). *Didactical phenomenology of mathematical structures*. Dordrecht: D. Reidel Publishing Company.
Heid, M. K., Choate, J., Sheets, C., & Zbiek, R. M. (1995). *Algebra in a technological world*. Reston, VA: NCTM.
Herscovics, N., & Kieran, C. (1980). Constructing meaning for the concept of equation. *Mathematics Teacher, 73*(8), 572–580.
Hill, H. C., Schilling, S. G., & Ball, D. L. (2004). Developing measures of teachers′ mathematics knowledge for teaching. *The Elementary School Journal, 105*(1), 11–30.
Knuth, E. (2000). Student understanding of the Cartesian connection: An exploratory study. *Journal for Research in Mathematics Education, 31*(4), 500–507.
Lakatos, I. (1976). In J. Worrall & E. Zahar (Eds.), *Proofs and refutations: The logic of mathematical discovery*. Cambridge: Cambridge University Press.
McConnell, J., Brown, S., Eddins, S., Hackwarth, M., Sachs, L., Woodward, E., et al. (1990). *The University of Chicago School Mathematics Project Algebra*. Glenview, IL: Scott, Foresman and Company.
Nathan, M., & Koedinger, K. (2000) An investigation of teachers' beliefs of students' algebra development. *Cognition and Instruction, 18*(2), 207–235.
Sierpinska, A. (1992). *Understanding in mathematics*. London: Falmer.
Tall, D. (2002). Continuities and discontinuities in long-term learning schemas. In D. Tall & M. Thomas (Eds.), *Intelligence, learning and understanding – A tribute to Richard Skemp* (pp. 151–177). Brisbane, Australia: PostPressed.
Usiskin, Z. (1988). Conceptions of school algebra and uses of variables. In A. F. Coxford & A. P. Schulte (Eds.), *The ideas of Algebra* (pp. 8–19). Reston, VA: National Council of Teachers of Mathematics.
VanDyke, F., & White, A. (2004). Examining students' reluctance to use graphs. *Mathematics Teacher, 98*(2), 110–117.
Vinner, S. (1983). Concept definition, concept image and the notion of function. *International Journal of Mathematics Education and Science Technology, 14*(3), 293–305.
Vinner, S., & Dreyfus, T. (1989). Images and definitions for the concept of function. *Journal for Research in Mathematics Education, 20*(4), 356–366.
Yerushalmy, M., & D. Chazan (2008). Technology and curriculum design: The ordering of discontinuities in school Algebra. In L. English (Ed.), *Second handbook of international research in mathematics education* (pp. 806–837). London: Taylor Francis.
Yerushalmy, M., & Schwartz, J. L. (1993). Seizing the opportunity to make algebra mathematically and pedagogically interesting. In T. Romberg, E. Fennema, & T. Carpenter (Eds.), *Integrating research on graphical representations of functions* (pp. 41–68). Hillsdale, NJ: Erlbaum

Part III
Examples of Learning Through Teaching: Mathematical Pedagogy

Interlude 2

As we already acknowledged, our initial idea when conceiving of this book was to focus on teachers' learning of mathematics. However, as the authors shared with us and the readers their own learning or the learning of teachers they observed, supervised, or collaborated with, many pedagogical issues surfaced.

As mentioned earlier, in order to explore the overlapping ideas related to both mathematical content and the teaching and learning of mathematics, we found the distinction between *mathematical pedagogy* and *pedagogical mathematics* as introduced by Mason (2007) useful (see also Introduction and the Interlude 1). To reiterate, mathematical pedagogy involves strategies for teaching mathematics and useful constructs, whereas pedagogical mathematics involves mathematical explorations useful for, and arising from, pedagogical considerations. We use this distinction as a lens for considering the chapters in this volume, focusing in Part III on mathematical pedagogy.

Mathematical Pedagogy

Mathematical pedagogy – as an outcome of teachers' learning – is explored in a variety of implementations. In several chapters it is conflated with what Mason (2007) denotes as mathematical didactics: Tactics for teaching specific topics or concepts.

Despite the emphasis on pedagogical mathematics in Part II we also find clear examples of mathematical pedagogy in those chapters. For example, a search for teaching strategies that results in improving students' learning is featured in the work of Marcus and Chazan. In their study, Ms. Alley and Ms. Lewis, who had acknowledged students' difficulty in dealing with linear equations of two unknowns, and the variety of interpretations that emerge from the instruction to "solve", developed a different set of instructions to enable the expressing of one variable in terms of another. What may initially appear as a "trimming" the idea of finding a solution, may in fact, as the authors acknowledge, serve as an appropriate bridge between students' initial understanding of "solving" and a broader mathematical interpretation.

In Part III we see more pedagogically focused examples: Liljedahl elaborates on sources of rapid changes in the strategies employed by the teachers in their mathematics classrooms: restructuring classroom learning from a traditional frontal presentation to collaborative group-work (Mary), exploring a series of interesting problem solving tasks (Mitchell), and communicating ideas (Danica). Alcock describes in detail the approaches she developed for teaching a Real Analysis course for a large class. These approaches are based on coordinating her personal experience and concern about students' learning with the current research literature. However, as Alcock acknowledges, her methods raise many yet-unanswered questions for further research.

Yerushalmy and Elikan demonstrate how a teacher who implements a reform-oriented inquiry-based mathematical curriculum develops her proficiency in managing whole class discussion. They also show reflections of two groups of teachers – one group involved with the reform Visual Math curriculum and another teaching using traditional methods – to the same video episodes. The differences in their foci of attention, from noting the nuances in interactions to being concerned with control and clarity, reflect differences in these teachers' learnt practices and their interpretation of the teacher's role.

Doerr and Lerman focus on the communicative practices learned by Cassie, one of the teachers participating in their research project. They describe Cassie's learning of the role of reading and writing, and the use of oral language, when teaching mathematics. They emphasize her appreciation of the centrality of these practices as one of the outcomes of her learning.

While some teaching strategies are developed by teachers for their own practice, there are cases where teaching strategies are suggested, or even imposed, by a third party. This party can be a researcher, as in the case of Michael in Kieran and Guzman's study, or an instructor/supervisor of student-teachers' practica, as in Hewitt's study. In the latter, several uncommon strategies – denoted as the "links lesson" and "silent lesson" – are demonstrated to student-teachers, with a requirement to implement these in their teaching. A search for appropriate responses to pupils, especially when novice teachers have to "think on their feet" and respond "in the moment," is featured in this chapter. Despite initial resistance, these lessons had a transformative effect on teachers, broadening their awareness of the notion of feedback and shifting their attention from providing clear explanations to listening to their students.

The common implicit theme in these chapters is the centrality of the teaching setting for teachers' learning. That is to say, the specific pedagogy learned by teachers could not have been developed by means other than teaching practice.

References

Mason, J. (2007, June). *Mathematical pedagogy and pedagogical mathematics. Two talks at the pacific institute for mathematical studies 10th anniversary*. Retrieved in March 2009 from http://mcs.open.ac.uk/jhm3/Presentations/%20Presentations%20Page.htm

Exploring Reform Ideas for Teaching Algebra: Analysis of Videotaped Episodes and of Conversations About Them

Michal Yerushalmy and Shulamit Elikan

The Challenge of Whole-Group Algebra Inquiry

In the past decade, school mathematics has moved to assimilate graphing technology into teaching and learning practices in both teacher demonstrations and student problem solving (NCTM, 1989, 2000; Nemirovsky, 1996; Stacey & Kendal, 2004; Kieran & Yerushalmy, 2004). Algebra reform has followed several approaches, some of which can be categorized as "a function approach to algebra," that organize the curriculum around the concept of function, emphasize and support concrete representations, and base learning on situations that appear realistic and are centered on modeling at different levels. In this type of curriculum, students make conjectures and perform actions with tools and representations in ways that were not possible in traditional algebra (Yerushalmy & Chazan, 2008). A large part of the learning is built on student ideas mediated by tools and activities. A key outcome of these innovations is changes in the sequence of learning and in the scope of traditional algebraic processes and objects. To this end, we have been developing and studying a guided inquiry technology-based algebra curriculum for grades 7–9 (Visual Math CET, 1995). The function approach we adopted and the curriculum we designed accordingly shift the emphasis from procedures to operations on functions. By its choice of sequence and the nature of the tasks, the "Visual Math" curriculum offers opportunities for students to raise questions and for teachers to offer tasks that promote inquiry in algebra. Each task has more than one solution, often demands considering aspects that have not been taught, and can be extended to support a compound problem-solving process.

In the inquiry algebra class, where students explore and conjecture, the teacher's primary role is to promote and organize discussions. Understanding this type of teaching is a major challenge for educators (Chazan & Ball, 1999). On one hand, studies suggest what teachers in reform classrooms should not do: for example, they should not teach solution methods, they should avoid telling, and emphasize

M. Yerushalmy (✉)
University of Haifa, Haifa, Israel
e-mail: michalyr@construct.haifa.ac.il

R. Leikin, R. Zazkis (eds.), *Learning Through Teaching Mathematics*, Mathematics Teacher Education 5, DOI 10.1007/978-90-481-3990-3_10,

listening. But as Chazan and Ball (ibid.) suggested in "Beyond being told not to tell," in normal classroom discussions teachers do a variety of things: give yes and no answers, ask questions, answer student comments, draw conclusions, ask for clarifications, comment on student answers, etc. What do we expect that teachers do differently in a reform classroom discussion?

In the published diary of her first year as a teacher of "Visual Math," Levenberg (1995) describes major difficulties and dilemmas relating to her conduct as a teacher, her accomplishments in guiding mathematical inquiry, in general, and when using the computer, in particular. Whole-group discussions were particularly different from what she was used to:

> Although it isn't entirely dark [when projecting computer screens], I miss the eyes of the students.
> I prepared an exposition introducing variables. A frontal lesson boosts my self-confidence. When I see the eyes of all the students then I have the feeling that I am teaching (Is that so?).
> The new trend is the class debate – I can create a debate that would last through the break... for example, a graphic description that suited a verbal description excited four students. I tried not to take sides nor to respond to the argument, but my body language denounced me (the children can read my facial expressions). (ibid., p. 73)

Beyond expressing her feelings as a teacher, Levenberg tried to clarify the difficulties inherent to whole-group discussions by analyzing elements that her students needed to learn:

> Debates disturb me. The voice tone (using vocal cords for persuading) may be disturbing and the self-expression is not always clear. But eventually they will learn to do it. It is interesting that such an ancient doctrine (Greece and Rome) is recapitulated in this curriculum and that discussions replace the ever-lasting paper assignments (ibid., p. 77).

In an interview held in May 1998 after three years of teaching "Visual Math," Gilead interviewed by Yerushalmy, Elikan, and Chazan (2000) talked about the nature of class assignments and team work that stimulate mathematical inquiry and lead to acquisition of knowledge. Gilead outlined the class events and queried the feasibility of whole-group discussions as a tool for acquiring shared knowledge. As Gilead described what happens in the classroom, she addressed the issue of class discussions and questioned how instructive whole-group discussions were.

> [Students'] team-work has to lead eventually to class discussion if you wish to distribute knowledge from the individual to the audience. Now the question remains if this is really instructive? If seven eloquent children talk, how does it affect the rest of the students?

Talking about her role as a teacher, Gilead emphasized the new, unplanned aspects and situations that occur during class discussion:

> I taught twice the seventh grade and each time new things emerge. This doesn't happen in traditional algebra class. Every lesson is a surprise. My challenge as a teacher in this kind of discussion is that there is no clear end to it, and it's hard to navigate this kind of class (Yerushalmy et al., 2000)

Discussions in algebra class are often considered difficult to initiate and maintain because school algebra is concerned mostly with symbols and procedures and not with discussions about big ideas, which seems more appropriate to do in geometry class. Teaching algebra for understanding, teachers must figure out how to open classrooms to questions about algebraic symbols, about meanings of expressions and equations, and about symbol manipulation. Carl, a mathematics educator who observed an algebra class at the Holt school (in Chazan, Callis, & Lehman, 2008) could not imagine that algebra beginners can be "so engaged in conversations about mathematics" (p. 166). Monitoring student knowledge through discussions produces various types of tension. Tensions are generated

- when the teacher must listen to individual students and at the same time keep the class active and instructive as a whole;
- when right or wrong ideas are offered by students;
- when teaching procedures, explaining rules, assigning problems, and introducing ways of finding the right answers must be reconciled with partnership with the students in resolving ambivalences;
- when it is necessary to find content that instigates discussions and at the same time teaches proficiency of basic skills and procedures that may not encourage classroom discussion.

Lampert (1990) described the unique relationship that must evolve in a class that seeks to function as an inquiring mathematical community where students constantly convey their observations eliciting ongoing and concluding arguments from peers and themselves. Previous studies about the professional development of teachers maintained that long practice is needed to embrace these change processes (Wood, 1995; Schifter, 1996, Chazan, Callis, & Lehman, 2008). The work of Leikin and colleagues about learning through teaching in reform secondary school suggested that teachers deepen and broaden their mathematical and pedagogical knowledge while teaching in the reform-oriented classroom. They learned new (for them) solutions to mathematical problems, new justifications to mathematical statements and they learned the art of problem posing (Leikin & Rota, 2006; Leikin, 2005).

In the present chapter, we attempt to understand what practitioners learn from their practice when adopting reform curricula that emphasize discussion-intensive lessons. First, we perform a comparative analysis of two class discussions with the same teacher and same students occurring two years apart. We analyze the discussion quantitatively and qualitatively and describe what the teacher learned from her practice. Then, we perform qualitative comparative analysis of two groups of teachers reflecting on the same two video episodes. Our analysis is based on the work of Lampert and Ball (1998), Santagata, Zannoni, and Stigler (2007), and the work of Herbst and Chazan (2003), who explored the practical rationality of mathematics teaching by analyzing conversations of videotaped episodes of teachers attempting to engage students in geometry proofs. Following Herbst and Chazan (ibid.), we examine the reality that viewers construct from the artifact we designed to learn about their teaching practices. By comparing the conversations of two groups of teachers who practice different algebra curricula, we attempt to gain a better understanding of which parts of the conversations are rooted in the episode as an artifact and which are personal constructions rooted in the reality of their practice.

Learning Through Teaching: First Round

Choosing the Video Episodes

As part of formative assessment of the "Visual Math" curriculum, we created a collection of video records focusing on investigative discussions, argumentation, and reasoning processes that would allow us to witness the construction of relevant mathematics knowledge. We used this collection to prepare episodes that can be shared by groups or individuals as practice cases to reflect upon (Yerushalmy et al., 2000). For the present chapter we selected two episodes from this collection. Each episode documents a conversation that takes place during a whole-group algebra class discussion and refers to a situation that does not usually occur in the traditional algebra class. In both episodes the discussion focuses on a similar task: understanding the meaning of a symbolic expression. In the first episode it is raised as an entirely open exploration: "find the expressions that represent a straight line." In the second one it was the class that offered, in a previous lesson, various expressions as solutions to a given problem in context, and in our episode the teacher chose to discuss interpretations of the meaning of one of the proposed expressions. The teacher and students are the same in both episodes; in the first episode the students are in the 7th grade, in the second episode, two years later, they are in the 9th grade.

Comparative Analysis of Interaction and Authority

One's almost immediate reaction watching the two video episodes is that in Episode 1 the teacher acted as expected in a traditional teaching situation. She surrendered her authority for short periods of time only, mainly when using the computer to present student suggestions. She acted under the constraints of time, of her original plans, and of the required curriculum, and guided the discussion in the direction of the planned answer. In Episode 2 the teacher acted differently: She circumscribed the boundaries of the discussion, listened to the students, and raised questions that were consistent with the themes chosen by her but more often by her students. The difference between the discussion routines is clear.

Comparative quantitative analysis using Cazden's (1988) model confirms the differences. Conversational interactions in a traditional class discussion are often

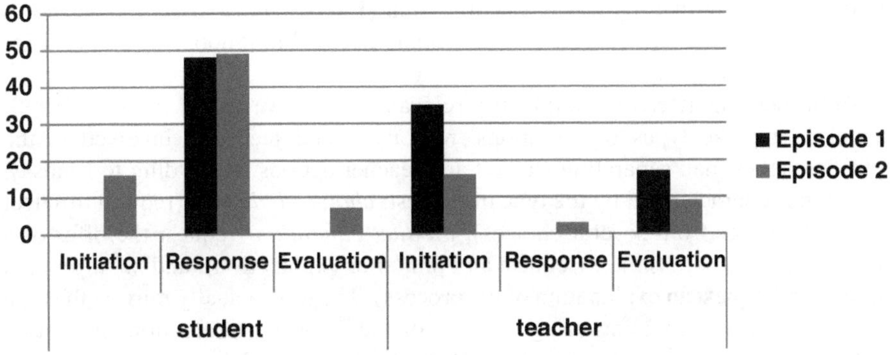

Fig. 1 Interaction analysis (percentages of interactions) of episode discussions based on Cazden's IRE Model

described as a process of initiation, response, and evaluation. Cazden's triad model trails the number of initiations, reactions, and evaluations, in which the teacher asks a question, the student answers, and the teacher evaluates the answer. Then another question follows, and so on. Students do not expect to contribute to initiations, and teachers dominate the reactions, focusing on assessing student answers. Figure 1 depicts a graphic view on the two episodes analyzed using Cazden's model:

In our analysis we extended the IRE model to document both student and teacher interactions. The interactions in Episode 1 resemble the traditional teaching procedures, where the discussion is dominated by teacher initiations, student responses, and teacher evaluations. This quantitative presentation does not paint a comprehensive picture. Viewing the episodes, we identified aspects that are not apparent in figure 1.

Episode 1 is divided into three parts. The first part was based on guided inquiry with a functions' graphing application that took place in a preceding lesson. The discussion began with the teacher's request to receive a list of rules the students had found for qualifying a linear function using the graphing software. The students offered answers and the teacher reconciled suggestions and conclusions, recalled student responses, at times rephrasing them, but did not take a stand or present a position. The students were barely capable of describing and explaining their computer lab results, and the teacher initiations and evaluations injected new relevant terms and descriptions. Two students disagreed about a rule that might describe a linear function. The teacher supported the discussion by asking the opinion of the rest of the students about the disagreement. In the second part the teacher provided examples to refute the hypotheses, using the application to let the students view the relevant examples. There were fewer teacher's initiations and evaluations and more student insertions based on onscreen examples. The teacher rephrased the task, asking students to provide expressions systematically (not from the list used to prepare for the discussion), and she mostly ignored suggestions that were not relevant to a formalization of the rule that qualifies the line $ax + b$. The third part began with a student proposing that the less used expression $(x + a) \cdot b$ can also be accepted as a linear rule. The teacher seemed surprised and decided to ignore this proposal for the rest of the discussion.

To further characterize and quantify relevant details, we used Lindsay's (1990) model to analyze types of initiations, responses, and feedback involved in the episodes rather than separating student and teacher actions. According to Lindsay, initiation is characterized by the type of request: *choice elicitation* (request to select the right answer from several choices); *product elicitation* (request the offer of a product); *process elicitation* (request an opinion or explanation); and *meta-process elicitation* (request an explanation of the process). The reply usually mirrors the type of initiation used, enabling categorization of the answers. Evaluations are based on three strategies that a teacher may adopt to improve and encourage responses: *involvement strategy* (asking for an explanation, example, or illustration); *recalling strategy* (recalling on the student's response to reinforce it); and *simplification* strategy (repeating the student's answer in the student's words or in different terms to validate the answer).

According to this analysis (see Table 1), the discussion in Episode 1 is characterized by an almost equal number of choice, process, and metaprocess initiations and by fewer questions that demand a fact as an answer. This distribution of initiations usually creates a similar distribution of types of responses. Over half the student responses consisted of voicing descriptions of findings and opinions about what they found, watching the graphing application in the previous lab-class and during the discussion. The more frequent strategy used for evaluation was recalling by repeating, rephrasing, and simplifying student answers without stating a position or judging. The students were engaged in making rules, but the teacher often interrupted and changed the discussion or repeated their statements to the class. Thus, although the IRE model described what may be interpreted as a "telling" teacher and "responding" students, in the subsequent analysis the episode appears as an arena

Table 1 Episodes 1 and 2: Percent of arguments based on subsequent classification

Elicitation Type		Episode 1	Episode 2
Initiation	Choice	26	0
	Product	19	14
	Process	26	21
	Metaprocess	30	64
Response	Choice	24	0
	Product	14	14
	Process	54	68
	Metaprocess	8	18
Evaluation	Promoting replies	20	29
	Recalling	47	71
	Simplifying	33	0

that provides students with ways to describe their own understandings, although the process is always initiated by the teacher.

The IRE analysis of Episode 2 (Fig. 1) shows that both students and teacher were involved in initiating, responding, and evaluating. The quantitative analysis reveals that student insertions outnumbered those by the teacher and initiations were equal in number. Students dominated the discussion, and their participation as initiators and responders overweighed the teacher's.

This episode took place two years after the first, involving the same teacher and more or less the same group of students. The mathematics was the algebra of quadratic functions. The discussion focused on one of the models that several groups of students arrived at while solving a problem during group work at a previous lesson: $f(x) = -1.25x(x - 80)$. The teacher wrote the function on the board and asked the students to consider it. The task was to reconstruct the story in the problem from this expression by explaining the structure of the expression and the meaning of the coefficients. Amit presented his arguments, which apparently enjoyed the approval of his team. The teacher did not respond. Inna, from another team, interrupted with a comment prompted by ambivalence in the story. Assuming that she can quickly solve Inna's misunderstanding, the teacher clarified the process described in the story. Inna was not satisfied with the answer and interrupted again. At this point other students attempted to answer Inna's question. Inna, a generally less involved student, was determined to obtain further clarifications and argued that the story and the model presented an illogical situation. Apparently her arguments were not due to what the teacher had thought were a minor misunderstanding but were rooted deeper in the challenge of describing a process that behaves as an increasing function with a decreasing rate of change. The teacher allowed the students to continue the discussion. Inna's questions caused Amit and his team to rephrase their arguments in order to strengthen them and to persuade others.

The subsequent analysis of Episode 2 (Table 1) shows a complete absence of choice elicitations: Most of the initiations, by both teacher and students, were of the metaprocess type, mainly requests asking for explanations of answers. There was a comparatively small number of evaluations (Fig. 1) dominated by students repeating their peers' answers and by the teacher repeating student statements without taking a stand, for example, "I see someone who doesn't agree yet. . .."

The primary and subsequent analyses shed light on changes in practices and norms that occurred within this setting. The first episode captured the teacher's first year of the guided inquiry teaching and the students' first year of algebra study; the second episode captured both teacher and students in the third year of this activity. Note that the comparative analysis is indicative not only of the learning through teaching that teachers underwent, but also of the new ways acquired by the students over the preceding two years. The teacher conducted discussions in both episodes, and in both episodes "telling" was involved in different forms. In the first episode, the teacher told students whether their answers were wrong or right but she also re-played student comments and rephrased and broadcast student descriptions to the class. This was one of the first attempts of the teacher to lead a non-authoritative discussion that allowed the students to discover, argue, and, thereby, comprehend the mathematical idea. Unlike in the traditional view of conversations about algebra, most teacher initiations and student responses were made to seek an opinion, and only one fourth of the statements were product (yes/no) answers. In Episode 1, the teacher reduced the immediacy of her evaluation and initiations when working with the graphing application, probably because she felt that her students were headed in the direction that she wanted to follow and were obtaining computer's feedback with fewer evaluations on her part. Two years later, probably owing to her own growing experience, the maturity of the students, and their more advanced knowledge of algebra, the teacher acknowledged the mathematical value of the conversation that proceeded almost without her intervention.

Disagreements appeared in both episodes. Based on Chazan and Ball (1999) analysis of the meaning and function of disagreements in mathematical discussions, we assume that stimulating and handling disagreements is a key concern in any class discussion. In both episodes the discussion was prompted by an important task that was suitable for producing disagreements. In the first episode the teacher presented an open task, and the students, who had no formal symbolic knowledge, brought a wide open collection of conjectures to the discussion based on observation of visual patterns while graphing with the software application. Throughout the episode, we learn that the teacher's goal was to reach a specific answer, central to her curricular agenda: the expression $ax+b$. Thus, in Skemp's terms (1976), there was a mismatch between the assigned task and what the students wanted to engage in (or what they assumed they were asked to engage in) and what the teacher attempted to achieve. Following Marti Schnepp distinctions (in Schnepp & Chazan, 2002), we would have described it as mixing two strategies of guided inquiry: in one, the teacher is a listener and an observer and in the second, the teacher is more involved in directing the discussion toward a planned curricular goal. In Episode 1, disagreement did not produce consensus, and the clock and curricular goals received higher priority. At the beginning of Episode 2, the teacher explicitly presented a clear goal: Interpreting the numerical coefficients that appear in the expression as part of the story. But after the teacher acknowledged the conceptual depth of the disagreement between Inna and the rest of the class, she seemed satisfied with the agenda and the productivity of the discussion.

The change in practice that appears in Episode 2 reflects also a change in the social and emotional atmosphere. In the first episode the discussion was not consolidated by the students, who kept providing answers of different types going in different directions. Students often did not listen to each other, did not comment on their peers' statements, and were anxious to offer their best answers to the teacher's questions. The teacher, concerned with losing her students' attention, controlled the diversity by her own initiations and by immediate evaluations. In the second episode, the students were greatly involved and sought to settle the conflict themselves. The teacher commented and stressed some points in the disagreement, but generally remained an observer, causing students to reevaluate and rethink their ideas.

Learning Through Teaching: Second Round

In editing an episode, we design an artifact because we isolate a short segment drawn from reality, but omit its context. Thus, the episode invites viewers to "reconstruct a possible reality from where that episode could have come" (Chazan & Herbst, 2003, p. 6). In creating such setting we hoped to study learning through teaching by examining constructions of teachers with different teaching practices. We use two comparable episodes to elicit specific elements we deemed important in reform and traditional algebra discussions. We asked two groups of teachers to watch the two episodes and we recorded their conversations or collected their written records. One group contained four teachers who taught "Visual Math" algebra but were not familiar with the school and the teacher appearing in the episodes. The second group contained five junior high and high school teachers who taught traditional algebra. The second group commented in writing.

Constructions by Teachers Who Practice Guided Inquiry

Four junior high school mathematics teachers, teaching "Visual Math" at the same school, met as part of their adoption of the reform curriculum to discuss issues related to whole-group inquiry. The meeting was led by an experienced educator who asked teachers to watch the two episodes and prepare to talk about issues that are meaningful to them. While watching the videos, the teachers were asked to consider the conditions that support or disrupt discussion, the teacher's involvement in leading the discussion and the students' involvement in conversations, and the quality of the mathematics being taught. The conversation that followed the viewing lasted hours. We summarized the discussed topics under the following five areas.

How Does One Learn to Discuss?

Noga: Did these students learn how to talk to each other? Here when they speak with each other, they also consult each other. I am trying to teach students to speak to each other, because usually when they want to answer other students' questions or to comment they look and talk only toward me (Episode 2).

Noga was surprised that students spoke at length and communicated with each other. She recognized the features she missed in her class discussion: students addressing each other, responding to each other's questions, and looking at each other. Noga spoke about her difficulties changing the traditional norm of always addressing the teacher. She understood that learning and teaching should address the social interaction, part of what Lampert (2001) described as the social complexities of practice. Noga pointed out elements distinguishing the two episodes that we had originally missed. For example, in Episode 1 a few students tried to turn around to face their peers, but they were stopped by the teacher who assumed that it was appropriate to work in a small group but not when participating in whole-class discussion. In Episode 2, the students sat in small groups, naturally turning to each other while speaking.

Who Convinces Whom? Is This the Teacher's or the Students' Job?

Noga: I didn't see them getting answers to their questions. The first child was left behind... She [the student] had very reasonable questions that would have helped her understand the question before getting the solution. And she didn't get answers (Episode 2).

Viki: Somehow in the end it is supposed to make sense. In the end things become clear, otherwise there is a problem and the students won't understand. The nice thing about the discussion is that each student can add something and in the end everything is summed up (Episode 2).

Noga: She [the teacher] doesn't give the answers... I noticed that she sometimes leads the students back to the original problem or repeats the student's arguments. When you repeat them it raises second thoughts and this can definitely help the discussion.

Nurit: She [the teacher] is not judgmental, she doesn't say right or wrong. It's important because otherwise she would close the discussion. If she says right or wrong there is no reason to continue the discussion (Episode 1).

The teachers exchanged ideas about the diverse views and put forward by the students. They were troubled by what seems to disturb every teacher involved in whole-group discussion: how to answer without terminating the discussion? Should the teacher answer, and when? Is it legitimate to deny the students' basic demand to receive an answer to a question? Noga realized that not telling does not mean not answering, and she identified in Episode 1 the indirect ways the teacher used by repeating the students' comments and leading them to the answer.

In Episode 2, no acceptable solution was reached, and the discussion did not end. The teachers' conversation went beyond what was shown in the video, and Viki envisioned a "happy end" where in the end the teacher summarized most of the students' arguments.

Why Is Mathematic Discussion Necessary?

Viki: Whole-group discussions can lead to new discussions or underline other important issues. And every team starts working on its own and develops new ideas. Maybe the discussion can raise other group ideas, and the rest of the class can talk about them and check them. And there is also a concluding discussion.

Nurit: First of all, it enhances reasoning, thinking together about a problem. Each student contributes something to the discussion. In our society it's possible to solve problems through teamwork. Different opinions are offered... but there are smart individualistic students who don't consult when they face a difficulty are stuck with it. Sometimes a weak student can provide a good idea that leads to the solution. So it's very important to think and work together.

This conversation reflects two modes of whole-group discussion. Viki thought that the discussion provides an opportunity for the teacher to introduce order in variability: She can open a discussion with one important aspect and acknowledge the work of individuals by bringing it to the knowledge of the class or attempting to reach a consensus. Nurit, however, spoke about the need to understand diversity of ideas and grant students the opportunity to clarify ambiguous issues. Nurit recognized that the discussion gives each student a chance to speak or listen, in contrast to the classic situation in which the best students offer their solutions. The discussion helps clarify certain issues, and a question asked by any of the students can generate a useful comment. Nurit emphasized that discussion was an opportunity for everyone to experience collaboration in problem solving.

What Does It Mean to "Be With Me" in a Discussion?

Nurit: We never have 100% of the students with us. But we try to reach out as much as possible. If I talk to all of them and I don't conduct a discussion with every group, or if I write things on the board and some children don't look at me, there is no chance of their being with me. Minimal conditions are needed for the student to know what we are talking about. This is my feeling. When he looks it doesn't necessarily mean that he understood, but if he doesn't look, it means he doesn't know.

Also, when children talk with each other during discussion, the teacher doesn't say that they are wrong but lets them continue. I guess that by the end of the lesson they know what's right and what's wrong. When I teach it's hard for me not to respond immediately, so sometimes I nod (Episode 2).

Constructing the reality from the episodes the viewers often ignored the fact that the episode spanned 2–5 minutes and tended to generalize them to a full class period, reaching the conclusion that only a fixed and small number of students participated in the discussion. This constructed reality apparently bothered teachers who asked whether students who did not speak in class and looked passive were participating

Nurit: It is my duty to make everyone talk; otherwise how do I know whether they understood? What is the best way to seat them for a discussion? Do all students need to look at the teacher or the board? Should the discussion involve the whole class or only small groups? What body language is expected from the teacher? They are all watching me, and if I move my head I disrupt the discussion.

What Is an Appropriate Task?

Ellen: A basic condition for a discussion is a suitable assignment. We learned the lesson. In the beginning we also used the questions from the "Visual Math" book "as is," and it didn't work, just like here. As Nurit said, not all students were with us (Episode 1).

Ellen discussed the need to match the assignment to the class situation as a condition for an appropriate discussion. She claimed that many students did not participate in Episode 1 because the assignment was not appropriate. Ellen referred to unsuccessful early attempts in her own teaching experience. Although teachers find it difficult not to adhere to the textbooks and to diversify (Pimm, 1996), Ellen argued that teachers should not adopt any textbook problem but rather examine how would the task appropriately serve the goals and the setting of their planned discussion.

Viki: I think it's very important that the task refers to a theme from daily life. Something that will make the children think harder. The student who did not get the quadratic equation and just experimented with the given numbers.... this is still a nice work (Episode 2).

Viki as well related the difference between the two episodes to the different nature of the tasks and argued that a question from a realistic context provides every student the opportunity to be involved and learn, thus creating a better discussion.

Constructions by Teachers Who Teach Algebra Traditionally

Five experienced junior high and high school teachers who teach traditional algebra showed interest in the "Visual Math" curriculum and asked to learn more about technology-based algebra taught as guided inquiry from practicing teachers and the development team. We suggested them to view and analyze the episodes and we asked them to answer some questions. The teachers received the video and were advised to watch the episodes several times. Teachers were given information about the classes and the climate, the age of the students, their mathematics background, and a short description of the lessons that preceded and followed the episode. The leading question we asked was, What resemblances do you find between the episodes and your work as a teacher? In addition, we asked questions similar to those posed to the "Visual Math" practitioners about the language of mathematic, the teacher's and students' roles, assignments, and discussions' principles. Often teachers chose not to answer the questions in order but to write a personal opinion and interpretation of each episode. Their responses were analyzed and compared with responses collected from mathematics educators. The results of this systematic analysis are described by Elikan (1999). Below is a selection of the most salient constructions of classroom reality as viewed by two sub-groups: first, the reflections of junior high teachers, followed by those of the senior high school teachers.

Junior high school teachers' main concern was securing students' confidence and motivation.

Mina: The discussion contributes to the students' understanding and helps build their self-confidence. The teacher asks the students to define the function and she repeats and clarifies the instructions. She could have been more involved (Episode 1).

Ravit: They seem to lack expertise in algebra. The inquiry starts with the students having very few clues about what they are looking for. The assignment is important for the students because it encourages insecure students who are afraid to speak. The teacher guides the class, trying to reach the goal and her [planned] conclusion. But in the end she reaches a dead end (Episode 1).

They were also concerned whether the discussion contributed to completing the assignment.

Ravit: Two students from two different groups argue and lead the discussion. The teacher expands the discussion, asks guiding questions, and gives others permission to speak. She gives other students the opportunity to clarify some issues [to their classmates]. I would rather have her answer the students' questions (Episode 2).

They recognized the strategies the teacher used to motivate student involvement and identified what they viewed as difficulties related to the minimal teacher involvement in providing answers.

Most of the responses of the senior high school teachers focused on the teacher's performance and on the way she led the class. Control is a dominant issue in their view:

Shery: The teacher lost control during the lesson. She didn't comment, she didn't interfere, didn't make clear what is right and what is wrong. The teacher must be more assertive and check the level of understanding. She was not focused enough. She should be more involved, directing, and conclusive, and ask more direct questions (Episode 2).

It was hard for them to understand how listening to students can help teach mathematics. They showed deep concern about how students feel in class:

Ron: There is a danger that the students will lose their self-confidence... Too many instructive techniques can confuse the students... The students look puzzled... (Episode 1)

They indicated confusion that was caused by the lack of discipline and the inability to balance the contributions among all the students. The viewers concluded that discussions are not suitable for large classes. They pointed out that only a few students participated in the discussion, and the more talkative ones sometimes dominate the discussion (not always on relevant issues).

Shery: The number of active students in class was small. Many students did not participate. Not all the students were given a chance to speak. Most of the students couldn't join the discussion and express their opinion. There were too many students in class. The discussion was disrupted because one student took control of it... two students from two groups argued with each other... (Episode 1)

Senior high school teachers considered the time frame of the discussion to be too long, a waste of time when the answer is simple and basic. They thought that the assignment in Episode 1, dealing with linear expressions, was simple and should not require a discussion, which is perceived as an instrument to present different options required to solve a mathematical problem.

Ron: The assignments required a simple technique that doesn't require a discussion (Episode 1).

Others thought that the task was not appropriate for discussion because it was too far from the students' knowledge.

Monny: The discussion didn't contribute at all. The question is too complicated for the students because they lack algebra expertise (Episode 1).

At the same time, the teachers thought that it was important to use mathematical problems in context, as in Episode 2, because the variety of methods available for reaching a solution in problems of this type makes the discussion worthwhile.

Monny: The question requires understanding, persuasion, and translation into mathematical language. A good assignment for discussion (Episode 2).

The discussion was perceived as an arena for the teacher to display various methods for a solution of more complex problems, mainly word-problems in context.

Shery: A realistic question highlights the use of mathematics in daily life. The question in not closely related to algebra and requires intuition and personal consideration. There are a few optional solutions, and reaching them will enrich the student (Episode 2).

Learning Through Teaching: Constructions That Uncover Teaching Practices

Summarizing the analysis of the two sources of data, we argue that the conversations about the two episodes are helpful in studying teaching practices and that they indicate processes of learning from teaching. Note, however, that the implications of the different settings in which we studied the two groups (as a result of technical constraints) may have had been greater than we had originally assumed. Colleagues within the same mathematics department, teaching the reform algebra program, share a common set of goals that can affect their conversation. Individuals watching a video and inquiring individually about it do not share these goals. But the answers documented in both settings, and the clear, consistent distinctions between the two groups support our suggestion that the constructions produced with the artifact indeed reflect commitments and requirements learned through different practices of teaching.

The group of the four "Visual Math" teachers participating in the conversation about the video episodes was going through a major change of their algebra curriculum and teaching. Watching the video episodes, they recognized practices similar to those taking place in their algebra classes, and they spoke about their own issues and tensions in ways that went beyond the specific actions in the episodes. In general, they directed their attention to Episode 2. Having recognized the reality they lived in the episodes, they probably considered Episode 2 as a more appropriate climate for discussion. They were not interested in the differences between the two episodes, in evaluating the changes, or in criticizing what may be viewed as the less mature reality of Episode 1. Instead, they focused on what provided them with an opportunity to describe their own practices.

Their reflections on Episode 2 illustrate their commitment to engage all students in the construction of knowledge through active learning and to guide whole-group discussions as a way to achieve this objective. It also became clear that they were committed to teaching that goes beyond the regular covering of the mathematics curriculum. They were concerned with what Lampert (2001, p. 265) described as "teaching students to be people who study in school." They made it clear that they are familiar with a classroom reality in which students talk and relate to each other, are ready to listen to others' ideas, learn to think together, and learn from the diversity of ideas. Because they identified differences between the two tasks being discussed, they spoke about their own sense and responsibility for the curriculum and the assigned tasks.

From their commitments we learn about their dilemmas. They identified in the episode activities apparently causing the tensions they faced. They appreciated the confrontations in the video as the ideal climate for learning, which they do not yet know how to achieve. Episode 2 helped Noga to describe her difficulties in achieving a desirable discussion. Noga indicated that she would like to learn how to teach her students to discuss. Watching the teacher attempt to guide without "telling," the viewers at first questioned whether the students really learned, but gradually, upon reflection, they were persuaded that even if students had difficulties during the few minutes shown in the video they would understand "by the end of the lesson." Nurit praised the teacher's habit not to provide judgmental feedback, and recalled her own failure to act in this way: "I nod my head." There is an implicit question whether "not telling" is consistent with the commitment to help students learn. Noga started by criticizing the practice of not providing answers to questions, but gradually identified several ways of "telling" that she was able to appreciate.

The second group consisted of five teachers that taught traditional algebra curriculum. The two junior high school teachers of this second group expressed commitment to constructing a secure environment for learning algebra. In their view, this environment requires a leading and guiding teacher. Although new for them, they appreciated the conversations taking place in both video episodes and were concerned about whether the discussion provided the teacher with the opportunity she needed to teach important mathematics. They did not extend their constructions to go beyond the specific episode, to speak explicitly about their own practices, or to discuss the great challenges involved in teaching this way. The three high school teachers expressed stronger commitment to teach the correct mathematics using an accurate vocabulary. The leading teaching practice exhibited in their reflections was "control." The teachers identified actions and norms that did not agree with their practices, in which the teacher controls the situation, asserts her views, determines who speaks and when, cuts short long conversations, points out exactly what is true or false, formulates conclusions, and asks questions to verify the students' comprehension. Engagement with problems in context through discussion made sense to them because it is not considered to be a purely mathematical conversation, and therefore requires argumentation skills that are different from the ones required in mathematical tasks such as the one in Episode 1. The ideal climate they envisioned for a discussion involved a task that can be discussed in a clear, simple, and univalent mathematical language and a teacher who knows when to interfere so that students are never lost. They did not address the possible challenges and complexities of the whole-class discussion, and although they did not explicitly refer to their own practices they did not hesitate to provide clear suggestions about tasks and teaching that would improve what they have seen.

Concluding Remark

The first group, which was undergoing a process of re-learning to teach algebra as a subject for guided inquiry, used the episodes as a *mirror*. Eventually, they recognized central components of their own teaching reality and focused on what provided them with an opportunity to describe their own practices, dilemmas, and tensions. The episodes opened for the group that taught traditional algebra a *window* to somewhat distant classroom reality. Watching these new views they implicitly uncover their own pedagogical principles, making an effort to "fix" the reality they viewed in ways that would coincide with their practices.

References

Cazden, C. B. (1988). *Classroom discourse. The language of teaching and learning*. Heinemann Portsmouth, NH.

Chazan, D., & Ball, D.(1999). Beyond being told not to tell. *For the Learning of Mathematics, 19*, 2–10.

Chazan, D., & M. Schnepp (2002). Methods, goals, beliefs, commitments and manner in teaching: Dialogue against a calculus backdrop. *Social Constructivist Teaching, 9*, 171–195.

Chazan, D., Callis, S., & Lehman, M. (2008). *Embracing reason: Egalitarian ideals and the teaching of high school mathematics*. New York, NY: Routledge.

Elikan, S. (1999). *Characterization of computers' supported inquiry discussions in mathematics classrooms*. M.A. Thesis. University of Haifa, Israel. (in Hebrew).

Herbst, P. G. & D. Chazan (2003). Exploring the practical rationality of mathematics teaching through coversations about videotaped episodes: the case of engaging students in proving. *For the Learning of Mathematics, 23*(1): 14–2.

Kieran, C., & Yerushalmy, M. (2004). Research on the role of technological environments in algebra learning and teaching. In K. Stacey, H. Shick, & M. Kendal (Eds.), *The future of the teaching and learning of Algebra. ICMI Study-12 Volume* (pp. 99–152). Dordrecht, the Netherlands: Kluwer.

Lampert, M., & Ball, D. (1998). *Teaching, multimedia, and mathematics: Investigations of real practice. The practitioner inquiry series*. New York, NY: Teachers College Press.

Lampert, M. (2001). *Teaching problems and the problems of teaching*. New Haven, CT: Yale University Press.

Lampert, M. (1990). When the problem is not the question and the solution is not the answer: Mathematical knowing and teaching. *American Educational Research Journal, 27*, 1, 29–63.

Leikin, R., & Rota, S. (2006). A case study on the development of teacher's proficiency through teaching. *Mathematics Education Research Journal, 18*(3), 44–68.

Leikin, R. (2005). Teachers' learning in teaching: Developing teachers' mathematical knowledge through instructional interactions. The paper presented at the 15th *ICMI Study: The Professional Education and Development of Teachers of Mathematics*. http://stwww.weizmann.ac.il/G-math/ICMI/log_in.html

Levenberg, I. (1995). Teacher's diary. *Journal for Mathematics Teachers* (ALEE') *17*, 73–77 (in Hebrew).

Lindsay, J. S. (1990). Classroom discourse analysis: A review of the literature with implications for educational evaluation. *Journal of Research and Development in Education, 23*, 107–116.

National Council of Teachers of Mathematics (1989, 2000). *Principles and standards for school mathematics*. Reston, VA: NCTM.

Nemirovsky, R. (1996). A functional approach to algebra: Two issues that emerge. In Bednarz, N., Kieran, C., & Lee, L. (eds.), *Approaches to algebra: Perspectives for research and teaching* (pp. 295–316). Dordrecht, the Netherlands: Kluwer.

Pimm, D. (1996) 'This is so': A text on texts. In A. J. Bishop et al. (Eds.), *International handbook of mathematics education* (pp. 371–409). Dordrecht, the Netherlands: Kluwer.

Santagata, R., Zannoni, C., & Stigler, J. W. (2007). The role of lesson analysis in pre-service teacher education: An empirical investigation of teacher learning from a virtual video-based field experience. *Journal of Mathematics Teacher Education, 10*, 123–140.

Schifter, D. (1996). *What's happening in math class? Envisioning new practices through teacher narratives*. New York, NY: Teachers College Press.

Skemp, R. (1976). Relational understanding and instrumental understanding. *Mathematics Teaching, 77*, 44–49.

Stacey, K., & Kendal, M. (2004). Algebra: A world of difference, In: K. Stacey, H. Shick, & M. Kendal (Eds.), *The future of the teaching and learning of algebra. The 12th ICMI Study* (pp. 329–346), Dordrecht, the Netherlands: Kluwer.

Visual Math: Algebra (1995). *Center for educational technology, Ramat-Aviv*. http://www.cet.ac.il/math-international/visualizing.htm

Wood, T. (1995). An emerging practice of teaching. In P. Cobb & Bauersfeld, H. (Eds), *The emergence of mathematical meaning. Interaction in classroom cultures* (pp. 203–228). Hillsdale, N.J: Erlbaum.

Yerushalmy, M., & Chazan, D. (2008). Technology and curriculum design: The ordering of discontinuities in school algebra. In L. English et al. (Eds.), *Handbook of international research in mathematics education* (2nd edn., pp. 806–837). New York, NY: Routledge.

Yerushalmy, M., Elikan, S., & Chazan, D. (2000). *Discussions in the Mathematics Classroom.* Multimedia package (Video and Documents). University of Haifa and Center for Educational Technology, Ramat-Aviv, IL.

On Rapid Professional Growth: Cases of Learning Through Teaching

Peter Liljedahl

Introduction

I do professional development. That is, I professionally develop inservice teachers in a number of different formal educational settings to achieve a number of different inservice goals. The sort of work that I am involved in is more than simply the delivery of workshops, it is often the creation and maintenance of a community of practice in which ideas are provisional, contextual, and tentative and are freely exchanged, discussed, and co-constructed (Little & Horn, 2007; McClain & Cobb, 2004; Wenger, 1998). Working within this context I am both a facilitator and a researcher. I facilitate the setting and I research my effectiveness. However, while it is true that as a researcher I am interested in the down-stream effects of the work that I am engaged in (improvement in students' experiences and performance, etc.), it is equally true that I am also interested in the effects on the teacher participants. There is much that happens in this regard.

Method

Working as both the facilitator and the researcher interested in the contextual and situational dynamics of the setting itself I find myself too embroiled in the situation to adopt the removed stance of observer. At the same time, my specific role as facilitator prevents me from adopting a stance of participant observer. As such, I have chosen to adopt a stance of *noticing* (Mason, 2006, 2002). This stance allows me to work within the inservice setting to achieve my professional development goals while at the same time being attuned to the experiences of the persons involved. I notice, first and foremost, myself. I attend to my choices of activities to engage in and the questions I choose to pose. I attend to my reactions to certain situations as well as my reflections on those reactions, both in the moment and after the session.

P. Liljedahl (✉)
Simon Fraser University, Burnaby, BC, Canada
e-mail: liljedahl@sfu.ca

R. Leikin, R. Zazkis (eds.), *Learning Through Teaching Mathematics*, Mathematics Teacher Education 5, DOI 10.1007/978-90-481-3990-3_11,

More importantly, however, I attend to the actions and reactions of the teacher participants both as individuals and as members of a community. And in so doing, from time to time I notice phenomena that warrant further observation and/or investigation. Often these are phenomena that occur in more than one setting and speak to invariance between individuals, settings, contexts, or behaviors. Once identified, these phenomena can be investigated using methodologies of practitioner inquiry that combine the role of educator with researcher – in this case teacher educator with researcher (Cochran-Smith & Lytle, 2004).[1]

Using this methodology of noticing within a diverse number of professional development settings – from workshops to learning teams to graduate programs – I have, from time to time, noticed teachers undergoing *rapid* and *profound*[2] changes in their beliefs and practices. This phenomenon is rare. Most teachers engaged in inservice work follow a trajectory of change that is much more pedestrian. At first such changes surprised me and I was immediately suspicious of the self-reported accounts of the almost instantaneous revision of practice that some teachers spoke about. But as more and more of these accounts accumulated I decided to investigate further. I stepped out of my role as facilitator and assumed a stance of researcher. I began to interview teachers, to visit their classrooms, to observe their teaching. I spoke with their colleagues, their administrators, and their students. I took field notes and I wrote narratives (Clandinin, 1992; Clandinin & Connelly, 1996).

Over time patterns began to emerge within the individual cases that I investigated. Using a grounded theory (Creswell, 2008) approach, the phenomenon of rapid and profound change began to fragment and converge into six different themes: *conceptual change, accommodating outliers, reification, leading belief change, exo/endospection*, and *critical questions*. Each of these six themes, although not entirely distinct, embodies a different mechanism for transformation of teaching practice that needs a unique theoretical framework to bring it into sharper focus.[3] The first three of these themes – *conceptual change, accommodating outliers*, and *reification* – are situated within what teachers have learned through their own teaching and are, thus, the three themes that I explore here. In what follows I present three

---

[1] It should be noted that the main distinction between a methodology of noticing and a methodology of practitioner inquiry is that noticing does not presuppose a research question. It is a methodology of attending to the unfolding of the situation while being attuned to the occurrence of phenomena of interest.

[2] Rapid and profound are relative terms. What is rapid for one person is not for another. Likewise, what is profound for one is not for another. In general, a transformation is deemed to be profound if a teacher's new practice is both visibly and invisibly different. That is, there are substantial differences evident from the perspective of an outside observer AND the teacher claims to have undergone a substantial change within how they view their teaching. In practice, this transformation is seen as being rapid if the bulk of the transformation occurs within a period of less than one month.

[3] The grounded theory approach that is used to spawn the six themes is not sufficient for analyzing the themes. At the same time, each theme, although not discrete, is distinct enough to be seen as its own phenomenon. As such, each theme is analyzed using the theoretical framework most relevant to its particular characteristics.

abbreviated cases[4] (and their analysis) each of which was selected for its ability to succinctly exemplify one of the three aforementioned themes.

The Case of Mary

Mary is an elementary school teacher with more than 20 years of experience. Although she has taught mostly every grade in the elementary curriculum, Mary has spent most of her years teaching the intermediate grades (4, 5, 6, 7). In recent years she has become disillusioned with the teaching of mathematics. In particular, she is questioning her ability to have a meaningful impact on students within the structures that she must work. Mary is unhappy with the current mathematics curriculum as well as the dispositions of her students with respect to mathematics and the learning of mathematics. For Mary, both the curriculum and her students expect the same things from her – to deliver mathematics in a piecemeal fashion with each dose carefully and precisely dispensed. She sees the curriculum as being a collection of disjoint topics each of which needs to be mastered in turn. There is little connection across these topics and there seems to be very little thinking required of the students. Mary reports that the teaching of mathematics is mundane. It is the one subject in her teaching where she feels that she lacks connection with her students as learners. She talks about the teaching of mathematics in very traditional ways – from the standard review-demonstrate-mimic-practice lesson plan to the use of homework, quizzes, and tests to leverage students' commitment to the acquisition of the taught skills. Ironically, she does not see herself as this type of teacher in the other subjects that she teaches. Most troubling to her, in this regard, is that she sees herself as bending to the will of her students. This is how they see mathematics and mathematics teaching and she is playing right into their expectations.

To revitalize herself, and to work with students in a more meaningful way in the context of mathematics, Mary decided to leave the classroom and work more closely with students within a learning support context. In particular, she works in supporting "at risk" students in the areas of mathematics and language arts. She finds this work to be rewarding in that she does not have to conform to the expectations of the students and the curriculum. However, she misses the dynamics of the classroom environment.

---

[4] In my research, I distinguish between cases and narratives. Although both generated in the tradition of narrative inquiry (Clandinin, 1992; Clandinin & Connelly, 1996) cases and narratives are written for different purposes. Narratives are written in the first person and are meant to capture some of the hidden aspects of a teacher's practice such as beliefs, anxieties, intentions, and goals. These are deeply personal aspects and require close relationships and deep trust in order to be produced. Cases, on the other hand, are written in the third person and deal more with the visible aspects of teaching such as lesson routines and assessment schemes. They may also include personal aspects such as goals, but these are usually aspects that are freely given or espoused. In all cases, both narratives and cases are shared with participants as a way to both enrich the descriptions and confirm their validity.

It was at this point in her career that Mary decided to enrol in an Elementary Mathematics Education Master's program. The program, which spans two years, is a collection of seven courses, each of which is designed to look at mathematics or mathematics education from a different perspective. Mary hoped that her participation in such a program would allow her to find what it was she felt she was missing in her practice. She did. Part way through her second course in the program, Mary returned to a regular classroom and completely reconstructed the teaching of mathematics for her new students. This was not a slow change.

As the instructor for the second course in the program I first met Mary in November when she was still very much disillusioned with the state of affairs in mathematics education, in general, and in her teaching, in particular. At the beginning of January, when she returned to the regular classroom she was a completely different mathematics teacher. Her classroom was now a place of inquiry and discovery, of meaning making and thinking. She had expectations of her students in the context of mathematics . . . and they lived up to them. Her classroom was transformed. Rows of desks were replaced by students working in groups of four at tables. There was no longer a well-defined "front" of the classroom as Mary now utilized the entire classroom for teaching and students were valued as partners in learning AND teaching. Mary was also re-examining the whole notion of assessment as a tool for learning as opposed to a way to control student behavior. Not only was her transformation as a mathematics teacher profound, it was fast when viewed in the context of professional growth in general, and even faster when viewed in the context of Mary's long teaching career.

Mary's Transformation as a Case of Conceptual Change

Mary reported that she had participated in many professional development workshops and district-based initiatives, from constructing assessments to piloting new textbooks and resources. But each of these failed to inform her practice in any significant way. In fact, many of them simply reinforced the very paradigm she was trying to escape. Based on interviews with Mary, I determined that what eventually initiated Mary's rapid transformation was the recasting of her view of mathematics curriculum from content to context. This resulted from an in-class exercise we did in which we looked at the mathematics curriculum documents from the perspective of mathematical processes. This view places the processes (such as estimation, problem solving, communication) as the primary goal of mathematics education and relegates traditional topics (such as geometry, fractions) to the context in which these processes are actualized. This view was what Mary was missing. In the mathematical processes she found the meaning that she felt was absent from her practice.

Mary had begun her transformation long before she entered the masters program, however. She had long ago rejected the beliefs upon which her practice was built as well as the beliefs upon which she saw the curriculum as being built. Outwardly there was no change, however. She still taught mathematics the same way as she always had ... and in the way she had been taught mathematics. And, although there seemed to be no difference in her teaching, Mary was persistently seeking a new way to teach.

The transformation that Mary underwent – belief rejection followed by belief replacement – can be seen as a special form of conceptual change, a theory that emerged out of Kuhn's (1970) interpretation of changes in scientific understanding through history. Kuhn proposed that progress in scientific understanding is not evolutionary, but rather a "series of peaceful interludes punctuated by intellectually violent revolutions", and in those revolutions "one conceptual world view is replaced by another" (p. 10). That is, progress in scientific understanding is marked more by theory replacement than theory evolution. Kuhn's ideas form the basis of the theory of conceptual change (Posner, Strike, Hewson, & Gertzog, 1982) which has been used to hypothesize about the teaching and learning of science. More recently, this theory of conceptual change has been applied to the learning of mathematics (Greer, 2004; Tirosh & Tsamir, 2004; Vosniadou, 2006; Vosniadou & Verschaffel, 2004). The theory has also been shown to be relevant to the meta-conceptual, motivational, affective, and socio-cultural factors of learning as well (Vosniadou, 2006).

In general, the theory of conceptual change starts with an assumption that in some cases people form misconceptions about phenomena based on lived experiences, that these misconceptions stand in stark contrast to the accepted theories that explain these phenomena, and that these misconceptions are robust. It is not a theory that applies to learning, in general. It is highly situated, requiring four primary criteria for relevance (Vosniadou, 2006) – (1) it is applicable only in those instances where misconceptions are formed through lived experiences and in the absence of formal instruction, (2) there is a phenomenon of concept rejection, (3) there is a phenomenon of concept replacement, and (4) there is the possibility of the formation of synthetic models. I propose that each of these criteria is equally relevant to the instance of Mary's rapid reformation of her beliefs about mathematics and the teaching and learning of mathematics as well as the rapid reformation of her teaching practice.

To begin with, Mary's relevant lived experiences occurred in her time as a student. As a learner of mathematics she has experienced both the learning of mathematics and the teaching of mathematics, and these experiences have impacted on her beliefs about the teaching and learning of mathematics (Chapman, 2002; Feiman-Nemser, McDiarmid, Melnick, & Parker, 1987; Lortie, 1975; Skott, 2001). The question is – can these experiences be viewed as having happened outside of a context of formal instruction (criteria #1)? Although her experiences as a learner of mathematics are situated within the formal instructional setting of a classroom, the object of focus of that instruction is on mathematics content. That is, while content is explicitly dealt with, within such a setting theories of learning, methodologies of teaching, and philosophical ideas about the nature of mathematics are not. Secondly, Mary had clearly rejected the paradigm in which she was working, and although this rejection did not manifest itself in her actual practice, it did exist within her belief structures around mathematics and the teaching and learning of mathematics (criteria #2). Mary then searched for something more meaningful around which she could construct her teaching practice. When she did eventually encounter the aforementioned view of mathematical processes as curriculum it instantly displaced the already rejected paradigm which she was still reluctantly using (criteria #3).

Finally, a "synthetic model" is a term reserved for the description of an incomplete or incorrect model. It is a middle ground between the initially rejected concept and the concept that is to be acquired. In many cases, it is a *synthesis* of the old and the new as the learner is making use of old resources to make sense of new ideas. In relation to Mary, a synthetic model manifested itself in the fact that although she made rapid and profound changes to her teaching practice she still continued to make changes and improvements as time went on (criteria #4). This is not unusual. Change begets change and as Mary settles into her new teaching practice she saw more and more details that required attention. This is no different than the context of a student's conceptual change around a mathematics or scientific concept. The synthetic models they develop are most often temporary and tentative, often giving way to more and more refined models as conceptual understanding is achieved.

Although Mary's case exemplifies rapid professional growth through a process of conceptual change she is only one of eight cases that I have encountered. In general, there are many teachers who engage in inservice education opportunities because they are extremely disillusioned with the paradigms that they are working in. They have learned, through their own teaching, that they do not work for them and they are looking for something better. Sometimes they find it, sometimes they do not. When they do find something better they may be able to transform their practice quickly . . . as Mary did.

The Case of Mitchell

Mitchell is a middle school teacher with eight years of teaching experience. He has always taught mathematics and he has a very clear sense of what is important for students to learn in mathematics and what his role as a teacher is in this context. For Mitchell, mathematics is really just a game – a game with set rules and very clear outcomes. Mathematics is a collection of skills and facts that need to be mastered before going on to the next level. As a teacher, he sees his job as assuring that each student learns these skills and facts – and to not let anyone advance to the next grade until they have done so. He also has a very clear idea of what the students' role is. Their job is to learn the material that is being taught and to be able to demonstrate mastery at the end of a unit . . . and at the end of the year. Mitchell is a traditional mathematics teacher in every sense of the word and he has no issues about stating so.

Mitchell's mathematics classroom is a pillar of traditional teaching. He adheres to a standard lesson of review-demonstrate-mimic-practice and students are expected to seek his help if they are stuck or do not understand. He only uses questions that are unambiguous and lead to closed-form single solutions. He feels that mathematics needs to be taught (and learned) in this fashion and that all of the problems that face mathematics education are due to deviation from this tradition.

Ironically, Mitchell is not this traditional in teaching his other subjects (science, language arts, and social studies) so there are some aspects of more reform oriented teaching that have seeped into his mathematics lessons. For example, he does allow his students to sit in pairs and to work together on in-class assignments. He also, from time to time, gives a problem solving activity to students, but he sees this as extracurricular and does not allow it to figure into his assessment and evaluation schemes.

Mitchell does not shun professional development opportunities, engaging in them with the expectation that he will "get something out of them." However, he openly admits that many of these opportunities turn out to be things that he already does, which further reinforces his conviction that his method of teaching is "on the right track." Occasionally he learns something that is interesting, which he then implements in his teaching. This is how he came to start doing some problem solving activities in his classroom, and as he states, "there are some really fun activities that I now do with my kids." Of course, there are things he sees in workshops that he also dismisses outright as being "completely pointless," such as a session on performance-based assessment that he once attended.

Although I had met Mitchell as a participant in a number of single session workshops, I did not begin to interact with Mitchell until he became a member of a district-based learning team that I was facilitating. This team was formed for the purpose of creating numeracy tasks for district wide assessment. Mitchell came to this learning team with the expressed purpose of offering some of his expertise in creating "really comprehensive final exams."

The first task of this team was to co-construct a definition of numeracy. Initial attempts to do this resulted in definitions that were more closely associated with fluency of arithmetic. In order to get past this initial definition, I suggested that they think of students that they had taught in the past who were very good in mathematics, and to further think what qualities they possessed that allowed them to be good in mathematics. This dramatically changed the discourse about numeracy and rather quickly a more sophisticated definition emerged – "Numeracy is not only an awareness that mathematical knowledge and understandings can be used to interpret, communicate, analyze, and solve a variety of novel problem solving situations, but also a willingness and ability to do so."

The team then set out to design a task that would measure some of the capacities embodied within this definition. Over the course of four additional meetings stretched out over approximately eight weeks the team went through three iterations of a design-test-refine process before they arrived at the final task (see Fig. 1). During this process I saw Mitchell undergo a tremendous transition in his teaching. After pilot testing the initial version of the task he was talking about things that needed to change in his classroom in order for his students to be successful. He was restructuring the way he thought about and facilitated group work; he was redefining his own notion of what constituted a good mathematics question . . . and a good mathematics answer; and he was trying to find ways to change the dispositions of

GIVING OUT BONUSES

You are the manager of the Teen-Talk Cell Phone company that employs a number of independent sales people to sell their phones **seven days a week**. These sales people work as much or as little as they want. As a sales manager you don't care how much they work, but you do care how much they sell. So, to motivate them to sell more you give out bonuses based on how productive they have been. There are two bonus plans:

- the top producing individual receives $500.
- the top producing team **shares $500 in a fair manner.**

However, there are also two problems:

- different people have different ways of reporting their productivity.
- the individual sales teams don't have the same number of people on them.

Based on the information provided in the table below, **who should get the bonuses this month, and how much do you think they should get? Justify your answers in writing.**

Sales Person	Team	Sales Reported for the Month of April (30 days)
John	A	300 cell phones sold this month
Peter	B	An average of 56 cell phones sold every 5 days
Lewis	A	An average of 10 1/3 cell phones sold each day
Amy	A	598 cell phones sold in the last 60 days
La Toya	A	An average of 98¾ cell phones sold every 10 days
Jennifer	B	An average of 11 4/15 cell phones sold each day
Steven	B	An average of 55 cell phones each week
Fantasia	C	4113 cell phone sold in the last year
Diana	C	An average of 10.05 cell phones each day
Matthew	D	An average of 10.87 cell phones each day
Camille	D	An average of 9 1/6 cell phones each day
Jasmine	C	267 cell phones this month

Fig. 1 An example of a refined numeracy task

the students in his classroom. In a very short period of time, Mitchell came to change most of what he held to be true about mathematics and mathematics teaching and learning.

This is not to say that Mitchell completely reconstructed his teaching practice overnight. He spent the remainder of that school year struggling to actualize some of his ideas as he swam against the current of already entrenched student expectations and dispositions. At the beginning of the following school year, however, his classroom was truly transformed. Lessons were now modeled on explorations initiated by interesting (often open ended) tasks which were worked on in groups and concluded with whole class discussion. His assessment practices were also completely redesigned, although still very much a work in progress.

Mitchell's Transformation as a Case of Accommodating Outliers

Unlike Mary, Mitchell did not come to the learning team looking for answers. He did not reject the teaching paradigm that he was working under, his teaching practice, or his world view of teaching mathematics. Quite opposite to Mary, Mitchell's teaching can be seen as impenetrable. He participated in a wide variety of professional development opportunities, but nothing had any effect on his practice. Invoking the discourse of adaptation á la Piaget, we might say that Mitchell was not accommodating new ideas into his existing schema of teaching mathematics (Piaget, 1968). He tended to deal with new ideas about the teaching of mathematics in one of the three ways. First, and most common, Mitchell would find within his professional development experience something specific that resonates with his current teaching. This point of commonality was then used to support his basic assumption that "I already do that." He was able to make this claim no matter how minute the point of commonality was. A nice example of this was Mitchell's insistence that he used effective questioning in his teaching because, when teaching from the front of the class, he asked a lot of questions and when he attended a workshop on effective questioning he learned that sometimes an effective question can be short and very directed, as his always were. Aside from ensuring that very little of substance penetrated his practice, this strategy of assumed commonality also served to entrench Mitchell's teaching practice as he was constantly reassured that he was "going in the right direction."

If Mitchell did not find any points of commonality, but the new ideas that presented to him caught his interest he tended to incorporate them into his practice. But he would do so without letting it impact on his conception of himself as a mathematics teacher, or on his general notions about mathematics and the teaching and learning of mathematics. As an example, Mitchell very easily introduced a program of problem solving into his practice. He had a small, but good, collection of problem solving tasks that he gave to his students from time to time. There was no attempt to assess their performance on these tasks or to pull some of the affordances such tasks could offer into the rest of his teaching. As such, he kept it very much as an extracurricular activity not allowing it to redefine his teaching or his conception of himself as a teacher. Mitchell was not only assimilating these problem solving experiences – he was actively not accommodating them.

Finally, if Mitchell found no points of commonality and no points of interest he would simply dismiss the new ideas presented to him as "pointless." As mentioned, this very effectively allowed him to deal with the complex nature of alternative assessment in general and the specific ideas around performance-based assessment in particular.

Ironically, as effective as Mitchell was at avoiding accommodation, it was accommodation that eventually led to the revision of his practice. From interviews with Mitchell it became clear that a real turning point for Mitchell was the construction of the definition of numeracy. The definition itself meant very little to him, it was more the consideration of the good students Mitchell had had in the past that lead to deeper changes. Mitchell had always been aware of these students, but he had effectively not allowed their existence to impact on his aggregated vision of a mathematics student. To him they were outliers, as were the capacities that they possessed. For Mitchell, a student was seen from a deficit perspective. They were children that lacked specific knowledge – knowledge that he possessed and would apportion out to them over the course of the school year. When he started to think of these outliers he not only saw them as capable, but he also saw a whole spectrum of skills that he had never really considered before. Problem solving abilities, divergent thinking, awareness of the mathematics inherent in a specific context, the ability to use mathematical concepts broadly in different contexts, etc. were suddenly seen as capacities that all students needed in order to be successful in mathematics.

As Mitchell struggled to reform his teaching for the remainder of that first year my work with him continued. In subsequent interviews Mitchell revealed that he was now beginning to make sense of why the problem solving tasks that he had previously been using as extracurricular were so effective at developing some of these aforementioned capacities. At the same time, Mitchell was beginning to see a new set of capacities requisite for students to be successful at the numeracy tasks that he had participated in designing. Group work, ability to articulate thinking, persistence, tolerance of ambiguity, and comfort with being stuck were now the deficiencies that he wanted to address.

In the consideration of both talented students and good problem solving tasks Mitchell was finally accommodating information into his practice. But this was not new information. Rather, it was information that he had previously incorporated into his schema by keeping it as outliers. That is, he had kept it compartmentalized and away from his normal constructs of what constituted a mathematics student and a mathematical activity, respectively. In the end, the reform of his teaching happened when he began to accommodate these outliers.

Mitchell is not the only teacher that I have encountered who was so effective at not accommodating new information. There have been many others. It is easy for a teacher to become entrenched in their practice, and it is easy for them to stay entrenched by using the "I already do that" strategy that Mitchell did. In my data there are five additional cases in which such teachers reformed their practice. In each case their change was initiated by an eventual accommodation of outlying information – information that was acquired through their own teaching.

The Case of Danica

Danica has been teaching middle school for 13 years during which time she has always enjoyed teaching mathematics. Her teaching, in general, and mathematics teaching, in particular, is dynamic and progressive. She is always willing to try something new and is a keen consumer of professional development opportunities.

Danica is unhappy with her mathematics teaching, however. She has a sense that what she is doing is correct, but her teaching "lacks cohesion." She feels that she is being pulled in so many directions by so many "new ideas" that she does not know who she is as a teacher any more. She is experimenting with formative assessment, pilot testing a new textbook, implementing problem solving in her classroom, and engaging students in her group work. Outwardly there is nothing really wrong with her teaching. But she feels she is lacking a holistic understanding of how it all fits together. She feels that everything she tries is sitting alongside every other thing she has tried and none of them are working in harmony with each other.

I first began working with Danica in the same learning team as Mitchell. And like Mitchell, this experience had a transformative effect on Danica. Almost immediately she gained confidence as she found cohesion in her teaching. This confidence continued to grow over the course of the rest of the meetings. Danica had found a thread that bound all of her previously disjoint ideas about teaching mathematics together – a thread of communication. Danica started to build all of her teaching around the idea that students needed to be able to articulate their thinking. This gave meaning to her efforts at group work, formative assessment, and problem solving. And, at the end of that school year Danica came to a surprising conclusion for herself and her students – students who were better able to articulate their thinking actually thought better.

Danica's Transformation as a Case of Reification

In interviewing Danica it became clear that for her, like Mitchell, the beginning of this transformation was the co-construction of the definition of numeracy. Unlike Mitchell, however, Danica was searching for a way to bring together her discrete experiences and the construction of the definition facilitated this. For Danica this was a reifying experience (Wenger, 1998).

For the most part, teaching has no concrete form. It exists in time and space in the relationships with students and the interactions between curriculum and learners. In essence, teaching is an experience. Occasionally this experience can be reified into some artifact that, at least for the creator, embodies the experience of teaching. For Wenger (1998), reification is "the process of giving form to our experiences by producing objects that congeal this experience into *thingness*" (p. 58).

For Danica the definition embodied teaching ... ideal teaching. But it wasn't her teaching ... at least not yet. The process of co-constructing the definition was an act of synthesis, drawing on the teaching experiences of eight different teachers. Danica contributed to this process, and as a result saw herself in part of the definition. However, the definition became so much more for her. This definition was the reification of the types of students she wanted to produce, and because she had contributed to its formation she felt she had access to all that it embodied.

That is not to say that all of Danica's transformation could be attributed to the reciprocating relationship she had with the definition ... but it was the start. The design and pilot testing of the numeracy task had a similar and, I would say, more profound effect on her practice. Again, the initial task that was created was a reification of aspects of all of the teachers' aspirations. For Danica, it carried within it the capacities that she wanted to develop within her students; capacities that she knew were important, and that she knew were absent within her students. The pilot testing of the task on her own students only served to solidify her resolve as with each iteration Danica became aware of more mathematical and pedagogical affordances embedded within the task (Liljedahl, Chernoff, & Zazkis, 2007) – each of which could now be used to achieve her goals.

Danica now possessed a definition and a task, each of which embodied the essence of the teacher she wanted to be. In these she found the thread of communication (or more specifically, the ability to articulate thinking) that bound her disjoint efforts together. Her assessment practices now had a purpose, as did her focus on group work and problem solving.

Danica is far from unique in this regard. Of the many teachers who I have worked with in the capacity of task design, 11 others have experienced a similar, although not as profound, transformation. The co-construction of tasks reifies teaching in ways that simple lesson plans, tests, and worksheets do not. They are both connected to individual teachers and idealized amalgamations of the best of teaching. However, other experiences tell me that it is not unique to the goal of task design. I am seeing similar effects working with a group of teachers in designing performance assessments and I have observed the same effects in teachers engaged in lesson study. The commonality in all these is teachers co-constructing some artifact of teaching ... reifying and amalgamating the best of their experiences and aspirations.

Discussion on Cases of Rapid Professional Development

There is an invariance that ties these three cases together. Although initiated by some disturbance (Mason, 2002) – a critical question, activity, or event – within the professional development setting the rapid and profound changes that followed burst from a basis of experience – experience in the classroom. These three teachers, and the 23 other teachers their cases represent, had already built up a large potential to change prior to entering into the professional development setting. In fact, in some cases it was this very build-up that brought them to the professional development setting to begin with. Both Mary and Danica came seeking something. They did not necessarily know what, but they knew that there was something missing from their practice. When they found what they were seeking they were instantly able to make sense of it and to utilize it in reformulating their practice. For Mary it was a viewpoint, a way to look at curriculum that allowed her to shed the oppressive paradigm that she was working in. For Danica it was an artifact, a definition and a task that embodied within it – and could act as a model for – the type of learning that she wanted to facilitate.

For Mitchell it is a little bit different. He was not really seeking anything in particular when he came to the learning team. However, like Mary and Danica he did come with a wealth of experience and a high potential for change. In his case, the critical activity that he encountered caused him to re-evaluate the outliers that he came in with and to accommodate them into his understanding of what it meant to teach and learn mathematics.

In general, Mary, Mitchell, and Danica (as well as the 23 other documented cases) came to the professional development setting with a potential for rapid professional growth already in place. From dissatisfaction with structures and paradigms to incongruencies in their teaching to unrealized gains from prior experiences, each of them had developed this potential through their own teaching, and in their own teaching experiences. In general, literature on the professional development of teachers does not do enough to acknowledge this.

Discussion on Professional Development Through Teaching

Current research on mathematics teachers and the professional development of mathematics teachers can be sorted into three main categories: *content*, *method*, and *effectiveness*. The first of these categories, *content*, is meant to capture all research pertaining to teachers' knowledge and beliefs including teachers' mathematical content knowledge, both as a discipline (Ball, 2002; Davis & Simmt, 2006) and as a practice (Hill, Ball, & Schilling, 2008). Recently, this research has been dominated by a focus on the mathematical knowledge teachers need for teaching (Ball & Bass, 2000; Ball, Hill, & Bass, 2005; Davis & Simmt, 2006; Delaney, Ball, Hill, Schilling, & Zopf, 2008; Stylianides & Ball, 2008) and how this knowledge can be developed within preservice and inservice teachers. Also included in this category is research on teachers' beliefs about mathematics and the teaching and learning of mathematics and how such beliefs can be changed within the preservice and inservice setting (Liljedahl, in press, 2007; Liljedahl, Rolka, & Rösken, 2007). Some of the conclusions from this research speak to the observed discontinuities between teachers' knowledge/beliefs and their practice (Cooney, 1985; Karaagac & Threlfall, 2004; Skott, 2001; Wilson & Cooney, 2002) and, as a result, call into question the robustness and authenticity of this knowledge/beliefs (Lerman & Zehetmeir, 2008).

The second category, *method*, is meant to capture the research that focuses on a specific professional development model such as action research (Jasper & Taube, 2004), lesson study (Stigler & Hiebert, 1999), communities of practice (Little & Horn, 2007; McClain & Cobb, 2004; Wenger, 1998), or more generally, collegial discourse about teaching (Lord, 1994). This research is "replete with the use of the term inquiry" (Kazemi, 2008, p. 213) and speaks very strongly of inquiry as one of the central contributors to teachers' professional growth. Also prominent in this research is the centrality of collaboration and collegiality in the professional development of teachers and has even led some researchers to conclude that reform is built by relationships (Middleton, Sawada, Judson, Bloom, & Turley, 2002).

> More accurately, reform emerges from relationships. No matter from which discipline your partners hail, no matter what financial or human resources are available, no matter what idiosyncratic barriers your project might face, it is the establishment of a structure of distributed competence, mutual respect, common activities (including deliverables), and personal commitment that puts the process of reform in the hands of the reformers and allows for the identification of transportable elements that can be brokered across partners, sites, and conditions. (ibid., p. 429).

Finally, work classified under *effectiveness* is meant to capture research that looks at changes in teachers' practice as a result of their participation in some form of a professional development program. Ever present in such research, explicitly or implicitly, is the question of the robustness of any such changes (Lerman & Zehetmeir, 2008).

As powerful and effective as this aforementioned research is, however, it can no longer ignore the growing disquiet that somehow the perspective is all wrong. In fact, it is from this very research that this disquiet emerges. The questions of robustness (Lerman & Zehetmeir, 2008) come from a realization that professional growth is a long term endeavor (Sztajn, 2003) and participation in preservice and inservice programs is brief in comparison. At the same time there is a growing realization that what is actually offered within these programs is often based on facilitators' (or administrators' or policy makers') perceptions of what teachers need as opposed to actual knowledge of what teachers really do need (Ball, 2002). Even the impact of relationships, one of the pillars of professional delivery methodologies, is coming under closer scrutiny. Questions about the complexity of relationships (Nickerson, 2008) are calling into question the exact nature of the collaborative and collegial interactions with distinctions being made between the public and private faces of practice (Little, 2002). That is, what components of their practice is a teacher actually willing to share and how do those choices affect the subsequent discourse and professional growth (Little, 2002)? The classification of teachers' knowledge into subcategories is also beginning to be seen as problematic (Askew, 2008, Davis & Simmt, 2006) in light of the complex, integrated, and situated nature of teaching, as are some of the long-held beliefs about teachers as learners (Ball, 2002) and the systemic practice of looking at teachers' knowledge as deficit (Askew, 2008). Finally, the very notions of our ability to change someone (in this case, a teacher) is being challenged (Mason, 1994).

This disquiet is leading, slowly and tentatively, toward the emergence of a new paradigm where the professional growth of teachers is seen as natural (Leikin, 2006; Liljedahl, in press; Perrin-Glorian, DeBlois, & Robert, 2008; Sztajn, 2003) and teachers are seen as agents in their own professional learning (Ball, 2002). But such a paradigm shift is going to require a complementary shift in how we look at certain stalwart traditions around teacher education. To begin with, we are going to have to challenge assumptions around what it means for teachers to be learners, and as such, relook at the effectiveness of inservice teaching methodologies such as *modeling good practice* and models of inservice delivery such as *single workshops* (Ball, 2002). We are going to have to change the way we look at teachers' knowledge, beliefs, goals, motivations, plans, and practice as well as what we consider as, and how we treat, evidence of these (Kazemi, 2008). We are even going to have to begin looking at discontinuities between teachers' knowledge, beliefs, goals, motivations, plans, and practice as sensible in the larger scheme of teachers' natural professional growth (Ball, 2002; Leatham, 2006; Liljedahl, 2008). Finally, we are going to have to rethink the structures of professional development programs and what it means to be responsive to the needs of teachers as they progress through their personal learning plans (Ball, 2002; Liljedahl, in press).

Conclusion

I began this chapter by stating that I do professional development. I wish to retract this statement. Instead I would like to state that I work with teachers in the context of their professional growth. Within this context I am sometimes privileged to observe their growth and sometimes I am even able to contribute to their professional growth. Sometimes the growth is planned and sometimes it is predictable. And occasionally the growth is neither planned nor predictable. But in all cases, my role in comparison to the breadth and depth of teachers' experiences, their agency, and their occasional intentionality, can be seen as ancillary at best. I am but a disturbance (Mason 2002).

References

Askew, M. (2008). Mathematical discipline knowledge requirements for prospective primary teachers, and the structure and teaching approaches of programs designed to develop that knowledge. In P. Sullivan & T. Wood (Eds.), *Knowledge and beliefs in mathematics teaching and teaching development* (pp. 13–35). Rotterdam, NL: Sense Publishers.

Ball, D. L. (2002). What do we believe about teacher learning and how can we learn with and from our beliefs? In D. S. Mewborn, P. Sztajn, D. Y. White, H. G. Wiegel, R. L. Bryant, & K. Nooney (Eds.), *Proceedings of the 24th international conference for psychology of mathematics education – North American chapter*. Athens, Georgia.

Ball, D. L., & Bass, H. (2000). Interweaving content and pedagogy in teaching and learning to teach: Knowing and using mathematics. In J. Boaler (Ed.), *Multiple perspectives on the teaching and learning of mathematics* (pp. 83–104). Westport, CT: Ablex.

Ball, D. L., Hill, H. C, & Bass, H. (2005). Knowing mathematics for teaching: Who knows mathematics well enough to teach third grade, and how can we decide? *American Educator, 2005*(Fall), 14–46.

Chapman, O. (2002). Belief structures and inservice high school mathematics teacher growth. In G. Leder, E. Pehkonen, & G. Törner (Eds.), *Beliefs: A hidden variable in mathematics education* (pp. 177–194). Dordrecht, The Netherlands: Kluwer Academic Publishing.

Clandinin, D. J. (1992). Narrative and story in teacher education. In T. Russell & H. Munby (Eds.), *Teachers and teaching from classroom to reflection* (pp. 124–137). Bristol, PA: Falmer.

Clandinin, D. J., & Connelly, F. M. (1996). Teachers' professional knowledge landscapes: Teacher stories – Stories of teachers – School stories – stories of schools. *Educational Researcher, 25*(3), 24–30.

Cochran-Smith, M., & Lytle, S. (2004). Practitioner inquiry, knowledge, and university culture. In J. Loughran, M. Hamilton, V. LaBoskey, & T. Russell (Eds.), *International handbook of research of self-study of teaching and teacher education practices* (pp. 602–649). Dordrecht, Netherlands: Kluwer.

Cooney, T. (1985). A beginning teacher's view of problem solving. *Journal for Research in Mathematics Education, 16*(5), 324–336.

Creswell, J. W. (2008). *Educational research: Planning, conducting, and evaluating quantitative and qualitative research* (3rd ed.). New Jersey: Pearson Education, Inc.

Davis, B., & Simmt, E. (2006). Mathematics-for-teaching: An ongoing investigation of the mathematics that teachers (need to) know. *Educational Studies in Mathematics, 61*(3), 293–319.

Delaney, S., Ball, D. L., Hill, H. C., Schilling, S. G., & Zopf, D. (2008). Mathematical Knowledge for teaching: Adapting U.S. measures for use in Ireland. *Journal of Mathematics Teacher Education, 11(3,)* 171–197.

Feiman-Nemser, S., McDiarmid, G., Melnick, S., & Parker, M. (1987). *Changing beginning teachers' conceptions: A description of an introductory teacher education course*. Paper presented at American Educational Research Association, Washington, DC.

Greer, B. (2004). The growth of mathematics through conceptual restructuring. *Learning and Instruction, 15*(4), 541–548.

Hill, H., Ball, D. L., & Schilling, S. (2008). Unpacking "pedagogical content knowledge": Conceptualizing and measuring teachers' topic-specific knowledge of students. *Journal for Research in Mathematics Education, 39*(4), 372–400.

Jasper, B., & Taube, S. (2004). Action research of elementary teachers' problem-solving skills before and after focused professional development. *Teacher Education and Practice, 17*(3), 299–310.

Karaagac, K., & Threlfall, J. (2004). The tension between teacher beliefs and teacher practice: The impact of the work setting. *Proceedings of 28th Annual Conference for the Psychology of Mathematics Education, 3*, 137–144.

Kazemi, E. (2008). School development as a means of improving mathematics teaching and learning: Towards multidirectional analysis of learning across contexts. In K. Krainer & T. Wood (Eds.), *Participants in mathematics teacher education* (pp. 209–230). Rotterdam, NL: Sense Publishers.

Kuhn, T. (1970). *The structure of scientific revolutions (second edition, enlarged)*. Chicago, IL: The University of Chicago Press.

Leatham, K. (2006). Viewing mathematics teachers' beliefs as sensible systems. *Journal of Mathematics Teacher Education, 9*(2), 91–102.

Leikin, R. (2006). Learning by teaching: The case of Sieve of Eratosthenes and one elementary school teacher. In R. Zazkis & S. Campbell (Eds.), *Number theory in mathematics education: perspectives and prospects* (pp. 115–140). Mahwah, NJ: Erlbaum.

Lerman, S., & Zehetmeir, S. (2008). Face-to-face communities and networks of practicing mathematics teachers: Studies on their professional growth. In K. Krainer & T. Wood (Eds.), *Participants in mathematics teacher education* (pp. 133–153). Rotterdam, NL: Sense Publishers.

Liljedahl, P. (in press). Changing beliefs as changing perspective. In *Proceedings of the 6th congress of the European society for research in mathematics education*. Lyon, France.

Liljedahl, P. (2008). Teachers' insights into the relationship between beliefs and practice. *Proceedings of the 14th international conference on mathematical views (MAVI)*. St. Wolfgang, Austria.

Liljedahl, P. (2007). Reification: Explicating teachers' tacit knowledge and beliefs. *Proceedings of the 13th international conference on mathematical views (MAVI)*. Gävle, Sweden.

Liljedahl, P., Chernoff, E., & Zazkis, R. (2007). Interweaving mathematics and pedagogy in Task Design: A tale of one task. *Journal for Mathematics Teacher Education – Special Issue on the Role and Nature of Mathematics-Related Tasks for Teacher Education, 10*(4–6), 239–249.

Liljedahl, P., Rolka, K., & Rösken, B. (2007). Affecting affect: The re-education of preservice teachers' beliefs about mathematics and mathematics learning and teaching. In M. Strutchens & W. Martin (Eds.), *69th NCTM yearbook – The learning of mathematics* (pp. 319–330). Reston, VA: National Council of Teachers of Mathematics.

Little, J. W. (2002). Locating learning in teachers' community of practice: Opening up problems of analysis in records of everyday work. *Teaching and Teacher Education, 18*, 917–946.

Little, J. W., & Horn, I. S. (2007). 'Normalizing' problems of practice: Converging routine conversation into a resource for learning in professional communities. In L. Stoll & K. S. Louis (Eds.), *Professional learning communities: Divergence, depth, and dilemmas* (pp. 79–92). Berkshire, UK: Open University Press.

Lord, B. (1994). Teachers' professional development: Critical colleagueship and the roles of professional communities. In N. Cobb (Ed.), *The future of education: Perspectives on national standards in America* (pp. 175–204). New York, NY: The College Board.

Lortie, D. (1975). *Schoolteacher: A sociological study*. Chicago, IL: University of Chicago Press.

Mason, J. (2006). What makes an example exemplary? Pedagogical and didactical issues in appreciating multiplicative structures. In R. Zazkis & S. Campbell (Eds.), *Number theory in mathematics education: Perspectives and prospects* (pp. 141–172). Mahwah, NJ: Lawrence Erlbaum Associates, Inc.

Mason, J. (2002). *Researching your own practice: The discipline of noticing*. New York, NY: Routledge.

Mason, J. (1994). Researching from the inside in mathematics education- locating an I-You relationship. In J. P. da Ponte and J. F. Matos (Eds), *Proceedings of the eighteenth international conference for the psychology of mathematics education* (pp. 269–283). Lisbon, Portugal: University of Lisbon.

McClain, K., & Cobb, P. (2004). The critical role of institutional context in teacher development. *Proceedings of 28th Annual Conference for the Psychology of Mathematics Education, 3*, 281–288.

Middleton, J. A., Sawada, D., Judson, E., Bloom, I., & Turley, J. (2002). Relationships build reform: Treating classroom research as emergent systems. In L. D. English (Ed.), *Handbook of international research in mathematics education* (pp. 409–431). Mahwah, NJ: Lawrence Erlbaum Associates.

Nickerson, S. (2008). Teams of practicing teachers: Developing teacher professionals. In K. Krainer & T. Wood (Eds.), *Participants in mathematics teacher education* (pp. 89–109). Rotterdam, NL: Sense Publishers.

Perrin-Glorian, M.-J., DeBlois, L., & Robert, A., (2008). Individual practising mathematics teachers: Studies on their professional growth. In K. Krainer & T. Wood (Eds.), *Participants in mathematics teacher education* (pp. 35–59). Rotterdam, NL: Sense Publishers.

Piaget, J. (1968). *Six psychological studies*, Anita Tenzer (Trans.), New York, NY: Vintage Books.

Posner, G., Strike, K., Hewson, P., & Gertzog, W., (1982). Accommodation of a scientific conception: Towards a theory of conceptual change. *Science Education, 66*, 211–227.

Skott, J. (2001). The emerging practices of novice teachers: The roles of his school mathematics images. *Journal of Mathematics Teacher Education, 4*(1), 3–28.

Stigler, J., & Hiebert, J. (1999). *The teaching gap. Best ideas from the world's teachers for improving education in the classroom.* New York, NY: The Free Press.

Stylianides, A. J., & Ball, D. L. (2008). Understanding and describing mathematical knowledge for teaching: Knowledge about proof for engaging students in the activity of proving. *Journal of Mathematics Teacher Education, 11*, 307–332.

Sztajn, P. (2003). Adapting reform ideas in different mathematics classrooms: Beliefs beyond mathematics. *Journal of Mathematics Teacher Education, 6*(1), 53–75.

Tirosh, D., & Tsamir, P. (2004). What can mathematics education gain from the conceptual change approach? And what can the conceptual change approach gain from its application to mathematics education? *Learning and Instruction, 15*(4), 535–540.

Vosniadou, S. (2006). Mathematics learning from a conceptual change point of view: Theoretical issues and educational implications. In J. Novotna, H. Moraova, M. Kratka, & N. Stehlikova (Eds.), *Proceedings of 30th Annual Conference for the Psychology of Mathematics Education, 1*, 163–165.

Vosniadou, S., & Verschaffel, L. (2004). Extending the conceptual change approach to mathematics learning and teaching. *Learning and Instruction, 15*(4), 445–451.

Wenger, E. (1998). *Communities of practice*. New York, NY: Cambridge University Press.

Wilson, S., & Cooney, T. (2002). Mathematics teacher change and development: The role of beliefs. In G. Leder, E. Pehkonen, & G. Törner (Eds.), *Beliefs: A hidden variable in mathematics education* (pp. 127–148). Boston, MA: Kluwer Academic Publishing.

Interactions Between Teaching and Research: Developing Pedagogical Content Knowledge for Real Analysis

Lara Alcock

Introduction

This chapter is about five of my current practices in teaching undergraduate mathematics. It is about my reasons for engaging in these practices and the way that they have been influenced by reflection on my teaching experience. I use the teaching of Real Analysis as a focus and, as such, the chapter is about aspects of my current pedagogical content knowledge for this subject. However, I also relate these practices to general issues in teaching and learning. One overarching theme is the need to help students develop skills on multiple levels, from understanding particular concepts to learning generally productive study habits. In all of these respects I write as a teacher.

I also write partly as a researcher, assessing what I do *not* know about the effectiveness of the five practices. As a researcher, I know that the vast majority of my teaching practices are unevaluated. My developing understanding of specific and general problems in learning Analysis leads me to invent, borrow, and use new ideas every time I teach the course. This means that I make modifications to my teaching faster than I can make a critical assessment of whether or not they work. I think that this is probably typical, and not necessarily bad – the creative work of making adjustments keeps me enthusiastic and my teaching "fresh." But it means that I use a number of practices for which I have a good theoretical rationale but little or no empirical evidence of effectiveness. Indeed, I know that if I experience personal satisfaction with a way of communicating an idea, I tend to conflate this with its effectiveness from a students' perspective.

With this in mind, this chapter constitutes an attempt to elucidate rationales for the five practices and to clarify what would be necessary in order to ascertain whether they genuinely aid learning. This discussion also leads to a second overarching theme – that of how best to use the lecture time available. As will become clear, the balance in my lecturing has shifted in recent years from explaining the

---

L. Alcock (✉)
Loughborough University, Loughborough, Leicestershire, UK
e-mail: L.J.Alcock@lboro.ac.uk

R. Leikin, R. Zazkis (eds.), *Learning Through Teaching Mathematics*, Mathematics Teacher Education 5, DOI 10.1007/978-90-481-3990-3_12,

mathematical content toward teaching how to interact with that content. This is in response to frequent conversations with students in which I have found myself giving explanations of how to study the material when I am not there. As with the specific practices, however, I do not know whether this shift has had a positive effect on my students' learning. Overall, therefore, a good alternative title for this chapter would be *Things I Don't Know About Teaching Real Analysis*.

My Current Approach to Lecturing

Before talking about the specific practices, I briefly lay out my overall views as a lecturer of Analysis. I have taught this subject in a variety of formats, including "supervisions" for groups of four students, co-operative classes of 20–30, and lecture classes of over 100. Most recently I have been working with large lecture classes in a UK university. Whatever the format, I see it as my role to explain the material in a way that allows students to: develop conceptual understanding consistent with the formal theory, follow the arguments, understand and appreciate the overall structure of the subject, and learn to reconstruct, adapt and apply the theorems and proofs. In the process, I discuss logical language and proof strategies, and I expect that my students will improve their ability to handle these. Students are expected to undertake exercises in which they make minor adaptations to given proofs or in which they are led through the steps of a proof. However, my approach is content-focused and I do not see it as my role to teach students how to construct proofs, in general. In most of the institutions where I have worked, other courses have dealt specifically with material on basic logic, quantifiers, conditional statements, proof types, etc.

My current approach involves giving lectures for which students are provided with "gappy" notes. These notes contain statements of definitions, theorems, etc., and tasks for students to work on together. There are gaps for students to respond to tasks and to fill in proofs and draw diagrams as I lecture. I use these notes to facilitate considerably more student–student discussion than might occur in a typical mathematics lecture. This is because, while I do not believe that there is anything wrong with traditional lecturing per se, I do believe that we often do a poor job of teaching students how to learn abstract material from lecture notes. Much of this is clarified in the rest of this chapter.

Five Current Practices

The practices that will be discussed are as follows:

1. Regular testing on definitions;
2. Tasks involving extending example spaces;
3. Tasks involving constructing and understanding diagrams;

4. Resources for improving proof comprehension;
5. Tasks involving mapping the structure of the whole course.

For each of these I do three things in the sections below. First, I describe how I came to focus on the practice, through teaching and research experience. Second, I describe the practice and explain what mathematical skills I hope it will help students to develop. Third, I discuss what I do and do not know about the effectiveness of the practice, and attempt to formulate specific research questions, answers to which would help me to make more informed decisions about how to use my lecturing time.

Regular Testing on Definitions

One of the first things I learned when I began my PhD in mathematics education was that students are often unaware of the status of mathematical definitions. They tend to interpret definitions as dictionary definitions, which describe a pre-existing concept, rather than as technical definitions that precisely specify the extension of a concept and should be used as a basis for deductions (Davis & Vinner, 1986; Tall & Vinner, 1981; Vinner, 1991). Thus new undergraduates often do not know that when mathematicians say, "Prove that every convergent sequence is bounded," what they mean is, "Prove that every sequence that satisfies the definition of convergence also satisfies the definition of bounded." Consequently, many reason about such statements by using concept images instead (Alcock & Simpson, 2002, 2004; Tall & Vinner, 1981; Vinner, 1991). This is particularly problematic in Analysis, in which students often hold conceptions that are at odds with the formal theory, due to a combination of the language of limits and convergence (Cornu, 1991; Davis & Vinner, 1986; Monaghan, 1991) and a variety of informal experiences with the ideas of limits and infinity (Przenioslo, 2004; Roh, 2008; Sierpinska, 1987; Williams, 1991).

This phenomenon was evident in my own teaching (of small groups of first year undergraduates) at the time, and obviously it is extremely important pedagogical content knowledge for a teacher of Analysis. Awareness of it means that throughout the course, I stress the importance of definitions and regularly explain their role in deciding whether an object has a property, in constructing general proofs and in mathematical theory as a whole. I also emphasize the importance of definitions by having regular 10-minute, 10-question tests in which students are required to state definitions and theorems from the course (they are also required to give examples, more of which is discussed in the next section). I like to do this every week, but in large lecture classes where this is not practical, I do it every two weeks instead.

I have reservations about this practice, because such testing may promote rote memorization. This is not something I want to encourage – I want my students to take a "deep" approach to understanding the course material and not to attempt to memorize it as a set of unconnected facts or statements (cf. Biggs, 1987, p. 15, cited in Kember, 1996). On the other hand, testing is possibly the most effective means that a teacher has of emphasizing the importance of an idea. I have learned, through my experience of teaching and marking examinations, that not having regular tests means that some students will not learn definitions as the course progresses *at all*. This puts them in a far worse position: They cannot hope to decide whether an object satisfies a definition or to recognize the use of a definition within a proof, unless they know what that definition is. So I have taken the view that in this case, the possibility of rote learning is the lesser of the two evils. I am inclined to agree with Bell (1993, p. 6), who focused on inquiry in mathematics learning but stated that, "This does not, of course, rule out important ancillary activities, such as the memorizing of important data or the practising of frequently needed skills."

What I do not have is empirical evidence for the effectiveness of this regular testing. This means that in my current course I am giving up an entire lecture's worth of time (out of 22 available lectures) to this practice without knowing whether it would be better for me to use that time simply to give another lecture. In fact, at my institution, the question is more interesting than this, because many comparable courses use a whole lecture for a single test approximately two thirds of the way through the course (much like an American mid-term). It could be that a single test would not promote regular study in the same way, but could involve longer, more in-depth questions. With all of this in mind, some questions pertinent to my teaching include the following:

- Does regular testing of definitions lead to better understanding, as measured by examination performance or in some other way?
- How much time do students spend studying for such tests and how would they otherwise have spent this time (do the tests promote more study or simply different study)?
- How do students study for such tests (do they attempt rote memorization or some other strategy)?
- Do regular small tests lead to better or worse understanding/performance than a single, more in-depth test?

Tasks Involving Extending Example Spaces

In response to knowledge about students' restricted concept images, I began varying the examples I use in my teaching. For instance, I often give something like [-253,0] instead of [0,1] as an example of closed interval and draw diagrams to show convergent sequences that are not monotonic, continuous functions that do not increase from the bottom left of the diagram and are not everywhere differentiable, etc. However, there is a limit to what one can do in this regard when time for giving illustrative examples is restricted, and I have also begun using more extensive example classification tasks in my teaching. This development was first prompted by teaching a version of an Analysis course that made use of the notions of open and closed sets and related concepts. The book I was using (as is typical) presented a definition, an example, and occasionally a non-example. In order to refamiliarize myself with the concepts, I found myself applying the definitions to many more sets. This helped me to become aware of the range of possible variation (cf. Marton, 2007; Mason, 2002) and to build myself a sort of "prototype" that I could then use as generic. As a result of this experience I set a task for students in which they would consider the application of definitions of open, closed, limit point, etc. to each of a list of sets of real numbers including [0,1], (0,1), [0,1), {0}, **N**, **Q**, **R**, and the Cantor set. This constituted my first concerted effort to avoid a situation in which students developed an inappropriately restricted concept image (e.g. Schwarz & Hershkowitz, 1999; Zazkis & Leikin, 2007) so that they would be less likely to treat an example as generic when, in fact, it incorporates unnecessary properties (Mason & Pimm, 1984). Teaching in this way was eye-opening for me in that it took far longer than I had expected for the students to complete the task. It was, however, consistent with research indicating that some students (the majority, in this case) did not spontaneously generate examples as a routine response to newly encountered definitions (cf. Alcock, in press; Dahlberg & Housman, 1997).

More recently, I have become familiar with further research on example generation and use. *Example spaces* (Watson & Mason, 2005), and tasks designed to promote their development, have been widely discussed (Mason, 2002; Watson & Mason, ibid.; Goldenberg & Mason, 2008). Particular attention has been paid to the role and structure of examples that promote cognitive conflict (Zazkis & Chernoff, 2008), to the way in which teachers and children make decisions about example classification (Tsamir, Tirosh, & Levenson, 2008), and to the relative value of different counterexamples in explaining the falsity of a general statement (Peled & Zaslavsky, 1997). At the undergraduate level, this is of particular relevance because we wish the student's personal example space for a concept to be consistent with the conventional example space as specified by the definition (using the terms in the sense of Watson & Mason, 2005). Further, one reason for students' poor proving capacity is a lack of familiarity with examples of concepts (Moore, 1994), which is important because at least some successful mathematicians do use examples in a substantive way in their own reasoning (e.g. Alcock & Inglis, 2008; Weber, 2008).

With this in mind, I have become progressively more systematic in my use of such activities. This year, I devoted the first two Analysis lectures to a "Chapter 0" for which I did very little lecturing and the students worked with each other on tasks that asked them to do the following:

1. Use number line diagrams to represent twelve different sets (including some written as intervals and some expressed in other ways such as $\{x:|x - 12| < 2\}$, $\{1 - 1/n:n \in \mathbf{N}\}$ and $\{[2n,2n + 1]:n \in \mathbf{N}\}$);
2. Use provided definitions to classify each set as bounded above or not, to give three upper bounds if possible, and to state the supremum if possible;
3. Sketch graphs of 16 different functions, including

$$f(x) = |\sin x|, \quad f(x) = ||x| - 1|, \quad f(x) = \begin{cases} 0 & x \neq 0 \\ 1 & x = 0 \end{cases}, \quad \textit{and } f(x) = \begin{cases} 0 & x \notin \mathbf{Q} \\ x & x \in \mathbf{Q} \end{cases};^{1}$$

4. Use provided definitions to decide whether each of the functions is bounded below, has a minimum at 0, and/or is decreasing.
5. Discuss (without definitions at this stage) whether they thought each of the sixteen functions has a limit at 0, is continuous at 0, and/or is differentiable at 0.

My intentions were to remind students about of a lot of previously-studied definitions for set and function properties, to expand their awareness of variation in the example space of real functions, and to allow them to become aware of uncertainty in their own knowledge about key concepts. For this reason, I chose to use example classification rather than generation tasks, because I wanted to introduce functions that the students had probably not encountered before and would probably not spontaneously invent, but that would later be useful for understanding boundaries of formal concepts. Of course, these tasks also allowed me to place early emphasis on the importance of definitions, and made it clear to the students that this would be a course in which active engagement was expected during lecture time.

This practice provided me with a useful set of experiences to refer back to in order to help the students see that they were making progress. As with the definition tests, however, from a research point of view I have no real measure of whether this was time well spent. With particular regard to the use of example classification tasks it would be useful for me as a teacher to know the following:

- To what extent are students able to independently classify examples as satisfying definitions or not (how much guidance do they need in interpreting quantified statements, for example)?
- Does correctness in such tasks improve if students are allowed to discuss the classifications in groups (should they be asked to work individually, or would discussion be better, even if it takes longer)?
- Do students have better recall of definitions when they have used these to classify examples than when they have listened to an explanation and/or seen these classifications demonstrated (perhaps with written proofs) by a lecturer?

Tasks Involving Constructing and Understanding Diagrams

Diagrams are, for me, a very important part of Analysis. Interestingly, and probably for historical printing reasons, many textbooks have very few diagrams. Like other lecturers, however, I draw many in my lectures and regularly encourage students to make use of them in supporting their own reasoning. Through conducting task-based research interviews, I have also become aware that giving a diagram alongside

[1]Obviously this function cannot be sketched accurately, but the intention was for the students to notice this and to think about how to give some reasonable representation.

a proof is probably not enough. Diagrams can provide insight (e.g. Gibson, 1998), but it is not always easy for students to make detailed links between what is in the diagram and what is in a formal proof. This means that the step between seeing that a result must be true and proving it can seem insurmountable (Alcock & Weber, in press; see also Raman, 2003). Through my small-class teaching, I have also learned that students often find it difficult to draw a diagram based on verbal and algebraic information. As a result, I now spend more time walking students through the process of drawing diagrams; for instance, noting explicitly that $|f(x) - f(a)| < \varepsilon$ can be thought of in terms of distances, and noting that it is "about $f(x)$ values" so that appropriate labels should go on the y-axis of a diagram.

I have also become progressively more systematic in designing tasks to develop students' diagram construction skills. I now have two tasks that I set during lecture time, each of which requires students to draw and think about their own diagram. The first of these is included in a set of instructions for proving the Mean Value Theorem,[2] which begins as follows:

1. Write down the assumptions.
2. Draw a nice big diagram representing the situation given by the assumptions. Make the graph curvy rather than straight, and make sure you make $f(a)$ and $f(b)$ different from each other.
3. Draw in the straight line that passes through the points (a,f(a)) and $(b, f(b))$.
4. Convince yourself that the equation of this line may be written as

$$y = f(a) + \left(\frac{f(b) - f(a)}{b - a}\right)(x - a).$$

5. We are going to consider a new function d defined on [a,b] by

$$d(x) = f(x) - \left[f(a) + \left(\frac{f(b) - f(a)}{b - a}\right)(x - a)\right].$$

   Don't panic. Label a point x on your x-axis and indicate vertically above it what $d(x)$ is measuring.

   ...

In writing tasks like this, which lead students through the reasoning needed to construct a proof, I have been very much inspired by Burn (1992). The additional focus on steps in drawing a diagram, however, is a more recent development for me.

The second task uses a similar set of instructions for a different purpose, that of developing understanding of the definitions of lower and upper sums in Riemann integration. The definitions state the following:

Suppose that f is bounded on $[a, b]$.

---

[2] If a function $f:[a, b] \to \mathbf{R}$ is continuous on $[a, b]$ and differentiable on (a, b), then there exists $c \in (a, b)$ such that $f'(c) = \frac{f(b)-f(a)}{b-a}$.

The *lower sum of f relative to the partition P* is

$$L(f;P)=\sum_{j=1}^{n} m_j(x_j-x_{j-1}),\ \textit{where}\ m_j=\inf\left\{f(x){:}x_{j-1}\le x\le x_j\right\}.$$

The *upper sum of f relative to the partition P* is

$$U(f;P)=\sum_{j=1}^{n} M_j(x_j-x_{j-1}),\quad \text{where}\quad M_j=\sup\left\{f(x):x_{j-1}\le x\le x_j\right\}.$$

I knew before I had taught this material that this would cause problems for the students, because it involves an off-putting amount of new notation and because recognizing that the idea is actually very simple involves being able to see clearly and accurately how this notation relates to a diagram. My gappy notes at this point, therefore, have the following set of instructions:

Consider the function f given by $f(x)=x^2$ on the interval [0,2].
Now consider the partition $P=\left\{0,\frac{1}{2},1,\frac{3}{2},2\right\}$. Draw a big graph of f on the relevant interval, and mark the points of this partition on the x-axis.
Now look at the definition of *lower sum.*
What is n in this case?
What are x_1, x_0 and m_1? Indicate m_1 on the y-axis in the diagram.
What are x_2, x_1 and m_2? Indicate m_2 on the y-axis in the diagram.
By making similar observations, write down an expression for $L(f\,;P_1)$, and indicate how we can "see" the area it represents on the diagram.
Write down an expression for $U(f\,;P_1)$ too.
Now consider the partition $P_2=\left\{0,\frac{1}{4},\frac{2}{4},\frac{3}{4},1,\frac{5}{4},\frac{6}{4},\frac{7}{4},2\right\}$. Without doing any calculations, answer the following questions:

1. Will L(f ; P_2) be greater than or less than L(f ; P_1)?
2. Will U(f ; P_2) be greater than or less than U(f ; P_1)?
3. Will U(f ; P_2) be greater than or less than L(f ; P_1)?
4. Which of L(f ; P_1), L(f ; P_2), U(f ; P_1) and U(f ; P_2) will give the best approximation to the area under the graph?
5. Which of these answers could be different if we had a different function?

Of course, there are other ways of teaching about this concept. One could begin with a diagram and develop the definition from this. One could also use dynamic demonstrations (a very nice geogebra demonstration is available, for instance[3]). However, in addition to understanding the concept, I want students to improve their ability to interpret formal notation and develop their own diagrams when I am not there to help. Again, this is an issue of balancing small-scale with large-scale learning goals. (I will use the geogebra demonstration after the students have completed the above exercise).

Research-wise, there are many interesting questions here, which I believe have been minimally investigated, at least at the tertiary level. Extensive work has been

[3] Under "Examples" at http://www.geogebra.org/cms/

done on providing students with diagram-based interactive materials designed to facilitate engagement with the formal limit concept (e.g. Tall, 1997) and with concepts and proofs in geometry (e.g. Marriotti, 2000), but I do not believe we know very much about students' capacity to construct and understand static diagrams such as one might see in a set of Analysis lecture notes. Also, my experience is that many students claim that they do not like diagrams. I have become convinced that a teacher should not necessarily be trying to change their minds about this, largely due to mathematicians' introspective claims about differences among expert practice (e.g. Burton, 2004; Hadamard, 1945) and to evidence showing that some successful mathematics students operate almost entirely syntactically in at least some proof situations (e.g. Alcock & Inglis, 2008; Alcock & Weber, in press). But I still encourage my students to engage seriously with diagrammatic representations before deciding whether or not this helps them, so it would be useful for me to have answers to questions such as those given below:

- To what extent can students construct diagrams to represent (Analysis) definitions or theorems?
- To what extent can students who have completed a course such as Analysis draw diagrams to illustrate concepts/definitions/theorems from the course?
- What proportion of students (perhaps at the end of such a course) believe that diagrams are important/indifferent/superfluous to their understanding of the course, and is this in any way correlated with their achievement?

Resources for Improving Proof Comprehension

Analysis involves a lot of proofs. I present and explain most of these in lectures; students construct some through Burn-inspired exercises (Burn, 1992, as above). Students are also advised to work on "the other case" in proofs in which only one case is covered in lectures.[4] They are then expected to write familiar proofs during their examination ("part (b)" of each of my examination questions usually requires giving a substantial proof from the course). A large part of their work, therefore, should be in understanding these proofs, but I know from my experience in marking examinations that many do not manage this well enough to perform well.

In fact, I know very little about whether and how my students go about studying proofs. I believe this is probably typical of undergraduate lecturers. We do not usually know how much time our students spend studying their lecture notes, nor how they spend this time. This is true in the case of research, too. Research on study habits has generally focused on subject areas other than mathematics (e.g. Entwistle & Ramsden, 1983; Kember, 1996; Vermunt, 2007) and studying a single proof in enough detail to understand it probably involves quite different skills from those

[4] For instance, in considering the Interior Extremum Theorem (if f is differentiable on (a, b) and attains a maximum or minimum at $c \in (a, b)$, then $f'(c) = 0$), I prove the maximum case and students are advised to write out a proof for the minimum case.

needed to assemble material from a variety of sources in order to write a coherent essay, for example.

Research in mathematics education does indicate that students find it difficult to *construct* proofs, and that they often behave as though their beliefs about the nature of proof were different from those of expert mathematicians (Harel & Sowder, 1998, 2007; Recio & Godino, 2001; Weber, 2001). Work more closely related to students' interpretations of written mathematics shows that there is often a distinction between what students find personally convincing and what they believe is acceptable as a proof (Healy & Hoyles, 2000), and that their ability to distinguish valid from invalid proofs is unreliable though may improve with ongoing mathematical education (Segal, 2000) or with prompts to think more carefully about what has been read (Alcock & Weber, 2005; Selden & Selden, 2003). Research is now beginning to reveal some of the processes mathematicians use to validate given proofs and check their own arguments (Inglis, Mejia-Ramos, & Simpson, 2007; Weber, 2008), though checking for correctness is not necessarily the same as reading for comprehension, the latter of which is a much more common task for students. It certainly seems that students often lack the ability to correctly "unpack the logic" of a mathematical statement and, therefore, establish whether the structure of a proof is such that it could conceivably prove that statement (Selden & Selden, 1995).

There is not a great deal of research on proof comprehension per se (Mejia-Ramos, 2008). Conradie and Frith (2000) discussed testing for comprehension of a presented proof by asking questions about its logical structure and about reasons for the validity of its statements. They suggested this as an alternative form of examination question that might avoid rewarding rote memorization. Lin and Yang studied proof comprehension with regard to geometry, first establishing comprehension criteria based on existing literature and on mathematicians' reflections, then using these to design a set of questions about a particular proof (Lin & Yang, 2007; Yang & Lin, 2008). They distinguished six facets of proof comprehension: basic knowledge, logical status, integration or summarization, generality, application or extension, and appreciation or evaluation.

I have not tried using proof comprehension tasks in tests, but in a bid to teach proof comprehension skills, I have designed a set of resources to demonstrate how one might go about understanding proofs by examining their internal logical relationships and overall structures. These resources are termed "e-Proofs" and I have constructed them for eight selected proofs in my Analysis course[5]. The e-Proofs are available to the students via the university's virtual learning environment (VLE). Each is presented as a sequence of screens with accompanying audio commentary, and each comes in three versions:

1. A *basic* version in which the proof appears one line at a time and the audio commentary simply reads that line;
2. A *line-by-line* version in which the whole proof is visible but grayed out, with each screen showing one line (or part of a line) fully visible, and arrows and

[5] With the support of a Loughborough University Academic Practice Award.

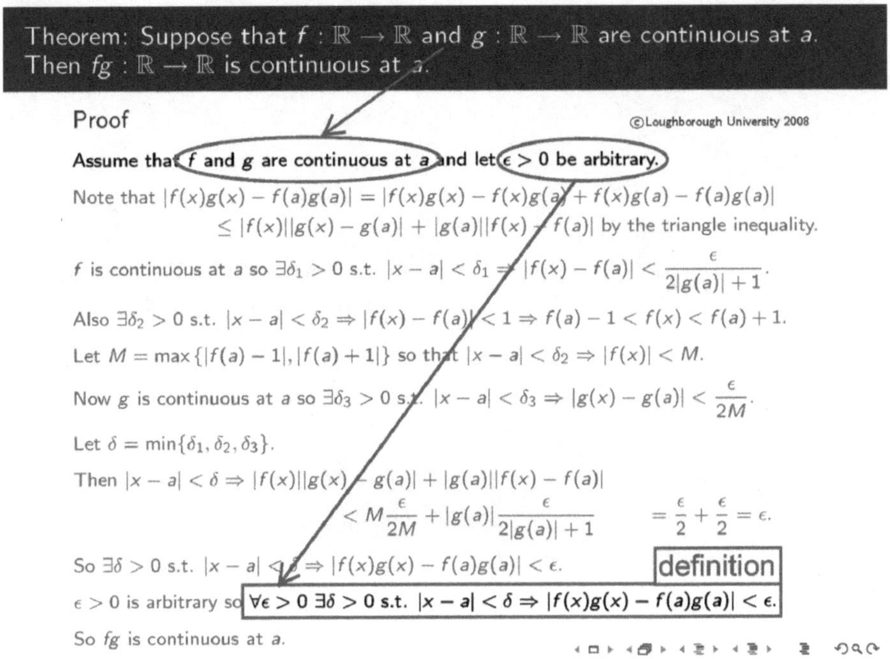

Fig. 1 A screen shot from the line-by-line version of an e-Proof

boxes indicating why that line is valid and its relationships to the other parts of the proof (a screen shot is shown in Fig. 1);

3. A *chunk* version, in which the whole proof is visible but grayed out, with each screen showing several lines fully visible and a box indicating what that section achieves (Fig. 2).

My intention is that the basic version should function as an opportunity to "watch the lecture" again (and might be particularly useful to second language English speakers who found it difficult to keep up), that the line-by-line version should direct attention to logical relationships among the lines (cf. Yang & Lin's *logical status*) and, in particular, to the process of inferring warrants to check each line's validity (cf. Alcock & Weber, 2005), and that the chunks version should help students to break a proof into parts and see it as having a relatively small number of main ideas (cf. Yang & Lin's *integration or summarization*). Overall, much of the information in the line-by-line and chunk versions is of the kind that would be discussed by a lecturer, but would be lost when the material is codified in static lecture notes.

I have been using the e-Proofs in lectures, usually giving out a copy of the whole proof, allowing students a few minutes to read and discuss it, then running through the line-by-line and chunk versions. It is difficult to quantify this experience, but my feeling is that using the e-Proofs has helped students to realize early on how

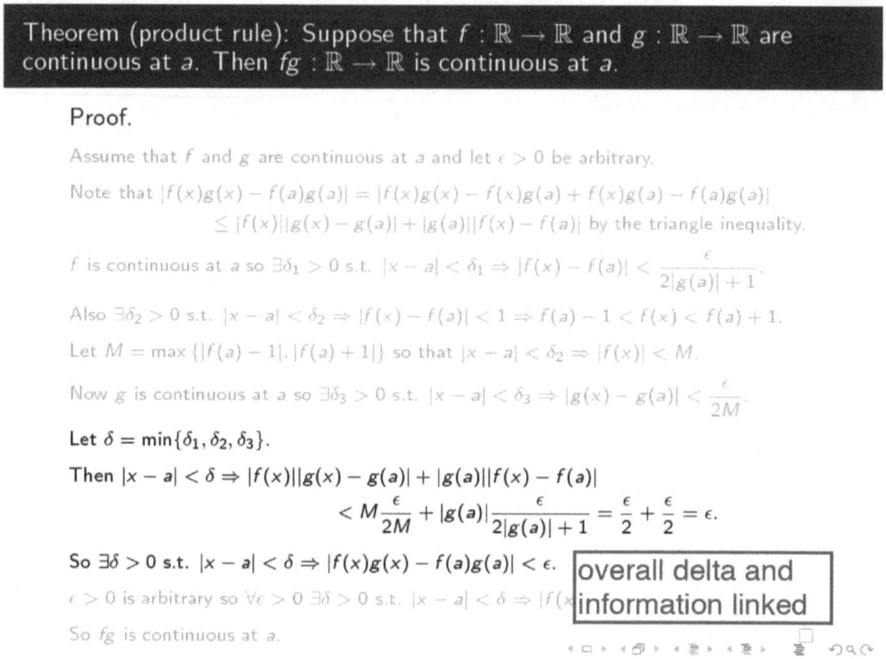

Fig. 2 A screen shot from the chunks version of an e-Proof

much work there can be in comprehending a proof, and that it has led to me making substantially more comments about this comprehension process.

For these reasons, I find the e-Proofs are useful, but once again I do not know whether they will actually improve comprehension of the proofs or learning in the course, in general. One particular issue is that I deliberately chose not to make an e-Proof for every proof in the course, partly because of time constraints, but primarily because I want students to develop their general proof comprehension skills and to transfer these to the other proofs in this course and beyond. So some questions I have are as follows:

- How do students usually go about studying proofs in lecture notes?
- Do students understand and remember a proof better if they spend time studying an e-Proof than if they spend time studying a standard written version?
- Are students who have studied e-Proofs better able to independently break down and understand a new proof than those who have not?

Tasks Involving Mapping the Structure of the Whole Course

The definitions tests, example classification tasks, diagram construction tasks, and e-Proofs are all things I use to get students actively thinking and doing mathematics during lecture time. They are all designed to address general skills in advanced mathematics, but at any given time they nonetheless focus on some particular definition/theorem/proof. Success in such a course, however, demands more than getting to grips with small elements of the content. It also depends upon the ability to manage one's learning efficiently, and to master a large amount of material in a restricted amount of time. Universities in the UK are increasingly providing general study skills advice through websites, and some institutions are going as far as instating modules dedicated to study skills and key employability skills (though I do not know of cases where this is happening in mathematics departments). However, as noted above, we do not know very much about mathematics students' study habits.

As a lecturer I do a number of things to help my students think of their learning in an integrated way and to be aware of their ongoing progress. One of these, which I did for the first time this year, is to make a very specific extended list of "measurable" course objectives/expectations to accompany the rather briefer and more ambiguous statements in the course specifications. The extended list (which could also be appropriate for other proof-based mathematics courses) reads as follows:

By the end of this course you should be able to

- fully and accurately state and explain the meaning of all the definitions;
- understand the role of definitions in providing a basis for systematic structuring of the Analysis topics of continuity, differentiability, and integrability;
- fully and accurately state and explain the meaning of all the theorems;
- fully and accurately write proofs of a large proportion of the theorems in the course (you should aim to be able to write proofs for all of them);
- explain line-by-line why the proofs are valid and describe their overall structure;
- use notation and logical language in a precise and unambiguous way;
- give a variety of examples of numbers, sets, and functions that satisfy definitions from the course and combinations of definitions from the course;
- sketch graphs and draw detailed diagrams to represent concepts, theorem statements, and reasoning in proofs;
- understand the relationships between the definitions and theorems across the whole course;
- with appropriate direction and hints, explore extensions and applications of theorems and of the reasoning used in proofs.

These are not elegant, and as with the definitions testing, there is a risk that some of them will be seen as an invitation to rote learning. However, my concern in the past has been that students are unaccustomed to this type of mathematics, and that they do not know at the outset what will be expected of them in the examination. This is my first real attempt to write objectives that are clear enough for what Biggs (2003) calls *aligned teaching*, in which "The curriculum is stated in the form of clear objectives, which state the level of understanding required rather than simply a list of topics to be covered. The teaching methods are chosen that are likely to realize those objectives; you get students to do the things that the objectives nominate. Finally, the assessment tasks address the objectives, so that you can test to see if the students have learned what the objectives state they should be learning." I find it useful to refer back to these objectives during "review" parts of lectures (I have about 20 minutes of review for each of the three major topics of continuity, differentiability, and integrability) and to invite the students to talk briefly to each other about whether they are progressing with respect to each one.

The remainder of each review is taken up with another learning-management activity based on the study strategies I developed as an undergraduate: Students each make their own one-column summary of the completed topic, with a view to ending up with a one-page course summary. In the final review session (or a longer revision lecture) I also use this list to help students decide how to focus their revision time. They are instructed to mark each item according to the following scheme:

- A tick (√) to indicate full knowledge and understanding;
- A question mark to indicate partial knowledge and understanding;
- A cross to indicate minimal or no knowledge and understanding.

I then suggest that they should work primarily on the items with question marks, since these are the ones on which they are likely to make the most progress (and since moving these items to the "ticks" pile will probably cause others to move from a cross to a question mark as they become accessible from the secure knowledge).

I give lecture time to these activities because I believe that they are important and because, while I hope that students would do them in their own time, I suspect that many would not: There is often a sense in the room that this is something new, or at least something that has not previously been done as a course progresses. As with my other practices, however, I do not have any actual evidence of this, and I do not know whether this represents lecture time well spent. It would be useful for me to have answers to questions such as given below:

- What summarizing activities do students typically do for mathematics courses, and when?
- How do students decide what to work on during their revision time?
- Do clear, detailed course objectives (in undergraduate mathematics courses) lead to better knowledge/understanding of mathematical material?

Discussion: Interactions Between Teaching and Research

There is a substantial two-way relationship between my teaching and research. Developments in my practice are often driven by my needs as a teacher but informed, or at least subsequently rationalized, with reference to mathematics education research. Knowledge of research literature also gives me advance warning of difficulties my students are likely to face, and conducting my own research has given me the opportunity to seriously *listen* to students' reasoning. The latter has given me insight that I would not have acquired as a teacher alone, especially in situations in which I am lecturing to large classes (even in small class or one-to-one situations, the tendency is to interrupt students, or at least to hear what I expect to hear). Conversely, my teaching puts me face-to-face with students' immediate questions and responses to tasks and alerts me to new areas of potential research interest.

The relationship between teaching and research is not unproblematic, however. I have never tried, for instance, to conduct systematic research on my own teaching or with my own current students[6]. I am not confident of my ability to remain objective in doing such work, and I do not want my students to become confused about my role. There is also the problem, mentioned in the introduction, that I do not make changes to my teaching one at a time, in a way that would allow me (or, preferably, someone else) to evaluate their effectiveness. Indeed, it may be meaningless to try to do that, for the reasons Schoenfeld points out:

> Imagine that one could construct a test fair to both old and new instruction. And suppose that students were randomly assigned to experimental and control groups, so that standard experimental procedures were followed. Nonetheless, there would still be serious potential problems. If different teachers taught the two groups of students, any differences in outcome might be attributable to differences in teaching. But even with the same teacher, there can be myriad differences. There might be a difference in energy or commitment: Teaching the "same old stuff" is not the same as trying out new ideas. Or students in one group might know they are getting something new and experimental. This alone might result in significant differences. (Schoenfeld, 2000, p. 645)

Where I do see possibilities, and where my primary interest currently lies, is in documenting students' responses to particular research-informed tasks, in order to provide information for lecturers about how their students are likely to interpret such tasks and what outcomes can be expected if they are used in or alongside lectures. While we wrestle with the less tractable questions, I think that there is considerable potential for design research at this scale. Its outcomes could provide mathematicians with a way to begin using more unusual, research-based instruction in a small-scale way as part of their existing teaching, in much the same way that teachers can use tasks from professional development experiences with their own classes.

Conclusion: Overarching Themes in Teaching

In this concluding section I return to the two overarching themes running through this chapter. The first of these is the need to help students develop skills on multiple levels. The practices discussed above vary in this respect. Some are about developing an understanding of individual concepts. Some are about what Hounsell and Hounsell (2007) might call *ways of thinking and practicing* in mathematics: knowing the status of definitions within mathematical theory, learning to translate

---

[6] Beyond collecting fairly basic, anonymous, opinion-based feedback.

between diagrams and formal language, learning to infer warrants in order to understand proofs, and so on. Some are about study habits, and what kind of thing a student ought to spend their study time doing. This variety, in my view, is part of what makes subjects at the transition-to-proof level so difficult: There are many, mutually supporting skills to be developed at once. I would stress, however, that I do not think this transition needs to be impossible; as discussed here, I believe that there are many practical things a teacher can do to help their students develop these skills.

This, however, brings us back to the main question of the second theme, which teachers negotiate all the time: how best to use the time available? As indicated above, my own teaching has become progressively less about the mathematical content and more about how to interact with this content. In this sense my focus has shifted from small-scale learning goals (that my students understand a particular concept or theorem) to larger-scale goals (that they develop the ability to relate examples to definitions, to draw diagrams, to break down proofs, and to make useful summaries of large amounts of material). This has contributed to a change in my overall lecturing style: whereas for the small-scale goals, a clear explanation might be best or at least quickest, for those at a larger scale, interactive lecture activities often seem more appropriate. I am content with this shift since it means that my students are more actively involved in learning and debating during lecture time, which also provides me with more opportunity for gaining insight into their current thinking. But it remains the case that there are all sorts of things I could do, in many different combinations, and I really have no evidence-based mechanism for deciding how much time to devote to what. This is quite a sobering thought, but also one that throws up many interesting sub-questions and challenges for those interested in research and development in learning and teaching.

References

Alcock, L. (in press). Mathematicians' perspectives on the teaching and learning of proof. In F. Hitt, D. A. Holton, & P. Thompson (Eds.), *Research in collegiate mathematics education.* (Vol. VII, pp. 73–100). Providence, RI: American Mathematical Society.

Alcock, L., & Inglis, M. (2008). Doctoral students' use of examples in evaluating and proving conjectures. *Educational Studies in Mathematics*, *69*, 111–129.

Alcock, L., & Simpson, A.P. (2002). Definitions: Dealing with categories mathematically. *For the Learning of Mathematics*, *22*(2), 28–34.

Alcock, L., & Simpson, A.P. (2004). Convergence of sequences and series: Interactions between visual reasoning and the learner's beliefs about their own role. *Educational Studies in Mathematics*, *57*, 1–32.

Alcock, L., & Weber, K. (2005). Proof validation in real analysis: Inferring and checking warrants. *Journal of Mathematical Behavior*, *24*, 125–134.

Alcock, L., & Weber, K. (in press). Referential and syntactic approaches to proof: Case studies from a transition-to-proof course. In F. Hitt, D. A. Holton & P. Thompson (Eds.), *Research in collegiate mathematics education* (Vol. VII, pp. 101–123). Providence, RI: American Mathematical Society.

Bell, A. (1993). Principles for the design of teaching. *Educational Studies in Mathematics, 24*, 5–34.

Biggs, J. (1987). *Student approaches to learning and studying*. Melbourne: Australian Council for Educational Research.

Biggs, J. (2003). *Teaching for quality learning at university*. Buckingham: SRHE and Open University Press.

Burn, R. P. (1992). *Numbers and functions: Steps into analysis*. Cambridge: Cambridge University Press.

Burton, L. (2004). *Mathematicians as enquirers: Learning about learning mathematics*. Dordrecht: Kluwer.

Conradie, J., & Frith, J. (2000). Comprehension tests in mathematics. *Educational Studies in Mathematics, 42*, 225–235.

Cornu, B. (1991). Limits. In D. O. Tall (Ed.), *Advanced mathematical thinking* (pp. 153–166). Dordrecht: Kluwer.

Dahlberg, R. P., & Housman, D. L. (1997). Facilitating learning events through example generation. *Educational Studies in Mathematics, 33*, 283–299.

Davis, R. B., & Vinner, S. (1986). The notion of limit: Some seemingly unavoidable misconception stages. *Journal of Mathematical Behavior, 5*, 281–303.

Entwistle, N., & Ramsden, P. (1983). *Understanding student learning*. London: Croom Helm.

Gibson, D. (1998). Students' use of diagrams to develop proofs in an introductory real analysis course. In A. H. Schoenfeld, J. Kaput, E. Dubinsky, & T. Dick (Eds.), *Research in collegiate mathematics education* (Vol. III, pp. 284–307). Providence, RI: American Mathematical Society.

Goldenberg, P., & Mason, J. (2008). Shedding light on and with example spaces. *Educational Studies in Mathematics, 69*, 183–194.

Hadamard, J. (1945). *The psychology of invention in the mathematical field* (1954 ed.). New York: Dover Publications.

Harel, G., & Sowder, L. (1998). Students' proof schemes: Results from exploratory studies. In A. H. Schoenfeld, J. Kaput, & E. Dubinsky (Eds.), *Research in collegiate mathematics education* (Vol. III, pp. 234–283). Providence, RI: American Mathematical Society.

Harel, G., & Sowder, L. (2007). Toward a comprehensive perspective on proof. In F. Lester (Ed.), *Second handbook of research on mathematics teaching and learning* (pp. 805–842). Charlotte, NC: National Council of Teachers of Mathematics.

Healy, L., & Hoyles, C. (2000). A study of proof conceptions in algebra. *Journal for Research in Mathematics Education, 31*, 396–428.

Hounsell, D., & Hounsell, J. (2007). Teaching-learning environments in contemporary mass higher education. In N. Entwistle & P. Tomlinson (Eds.), *Student learning and university teaching* (pp. 91–111). Leicester: The British Psychological Society.

Inglis, M., Mejia-Ramos, J. P., & Simpson, A. (2007). Modelling mathematical argumentation: The importance of qualification. *Educational Studies in Mathematics, 66*, 3–21.

Kember, D. (1996). The intention to both memorize and understand: Another approach to learning? *Higher Education, 31*, 341–354.

Lin, F.-L., & Yang, K.-L. (2007). The reading comprehension of geometric proofs: The contribution of knowledge and reasoning. *International Journal of Science and Mathematics Education, 5*, 729–754.

Marriotti, M. A. (2000). Introduction to proof: The mediation of a dynamic software environment. *Educational Studies in Mathematics, 44*, 25–53.

Marton, F. (2007). Towards a pedagogical theory of learning. In N. Entwistle & P. Tomlinson (Eds.), *Student learning and university teaching* (pp. 19–30). Leicester: The British Psychological Society.

Mason, J. H. (2002). *Mathematics teaching practice: A guide for university and college lecturers.* Chichester: Horwood Publishing.
Mason, J., & Pimm, D. (1984). Generic examples: Seeing the general in the particular. *Educational Studies in Mathematics, 15*, 277–289.
Mejia-Ramos, J. P. (2008). The construction and evaluation of arguments in undergraduate mathematics: A theoretical and a longitudinal multiple case study. Unpublished PhD Thesis, University of Warwick, UK.
Monaghan, J. (1991). Problems with the language of limits. *For the Learning of Mathematics, 11 (3)*, 20–24.
Moore, R. C. (1994). Making the transition to formal proof. *Educational Studies in Mathematics, 27*, 249–266.
Peled, I., & Zaslavsky, O. (1997). Counter-examples that (only) prove and counter-examples that (also) explain. *Focus on Learning Problems in Mathematics, 19 (3)*, 49–61.
Przenioslo, M. (2004). Images of the limit of function formed in the course of mathematical studies at the university. *Educational Studies in Mathematics, 55*, 102–132.
Raman, M. (2003). Key ideas: What are they and how can they help us to understand how people view proof? *Educational Studies in Mathematics, 52*, 319–325.
Recio, A. M., & Godino J. D. (2001). Institutional and personal meanings of mathematical proof. *Educational Studies in Mathematics, 52*, 319–325.
Roh, K. (2008). Students' images and their understanding of definitions of the limit of a sequence. *Educational Studies in Mathematics, 69*, 217–233.
Schwarz, B., & Hershkowitz, B. (1999). Prototypes: Brakes or levers in learning the function concept? *Journal for Research in Mathematics Education, 30*, 362–389.
Schoenfeld, A. H. (2000). Purposes and methods of research in mathematics education. *Notices of the American Mathematical Society, 47*, 641–649.
Segal, J. (2000). Learning about mathematical proof: Conviction and validity. *Journal of Mathematical Behavior, 18*, 191–210.
Selden, J., & Selden, A., (1995). Unpacking the logic of mathematical statements. *Educational Studies in Mathematics, 29*, 123–151.
Selden, J., & Selden, A., (2003). Validation of proofs considered as texts: Can undergraduates tell whether an argument proves a theorem? *Journal for Research in Mathematics Education, 34*, 4–36.
Sierpinska, A. (1987). Humanities students and epistemological obstacles related to limits. *Educational Studies in Mathematics, 18*, 371–397.
Tall, D. O. (1997). Functions and calculus. In A. J. Bishop et al. (Eds.), *International handbook of mathematics education* (pp. 289–325). Dordrecht: Kluwer.
Tall, D. O., & Vinner, S. (1981). Concept image and concept definition in mathematics with particular reference to limits and continuity. *Educational Studies in Mathematics, 12*, 151–169.
Tsamir, P, Tirosh, D., & Levenson, E. (2008). Intuitive nonexamples: The case of triangles. *Educational Studies in Mathematics, 69*, 81–95.
Vinner, S. (1991). The role of definitions in teaching and learning mathematics. In D. O. Tall (Ed.), *Advanced mathematical thinking* (pp. 65–81). Dordrecht: Kluwer.
Vermunt, J. D. (2007). The power of teaching-learning environments to influence student learning. In N. Entwistle & P. Tomlinson (Eds.) *Student learning and university teaching* (pp. 73–90). Leicester: The British Psychological Society.
Watson, A., & Mason, J. (2005). *Mathematics as a constructive activity: Learners generating examples.* Mahwah, NJ: Lawrence Erlbaum Associates.
Weber, K. (2001). Student difficulty in constructing proofs: The need for strategic knowledge. *Educational Studies in Mathematics, 48*, 101–119.
Weber, K. (2008). How mathematicians determine if an argument is a valid proof. *Journal for Research in Mathematics Education, 39*, 431–459.
Williams, S. (1991). Models of the limit held by college calculus students. *Journal for Research in Mathematics Education, 22*, 219–236.

Yang, K.-L., & Lin, F.-L. (2008). A model of reading comprehension of geometry proof. *Educational Studies in Mathematics*, *67*, 59–76.

Zazkis, R., & Chernoff, E. J. (2008). What makes a counterexample exemplary? *Educational Studies in Mathematics*, *69*, 195–208.

Zazkis, R., & Leikin, R. (2007). Generating examples: From pedagogical tool to a research tool. *For the Learning of Mathematics*, *27*, 11–17.

Teachers Learning from Their Teaching: The Case of Communicative Practices

Helen M. Doerr and Stephen Lerman

Introduction

The nature of the professional knowledge of teachers, in our case teachers of mathematics, has been and continues to be of major interest in research on teaching. Research has focused on three major questions: What does the professional knowledge base for teaching mathematics look like; how do practicing teachers acquire it; and does it develop from practice and if so how? In this chapter, we will address these questions, and in particular the third question, by examining some of the data from a four-year research project carried out by the first author and her colleagues.[1] The focus of the research project was to investigate the role of literacy in mathematics teaching, taken to include speaking, writing, and reading. Broadly speaking, literacy is driven by the need to communicate, a driving force that encompasses social development, in general, and all learning, in particular. The shared question for the researchers and teachers was how to develop students' abilities and skills for communication in the mathematics classroom. The researchers also focused on how experienced and competent teachers learned from their teaching, in this case about the development of students' abilities to communicate mathematically.

In this chapter, we wish to distinguish between the nature of teachers' knowledge and the development of that knowledge. Our primary interest is in the latter: how that knowledge develops as teachers interact with other teachers and researchers and experiment with new ideas that become absorbed into their repertoire of professional knowledge. For us, watching what teachers *do* as they engage in practice leads to a focus on pedagogical strategies or actions in the classroom. Going further

H.M. Doerr (✉)
Syracuse University, Syracuse, NY, USA
e-mail: hmdoerr@syr.edu

[1]This material is based upon work carried out by the first author, K. Chandler-Olcott, K. Hinchman, and J. Masingila and supported by the National Science Foundation under Grant Number 0231807. Any opinions, findings, conclusions, or recommendations expressed in this material are those of the authors and do not necessarily reflect the views of the NSF.

R. Leikin, R. Zazkis (eds.), *Learning Through Teaching Mathematics*, Mathematics Teacher Education 5, DOI 10.1007/978-90-481-3990-3_13,

and unpacking *why* teachers do what they do leads us and them to a focus on interpretation, reasoning, and explanations about their teaching of mathematics. Several years ago, Hiebert and colleagues (Hiebert, Gallimore, & Stigler, 2002) posed the question of what does the professional knowledge base for teaching mathematics look like? In addressing this question, we find it useful to work with a distinction between local and global theories, drawing on arguments that have been put forth to describe the role of design research in the field of education (Cobb, Confrey, diSessa, Lehrer, & Schauble, 2003; Design-Based Research Collective, 2003). We also draw on researchers (diSessa & Cobb, 2004; Lewis, Perry, & Murata, 2006) who have described how local theories grounded in classroom practice can contribute to a theoretically grounded knowledge base for teaching. Teachers, such as those in this study, often develop strategies to deal with the very specific issues that arise for them in practice. Such strategies are specific responses to problem situations involving particular students, materials, and mathematical goals. But at the same time, some of these strategies are *local theories* or *principles for action* that can cut across contexts and can be adapted to other problem situations. For example, the teachers in this study developed, over a lengthy period of time, a way of working with the development of students' mathematical writing through a sophisticated model for thinking about writing and a resulting set of rich resources (Doerr & Chandler-Olcott, 2009). The teachers realised that they could use the same way of thinking to approach the development of students' mathematical reading skills. In this sense, local theories are more than "mere" strategies that are relevant to a particular time and place and no more. These local theories are principles for action that can be shared among teachers and across contexts and problem situations. Our focus in this chapter is on how these aspects of teachers' knowledge develop in practice and potentially contribute to the professional knowledge base for teaching.

This raises the question of whether there are global theories about the professional knowledge base for teaching. Can we identify *generally accepted theories* about teaching mathematics, taking pedagogy and mathematical knowledge together, as inseparable? We know that mathematical knowledge has some kind of universal certainty, but we are much more cautious about the certainty of mathematical pedagogic knowledge. The former exhibits a strong grammar, enabling one to be quite specific about what is being claimed, and there are deductive and inductive methods for verification or refutation. The latter exhibits a much weaker grammar, leading to different and in many senses incommensurable discourses. However, our experience leads us to believe that there are some elements of a global mathematical pedagogic knowledge, as we are calling it. For example, we find that teachers reading Skemp's (1976) article about instrumental and relational thinking for the first time are almost always very persuaded and take the ideas into their thinking about teaching. Similarly, many teachers respond positively to Wood's (1998) descriptions of funneling and focusing classroom conversations, which was very significant for the teachers in this study, as we will report below. But neither of these elements of mathematical pedagogic knowledge is as universally accepted as, say, the Pythagorean theorem or the Euclidean algorithm.

Turning now to our primary interest, how the professional knowledge base for teaching mathematics develops, we hypothesize that one way is through teachers learning from their teaching as they work together, with the support of university researchers, and sharing their ideas with others through dissemination at conferences and in journals. Our concern is to elaborate on the process whereby teachers learn from their teaching and, in so doing, potentially contribute to the development of the professional knowledge base for teaching mathematics.

Background of the Study

The data reported in this chapter are from a four-year research project on mathematics and literacy. The research was carried out by a team of university-based researchers in mathematics education and literacy education, working in concert with mathematics teachers in a mid-sized urban district in the United States. The district had recently adopted what are known as Standards-based curriculum materials, namely, *Connected Mathematics Project (CMP)*, (Lappan, Fey, Fitzgerald, Friel, & Phillips, 1998). These particular Standards-based materials, along with several others, were developed with the support of the National Science Foundation in the 1990s to align with curriculum standards that had just been put forward by the National Council of Teachers of Mathematics (NCTM, 1989). These materials represent a significant shift from traditional textbook materials in that they are structured around a sequence of "investigations" that require the students to engage with mathematical tasks that need to be interpreted through the stories, pictures, diagrams, and charts in the texts. In addition, the students are expected to provide descriptions, explanations, and justifications about their work with the tasks, both orally and in writing. As such, these curricular materials presented new challenges for the teachers who now needed to learn to support the development of students' abilities to communicate mathematically. In this chapter, we report on the learning that occurred for one of the teachers as she learned to address the literacy demands of these mathematically rich and contextually complex curricular materials.

The teachers participating in this project were from one high school and its three feeder schools. The results reported here are from one of the teachers, Cassie, in the feeder school where the first author worked with the teachers. Belmont School had approximately 860 students and 45 teachers and support staff and was considered high poverty with over 80% of the students qualifying for free or reduced-fee lunch. The school population was quite diverse with approximately 31% African-American, 21% Asian, 35% Caucasian, and 11% Latino/a students. Approximately 20% of these students were English language learners, and 25% were identified as having special needs. As we began the project, it was the second year using the *CMP* materials for most of the teachers. There were five participating teachers at this school, who taught grades 6 through 8 (student ages 11–13). Four of the teachers were very experienced; one was in her first year teaching as we began this project. Cassie was in her 11th year teaching; she was enthusiastic about the use of the *CMP* materials. Cassie held multiple teaching certifications as she was certified to teach middle grades mathematics, special education, and elementary education. She was teaching students in both grade 7 and grade 8 as we began our work together.

The work with the teachers consisted primarily of four on-going activities: summer workshops, quarterly project meetings with teams from other project schools, bi-weekly team meetings, and "lesson cycles" (described more fully below). The first week-long summer workshop provided an introduction to the project for all the teachers as we began our collaborative work in addressing the literacy demands of the Standards-based texts. As the project progressed, the teachers at Belmont School chose to work together for three weeks each summer to develop specific instructional goals and plans related first to mathematical writing and later to mathematical reading. During the school year, the quarterly project meetings and the bi-weekly team meetings provided forums for the sharing and continued discussion of instructional strategies.

Since our primary research questions concerned teachers' learning about mathematical communication in their own practices, we used "lesson cycles" to work jointly on planning, implementing, and debriefing lessons for supporting literacy opportunities for students (Doerr & Chandler-Olcott, 2009). Each lesson cycle consisted of three elements: (1) A planning session that followed the overall *CMP* guidelines for the investigations, but asked specifically the question "what are the literacy opportunities in this lesson?" In planning with this focus, the teacher discussed her ideas for reading the text, described opportunities for students to speak with each other, and identified prompts for student writing that would be used in the lesson. (2) The implementation of the lesson, where a member of the research team would observe the lesson, take extensive field notes, and generate questions for discussion that arose during the observation related to the literacy opportunities in the lesson. (3) A de-briefing session with the teacher, where the intent of the session was to collaboratively gain insight into the teachers' thinking about the literacy opportunities of the lesson and to collect shareable artifacts from the lesson, such as insights gained or tools used to support students' learning. The de-briefing session often centered on a discussion of the students' written work and how that might be used to inform subsequent lessons. The planning and debriefing sessions were audio-taped and later transcribed. Brief memos were written based on notes taken and the artifacts of the session. The lesson cycles began halfway through the first year of the project and continued through the fourth year of the project. Each teacher participated in a lesson cycle approximately once every three to four weeks with a member of the research team.

The observation of the lesson allowed us to focus on the teachers' pedagogic strategies enacted during the lesson. The planning and debriefing sessions gave us a focus on how the teacher interpreted and reasoned about the events that occurred in the classroom. Taken together, these sessions and the observations led to our developing insights into the development of students' abilities to communicate their mathematical thinking.

From the beginning of the project, the teachers at Belmont School shared a common focus on the need for their students to become better mathematical writers.

This was, in part, driven by the high-stakes testing that took place at the end of grade eight, where students were asked to explain their reasoning or solution strategies in writing. All the teachers felt a school-level shared responsibility for preparing students for this exam. The focus on writing was also driven, in part, by the curricular materials that included many tasks that asked students to explain their reasoning or solution strategies. The teachers valued these tasks, since they required elaborated descriptions and explanations. Finally, the focus on writing reflected the teachers' concerns for many of their students who were not at grade level in reading or writing for reasons of second language learning, learning disabilities, or special needs. This shared interest in student writing became the focus of the discussions at the bi-weekly team meetings, during the lesson cycles, and at the summer workshops.

At the end of the third year of the project, the teachers wanted to give increased attention to students' mathematical reading. In part, this shift occurred as a consequence of the progress that the teachers had made in planning for the development of students' written expression. They wanted to take what they had learned from their success in improving students' writing and address students' development as mathematical readers. In part, this shift was influenced by the teachers' increasing sense that reading was a critical barrier to students' performance on the state exams. At Belmont School, we never had an explicit, sustained focus on oral language within the project, although other schools in the project did. However, this is not to suggest that either the researchers or the teachers saw reading, writing and speaking as unrelated or disconnected practices. Indeed, it was quite the contrary! There were many instances of the teachers explicitly and specifically using writing to support speaking, using speaking to support reading, and so on. These teachers developed in their perspectives and practices on the relationships or interplay among reading, writing, and speaking. But that interplay is not the focus of this chapter.

Our data sources included field notes from the summer work sessions, the quarterly project-wide meetings and the bi-weekly school-level team meetings and the field notes and transcripts from the lesson cycles. In addition, each teacher was interviewed five times: once prior to the start of the project and at the end of each year of the four-year project. At the final interview, each teacher was asked to reflect back on her experiences in the project and to share the stories and events that had been most salient from her perspective and the insights that she had into teaching through participating in the project. This final interview is the starting point for the results reported in this chapter. We are using our shared analysis of this final interview to frame the learning that took place for Cassie by using how she told the story of her own learning about mathematical communication. We wish to point out that Cassie's perspectives on her learning occurred through her sustained interactions with her colleagues, her students, the researchers, and the curriculum. Taken together, these interactions provided a focus on the role of communicative practices in the classroom and this sustained focus appears to have been a critical factor in supporting the development of Cassie's knowledge of teaching. We begin with Cassie's insights into the role of oral language, writing, and reading in teaching mathematics.

Insights Related to Oral Language

I Used to Be a Big Funneler

An important area of learning for Cassie occurred as she came to recognize that her patterns of questioning were largely dominated by the kinds of initiate-respond-evaluate (Mehan, 1979) patterns that Wood (1998) described as "funneling." We referred to this above when discussing teachers' acquisition of the professional knowledge base for teaching mathematics. Cassie saw this as a problematic area in her own practice in that it created the "illusion of learning." She explained that she had taught material, but when she tested students, she was finding that they had not mastered it. She now sees her funneling as a cause of the difficulty and something that she could change. Cassie explained the following:

> Deep down inside I'd say, you know, I taught this stuff. I thought they had it. I gave a test, and they didn't have it. Where are they losing it? It's like you're looking at the kids and wondering where they're losing it when, in fact, it turned out, it was the way I was doing things. It was me. Because they weren't getting it. They were just regurgitating what they thought I wanted to hear.

Cassie was very self critical as she says that "I think I had a lot of kids who had the illusion of learning because I'm saying they had this [understanding]. No, I had this [understanding] and I was feeding it to them. And that was not very helpful." She realized that funneling created the illusion that the students had learned the mathematics she was trying to teach, but in the end this strategy did not help her students learn.

Cassie attributed her own learning about this to two factors. The first factor was a set of articles by Wood (1998) and by Herbel-Eisenmann and Breyfogle (2005) that draws on Wood. The team of teachers read and discussed these articles during one of their summer team meetings. We also looked at some video clips of teachers from professional development materials with these articles in mind. Together these articles describe funneling conversations and an alternative approach called "focusing." Cassie pointed out that "it's hard not to funnel" and that it still occurred in her practice: "not that I'm still not guilty of funneling now and then, because I am. But it's something you watch out for now." Cassie pointed out that this is a difficult practice to change; working on alternatives is not easy. She was now much more cognizant of when she was falling into the funneling trap. As Chazan and Ball (1999) point out about this changed role for the teacher: When the role of the teacher is not telling, then what is it? At the end of the project, Cassie was still working on this new role for herself.

The second factor that influenced Cassie's learning was a conversation that occurred during one of the lesson cycles in the third year of the project. During the observation of the class, Cassie had a conversation with one of her students, Kiesha, who was having difficulty in solving a particular problem that drew on understanding the area of a rectangle. This student was low achieving in mathematics and her school attendance was poor, thus contributing to her achievement. Cassie's conversation with Kiesha highlighted the extreme difficulties that Cassie experienced in the moment of teaching as she was struggling to not funnel Kiesha into the solution to the problem. Cassie's distress in this episode was clearly visible as she turned to another student for help in her conversation with Kiesha and then later to the researcher who was observing the lesson. Cassie's conversation with Kiesha was a very long one and did not come to a "successful" ending.

During the debriefing of the lesson, Cassie attributed part of the difficulty to a mathematical issue. Kiesha did not appear to have had prior experience with *CMP* in grade six where the concepts of area and perimeter are developed; she had transferred to this school in grade seven. Cassie is familiar with lessons from grade six, both because of her prior teaching experience at that grade level and because of the work of the project. She referred to the "bumper cars" which is the context of students' initial investigation into area and perimeter in grade 6 (where the problem is cast in terms of the area that the bumper cars drive in and edges that are needed to fence in the area). Cassie saw that this context might have helped Kiesha by giving her something to draw on in reasoning about the current problem: "I think that makes a difference because she didn't have that context to fall back on." The student's familiarity with that context might have provided Cassie with a language to draw on as an alternative to funneling. Cassie later shared this conversation with her colleagues and for Cassie it became known as the "Kiesha conversation." Cassie saw this as the increasingly important role of context in developing mathematical connections to support students' learning.

We see Cassie's struggle over the Kiesha conversation as the local enactment of a principle for action: Kiesha's having missed an important context had an adverse impact on her consequent ability to draw on the relevant mathematics of area and perimeter. Cassie was operating under her emerging and more global theory of resisting funneling, thus leading to a frustrating interchange with the student. This interchange sharpened awareness on Cassie's part that what she wanted to elicit from Kiesha was her reasoning about the problem, while not feeding her with an answer.

Letting Kids Develop Their Own Words for Concepts

Cassie recognized that students needed to have opportunities to discuss mathematics, but it was not easy for Cassie to engage students in having discussions. Cassie had classes with significant numbers of students with behavioral problems and many students whose basic skill levels were very low. Cassie had used an activity structure for paired interactions called "Rally Coach" that had been introduced through a project workshop on Kagan structures (www.kaganonline.com). In this activity structure, students work in pairs on a sequence of problems, where each partner checks the previous work and coaches the other student if need be. This structure enabled Cassie "to let go a little bit" so that the students' interactions could happen. Cassie saw this paired coaching as giving students an opportunity to teach each other. Cassie explained the following:

> Teaching it makes it so much clearer and that's why I like the Rally Coach. And I think that's something I began this year and I don't know if I should have. Skill levels are kind of weird to do that.... When I gave them more time at the end of the year and I was willing to let go a little bit, I think the kids did better [working in pairs] on that information than if I just let them work independently [by themselves]. They really need that [paired] discussion.

But Cassie's view of oral language was not simply that students need opportunities to talk. Rather, Cassie had developed a view of the different kinds of talk that needed to happen so that that students' language for mathematical ideas would develop over time. Her framing of this development (an example of a local theory or principle for action) was influenced by the work of Herbel-Eisenmann (2002), described in an article that was read and discussed by the team of teachers during their summer work together. In her work, Herbel-Eisenmann frames the developing language of students in three main categories: "contextual language," "bridging language," and "official mathematical language." The language that depends on specific contexts or problem situations is referred to as "contextual language." This category often occurs when using *CMP*, since instruction is organized around and driven by the solution of mathematical problems in specific contexts (such as the bumper cars or a walking race). Herbel-Eisenmann delineates two forms of bridging language: that which is idiosyncratic to the student- or teacher-generated language in a particular classroom and that which is transitional mathematical language in that it refers to a particular process or representation without a contextual reference. These forms of bridging language help students move from less mathematical ways of talking about ideas to ways that are more mathematically precise. Herbel-Eisenmann argues that bridging language "offers access to a larger range of students because classroom discussion can include more levels of entry" (p. 101). This argument resonated with Cassie and the other teachers in the project.

The third category of the framework is that of "official mathematical language." This refers to the language that is "part of the mathematical register and would be recognized by anyone in the mathematical community" (p. 102). The teachers took up these three descriptors for students' oral language, as well as for students' written language. However, in so doing, the teachers renamed Herbel-Eisenmann's category of "contextual language" as "everyday language." This was, at least in part, a recognition on their part that in many cases they needed to address the "everyday" meanings students had for words that might be unfamiliar to them by reason of context, life experiences, and/or second language learning. Herbel-Eisenmann's framing of students' language development became a local theory of mathematical pedagogic knowledge used by Cassie and the other teachers to guide their actions in planning and in teaching lessons.

Cassie articulated these principles as the need for students to "develop their own words for concepts." She saw this as essential for building their understanding of concepts and she doubted the efficacy of giving students the official mathematical language. Cassie explained her views: "I think if they develop their own words for concepts, they're going to understand those concepts more clearly than if we give them the official math language. I think that's really important." She recalled an episode in her classroom that occurred near the end of the fourth year of the project. In this episode, the students were explaining how they found various equivalent expressions by combining like terms. When one student gave his description, the special education provider, who was with the class at the time, interjected and pressed the student to say that two terms could not be combined because they were "not like terms." Cassie was unhappy with this interjection and felt it took away the student's voice. She argued: "We're not giving kids enough time to develop that everyday math language, the bridging [language]. I think the students had a good description in their own words about what was going on. Instead, we're giving them the words they really have no connection to." She was adamant that the student's description, given in bridging language, was satisfactory for now and that the student needed to express his ideas in his own words. She further commented:

> Because what he said was perfectly okay. It made perfect sense. ... He said it wasn't the same [the terms were not alike] ... His explanation was perfectly fine. I think a lot of kids understood his explanation. And when the words "like terms" came in, I think a lot of kids said, 'okay, that's too much. I'm done.' I think we do that quite a bit.

Cassie recognized that the student's description was a temporary bridge to official mathematics language. The description made sense to the student who gave the description, and other students were able to gain entry into this student's way of expressing his idea about adding like terms.

Cassie was both self-critical and reflective on her own learning from this episode. Recognizing the futility of giving students words that they have no connection to, she saw her role as helping students make connections. Later, she talked about students needing to connect the words "slope" and "rate of change" since they refer to the same underlying concept. Cassie said: "We're giving them the words they really have no connection to. I mean, we're not connecting it. And I think that needs to be done. And that really surprised me because I used to do stuff like that [not helping with connections]." She articulated a new role for herself in wanting to hear what students have to say, recognizing from Herbel-Eisenmann's framework that what one student has to say may help others understand. Cassie said: "And now I wanted to know what they have to say. Because sometimes what they have to say will sometime trigger something in someone who doesn't understand." Cassie's principles for action and a new role for herself in the classroom emerged from her reading, her interactions with the researchers and her fellow teachers, in the context of addressing issues of concern in her practice.

Insights Related to Mathematical Writing

Writing Over Time ... Because I Actually Saw Growth in the Students

This comment by Cassie is a reference to an episode that occurred early in the project where she had given her students the same writing prompt ("what makes two figures similar") at the beginning, middle, and end of a unit of instruction on similar figures. Cassie and the researcher (first author) had an extended conversation about this writing, and Cassie spent some time classifying various examples of students' work. There were three categories that Cassie used in this classification: weak, average and strong. These represented her assessment of the level of student understanding that was evident in the writing. In this process, Cassie was often tentative with her judgments about what students might understand, especially for those students with special needs (perhaps diagnosed with a learning difficulty related to written expression) or second language learners. There were many students whose growth she could see over time and this was exciting. Cassie said: "I actually saw the growth in students . . . It's just like you really get a picture of what they're thinking and what they know about math." These insights into students' thinking were useful for her instruction. However, what was most striking about her analyses of this student work was her realization that there were some students who started out weak and stayed weak. This greatly concerned Cassie and provided her with compelling evidence that she needed to change her instructional strategies to address the needs of these students. This analysis of student writing furthered Cassie's more general commitment to using writing for insight into student thinking and student growth. Later in the project, Cassie selected an example of this "writing over time" and shared it with her colleagues and at a national conference. This example of writing over time became a local theory (or principle for action) for Cassie and the other teachers to draw on in their teaching practices.

Insights Related to Mathematical Reading

The Reading "Is Where We Lose a Lot of Kids"

From the beginning of the project, the teachers had identified that their students struggled with reading the *CMP* investigations. As with oral language and writing, the teachers were concerned with the learning of students who were below grade level in reading, students with special needs, and second language learners. The teachers wanted to improve the students' abilities to read and interpret the questions on the ever-present state assessments. Cassie reflected the concerns of all the teachers when she commented: "I was struggling with the groups [of students] I've had the last couple of years and I was trying to figure out what can I do. And I've had lower readers. But the group that really sticks out is the one I had this [last] year. They were very low readers. Most of them were special ed students." To help these students gain access to the investigations, Cassie developed what she called "guided notes." This was influenced, in part, by her collaboration with the reading and language arts teacher who was a member of Cassie's grade level team.

Over the course of the project, Cassie developed her expertise in using these guided notes. One critical episode occurred during a summer workshop when she articulated to the other teachers the rationale behind the construction of these "guided notes" sheets and how they are used in her teaching. Cassie summarized the three components of the guided notes structure: "First, have the students read a passage independently and high-light important information. Second, give the students the guided notes sheets. Third, review the notes as a class." Cassie explained that the guided notes sheets required the students to re-read the text (since the sheet of questions is given out to the students after they have initially read the text). The questions focus on the important information in the problem, supporting students in internalizing and applying information from the text. In sharing this pedagogic knowledge with her colleagues during a summer workshop, Cassie offered explicit guidelines (or principles for action) for creating effective questions: "Create 'right there' questions, create questions that allow students to make predictions, create questions that allow students to apply their knowledge." Each of these categories served specific instructional purposes: The "right there" question had the answer "right there" in the text, providing easy access for all students, while focusing on the most important information in the problem. The prediction question engaged students with interpreting and making inferences from the text; the application question engaged students with specific mathematical elements in the problem. The rationales for these categories are the underpinnings of Cassie's local theory for the efficacy of the guided notes strategy.

Cassie described how the guided notes strategy gave even her very low readers (some of whom were at grade two reading level) access to an investigation on linear relationships that centered on a walking race (and hence walking rates) and a head start for one of the racers. Cassie commented about the reading of the story for that investigation:

> There's a lot of information in that little bit. Not a lot of reading, but a lot of information about the head starts and the walking rates. And so doing the reading and the discussion [with the guided notes], everyone knew what was going on [in the story]. And that investigation was open, then, for everyone to do the math. The reading wasn't slowing them down. And we need to do more of that, because that's where we lose a lot of kids.

The guided notes strategy pushed the students to go back and re-read the text. This re-reading of the math text was a crucial idea for Cassie and the other teachers as they had realized (through our summer workshop) the full extent to which they re-read mathematical texts. Indeed, in many lessons, Cassie would say to her students "This is math reading. How many times do we read it?" It was the guided notes that enabled Cassie to help the students have a more focused purpose for going back and doing this re-reading. Cassie described how spending the time in re-reading and discussing these central stories helped her students remember the context and helped them make mathematical connections. The design of these guided notes was driven by clear principles that served explicit instructional purposes. As such, this became a local theory for Cassie and the teachers in this project.

Some Broader Changes

In addition to the learning that occurred relative to specific classroom practices of oral language, reading and writing, Cassie articulated two changes that addressed broader aspects of her practice. The first change was related to her perspective on the centrality of communicative practices in learning mathematics. The second change occurred in relationship to her use of the curricular materials. We discuss each change in turn.

Reading, Writing, and Speaking Really Are Central to Math

The first (and perhaps the most sweeping) insight was that "reading, writing, and speaking really are central to math." Cassie reflected that she had "always wanted to think, oh, that's just the math that's the important part." She now saw the necessity of directly addressing students' skills and abilities to read, write and speak as central to students' mathematical learning. She quickly acknowledged that students would have difficulty with these practices, but now saw that she had a role to play in supporting students' development as they acquired these practices. Her work with her colleagues and her developing repertoire of instructional strategies related to communicative practices began to shape this role for her. But Cassie continued to acknowledge the difficulties in forging this new role. She commented that "organized chaos isn't my thing" as she wanted her classroom to be well-organized and structured and that this worked against her giving up some control when the consequence of that might be more "chaos" in the room – often reflected in off-task behaviors and conversations by students. As the project progressed, Cassie continued to experiment with forms of group work and paired work that became much more common events in her classroom, as students engaged in various communicative practices.

Making the Program "A Little More Traditional. That's What I Was Doing."

Cassie's perspectives and responses to the curricular materials changed over time. She directly attributed this change to the project's focus on communicative practices. She pointed out that she initially "like[d] *CMP*, but I was still very traditional with it. I took a program that was not traditional and made it a little more traditional. That's what I was doing." In terms of the literature on curriculum use (Remillard, 2005), this could be interpreted as her taking a new set of curricular materials and adapting and enacting the materials in ways that largely left her core practices intact. But her core practices began to change. Cassie pointed out that because of the math and literacy project, she began to experiment in her practice; because the teachers had chosen writing as a focus for the team, this experimentation was largely one of trying writing with her students. The results of this writing often gave Cassie new insight into what her students were thinking and into how she might want to plan for the next day's instruction. These insights into students' thinking and instructional decisions influenced changes in Cassie's practice. As Cassie said "you start to do little things here and there. And you start to see the value of it. And then you do it more and more and more." These were not all-at-once or "aha" moment changes in Cassie's teaching, but rather these changes occurred incrementally, as they were tried and revised in practice. In looking back, Cassie now characterized these changes as substantial: "It's just amazing how much it changes. Because you're just looking out for the literacy and where can I do this. And what can I have the kids write about to see what they really understand? Can I get a Rally Coach out of this? . . . Where can I get them talking? And that's how you think." This new perspective is closely related to Cassie's first insight, namely, that reading, writing and speaking are central to mathematics learning. Attending to these practices had become central to her teaching and this supported her shift from a traditional teaching approach to one that was more nearly aligned with the Standards-based vision that under-girded the conceptualization of the *CMP* curricular materials. Cassie pointed out how much of a change this has been for her: "And then it becomes part of your teaching, [so] that you don't even remember what you were like before. . . . I used to teach like that. And I know I did. I taught the traditional way. And, now, I'm like good God, I'd rather gouge my own eyes out than teach [like] that."

Discussion

We will address each of our three major questions in turn, indicating what we have learned from the project in terms of each of the questions, in the hopeful expectation that we are speaking beyond the specifics of this project to mathematics teaching in general. As we noted in our opening paragraphs, the main emphasis in this chapter is on the third question.

Our first question addresses the nature of the professional knowledge base for teaching mathematics. This knowledge base differs from the mathematics content knowledge base in the lack of expectation of the same degree of certainty in the former as compared to the latter. This is not surprising given that education is a social science, not a physical science. What becomes accepted in a research field depends on the gate-keeping procedures of the research community, which certainly change over time.

> Educational research is located in a knowledge-producing *community*... Of course, communities will display a great deal of variation in their cohesiveness, the strength of their 'disciplinary matrix', and the flexibility of the procedures by which they validate knowledge claims. (Scott & Usher, 1996, p. 34)

In terms of the "knowledge producing community" argument by Scott and Usher (1996), and others, we understand that there is something we can call the professional knowledge base for teaching mathematics whose validity lies only in its acceptance by members of the community. This is a weak claim, we recognize, given especially that what is accepted by one group of researchers may not be accepted by others (witness the debates around constructivism). Perhaps the most fruitful engagement with theory is the resonance with and local adoption, and therefore testing, of accepted theories, such as funneling, and the generation of local theories that are meaningful and powerful in that context but, with sharing and dissemination, find resonance more widely.

Regarding our second question, and moving towards our third, we argue that we have seen the acceptance by the teachers of theories from well-known research in the field, such as Wood (1998) on funneling and Herbel-Eisenmann and Breyfogle (2005) on contextual language, bridging language and official mathematical language. We have also seen the development of pedagogic theories that are at least initially local, as for example when Cassie said that "reading, writing and speaking really are central to math." This insight initially developed from a perceived need to support students in developing their writing so as to evidence their reasoning and solution strategies and then subsequently to support their reading the texts of mathematical investigations. A focus on writing led to *principles for action* embodied in sophisticated planning tools and a set of shared pedagogical strategies, which carried across to reading. At a later stage, the teachers presented their work at a meeting of the National Council of Teachers of Mathematics (NCTM) and their experiences and, in particular, the theories that they had developed about their work (such as Cassie's statement above) were communicated to the audience and resonated strongly with them. We would want to say that we are now seeing the locally initiated theories becoming more global as "generally accepted theory."

The study has pointed to some key factors in that acquisition and potential development of a professional knowledge base for teaching: Working with colleagues; collaborating with researchers who come into the school; having a local focus that is important to the teachers; engaging in a longitudinal process. These factors interact with each other, as we will point out below.

In the mathematics education community we have long used the notion of reflective practice as a potential tool from which teachers can gain insight into their teaching and hence learn. We would argue, both in principle and from the experience of the project, that reflective practice requires an interlocutor to be meaningful. Reflection calls to mind looking at oneself in a mirror, where what one "reads" there is entirely within the sphere of one's own interpretation. The need to explain what one is doing, examine what one is doing with others with the aim of change and improvement is what makes the idea of learning from reflection potentially a reality. Cassie's experience of the interjection of the special education provider concerning combining like terms focused her attention on the need to give students "time to develop their everyday math language" in a way that, working on her own and not having the other teachers and the researchers with whom to communicate, would have been far less fruitful and perhaps would not have happened at all.

The interaction with researchers was vital in a number of ways. First, we feel sure that the extended engagement of the teachers with issues of reading, writing and speaking mathematics would not have happened without the researchers. Second, the researchers provided informed input throughout the project and research literature played a very important part. Cassie was able to see her "illusion of learning" as resulting from her unintentional funneling, when the teachers encountered Wood's work in a summer team meeting. Not only did the notion give a language for what was happening, it provided also the possibility for change, as we see in Cassie's story. We have little evidence, regrettably, that such intensive teamwork in researching and developing the teaching of mathematics continues beyond engagement with researchers in projects or beyond higher degree study, for which the pressures on teachers are certainly to blame.

Equally, we feel sure that the local focus, on issues that arose in their classrooms, was an essential factor in the teachers' learning. They knew that the Standards-based curricular materials provided new challenges for their students and that they needed to respond in ways that would support their students' learning. There may even have been some skepticism about how much of an effect there would be on students' understanding and performance with a focus on writing, then reading and speaking, but it is clear that the teachers learned a great deal about the importance of these aspects of mathematical thinking. As Cassie said, she saw the importance of "writing over time . . . because I actually saw growth in the students."

We return to our third question, "does the professional knowledge base for teaching mathematics develop from practice and if so how," which has been our major concern and interest in this chapter. We have already suggested that the teachers in the study adopted theories that came from the literature, thus were largely researchers' productions, and subsequently developed their own theories in response to locally perceived needs (nascent research questions) into more global theories as they were disseminated and were found to resonate with others. At the same time, we as a community need to continue to work with the differences in the strengths of the grammars of mathematics and of mathematics education, as we mentioned in the early part of this chapter.

Teachers are both productive and reproductive in this knowledge domain, that is, they both adapt and adopt theories from the field of professional knowledge for teaching mathematics and produce their own. In the transition from local theories to more global theories, teachers potentially produce new elements of the professional knowledge base. This latter stage depends largely, of course, on the possibility of dissemination amongst the mathematics education community, both those engaged in school practice and the research community. We would argue that we have presented strong evidence, in Cassie's story, of both the teachers' potential contribution to the professional knowledge base in the guided notes and the insight that reading, writing, and speaking are central to mathematics, and the actual contribution, as presented at the national conference, of the effects of writing over time.

References

Chazan, D., & Ball, D. (1999). Beyond being told not to tell. *For the Learning of Mathematics, 19*, 2–10.

Cobb, P., Confrey, J., diSessa, A., Lehrer, R., & Schauble, L. (2003). Design experiments in educational research. *Educational Researcher, 32*, 9–13.

Design-Based Research Collective. (2003). Design-based research: An emerging paradigm for educational inquiry. *Educational Researcher 32*, 5–8.

diSessa, A., & Cobb, P. (2004). Ontological innovation and the role of theory in design experiments. *Journal of the Learning Sciences, 13*, 77–103.
Doerr, H. M., & Chandler-Olcott, K. (2009). Negotiating the literacy demands of Standards-based curriculum materials: A site for teachers' learning. In J. Remillard, B. Herbel-Eisenmann, & G. Lloyd, (Eds.), *Mathematics teachers at work: Connecting curriculum materials and classroom instruction* (pp. 283–301). New York: Routledge.
Herbel-Eisenmann, B. (2002). Using student contributions and multiple representations to develop mathematical language. *Mathematics Teaching in the Middle School, 8*, 100–105.
Herbel-Eisenmann, B., & Breyfogle, M. L. (2005). Questioning our patterns of questioning. *Mathematics Teaching in the Middle School, 10*(9), 484–489.
Hiebert, J., Gallimore, R., & Stigler, J. (2002). A knowledge base for the teaching profession: What would it look like and how can we get one? *Educational Researcher, 31*, 3–15.
Lewis, C., Perry, R., & Murata, A. (2006). How should research contribute to instructional improvement? The case of lesson study. *Educational Researcher, 35*, 3–14.
Lappan, G., Fey, J., Fitzgerald, W., Friel, S., & Phillips, E. (1998). *Connected mathematics.* Menlo Park, CA: Dale Seymour Publications.
Mehan, H. (1979). *Learning Lessons.* Cambridge, MA: Harvard University Press.
National Council of Teachers of Mathematics (NCTM). (1989). *Curriculum and evaluation standards for school mathematics.* Reston, VA: NCTM.
Remillard, J. T. (2005). Examining key concepts in research on teachers' use of mathematics curricula. *Review of Educational Research, 75*, 211–246.
Scott, D., & Usher, R. (Eds.) (1996). *Understanding educational research* London: Routledge.
Skemp, R. R. (1976). Relational understanding and instrumental understanding. *Mathematics Teaching, 77*, 20–26.
Wood, T. (1998). Alternative patterns of communication in mathematics classes: Funneling or focusing? In H. Steinbring, M. G. Bartolini Bussi, & A. Sierpinska (Eds.), *Language and communication in the mathematics classroom* (pp. 167–178). Reston, VA: National Council of Teachers of Mathematics.

Feedback: Expanding a Repertoire and Making Choices

Dave Hewitt

Teaching is a complex activity and yet it can be viewed by some from the outside as a relatively simple process of passing on information in a clear way. My own experience of learning to teach – something which I feel I am still involved with 30 years after teaching my first lesson – is that my awareness of the complexity involved in teaching has increased over time. The more I fool myself into thinking I am sorting one aspect out, the more I become aware of issues which I had not considered before. In one way it is precisely this which makes teaching such a fascinating and engaging activity, in another way it can become frustrating for those who seek definite answers to the question of how to teach. Wheeler (1998, p. 98) suggested that rather than thinking of teaching as an art or a science we might "settle for teaching as essentially a technical matter – not in the sense of a full-fledged technology but as a set of knowhows, a sort of kitbag for dealing with the practical demands of the classroom, a kind of *bricolage*." In this spirit I explore here one aspect of such a kitbag, that of how a teacher might respond to contributions which pupils make in the classroom. These are not planned contributions but more responses to the in-the-moment incidents, many of which are unexceptional but all of which require decisions to be taken as to whether and how to respond. The issue of giving feedback to pupils' comments is, of course, a complex issue and here I am concentrating on particular aspects of such feedback – that of whether to act as judge when giving feedback and whether to offer explanations. The effects of such feedback are not my concern here (see Kluger & DeNisi, 1996, for a general review of effects of feedback) but instead I am interested in the beliefs which might lay behind such decisions. In this case my interest is with student teachers, some of whom are on a one year Post-Graduate Certificate in Education (PGCE) course and some of whom are in the second year of a two-year PGCE course (due to their degrees not having sufficient mathematics content).

D. Hewitt (✉)
University of Birmingham, Birmingham, UK
e-mail: d.p.hewitt@bham.ac.uk

R. Leikin, R. Zazkis (eds.), *Learning Through Teaching Mathematics*, Mathematics Teacher Education 5, DOI 10.1007/978-90-481-3990-3_14,

Calderhead (1984) classified different types of decisions that teachers make as reflective, immediate, and routine. For student teachers, who have little or no experience of teaching in a classroom, relatively little is routine and ways to respond to pupils in the classroom can be based upon what is familiar such as their personal mathematical knowledge. They can judge whether an answer is correct or not – this requires subject matter knowledge rather than pedagogic content knowledge (Shulman, 1986). So acting as judge to pupil comments can be initially a common form of feedback to pupil comments. Offering explanations also requires subject matter knowledge but in addition there is a sense of becoming aware of audience and trying to offer "clear" explanations. Explaining stresses one's own personal knowledge and as such it is possible to try to explain something in response to what a pupil has done without needing to try to understand what that pupil was thinking. A shift away from explaining can require a corresponding shift from attending to oneself to attending to the pupils. Iannone and Nardi (2005) offer an example of a lecturer shifting from providing model answers to questions ahead of a lecture (demonstrating the lecturer's own knowledge) to waiting until after receiving work from the students and responding to their collective errors. As part of the PGCE course we want to shift their attention away from their own performance and onto what the pupils are doing. We also want to widen the student teachers' awareness of possible ways of responding so that they can make informed and reasoned choices rather than following a limited range of already known possibilities. Scherer and Steinbring (2006, p. 171) spoke of experienced teachers when they said that "The communicative patterns and reciprocal actions between teacher and students are usually stable, unconscious *habits*, which have grown over the years, and which make it possible to cope spontaneously with the complex teaching events in which all the participants are involved" (their emphasis). The need to cope with such complexities as exist in a classroom drives certain responses into spontaneous unconscious actions as there is a need to respond quickly. Student teachers also begin teaching with existing images of what it is to be a teacher and have a certain set of spontaneous actions and responses within the classroom. Our desire during their PGCE year is to help them become aware of a greater range of ways to respond and also to develop a personal pedagogy which can inform their choice in how to respond. During a lesson, decisions need to be made in the moment and afterwards there is the luxury to reflect upon that decision and to consider alternative ways of acting. One way to work on practice is to reflect on a moment when an action was taken, perhaps out of habit or lack of awareness of alternative possible actions. This reflection classically takes place after the event at the end of a lesson. Over time such reflection becomes particularly powerful if the ability to reflect and consider alternative actions occurs closer and closer to the event itself. Eventually, the awareness of issues and of alternative actions can occur in the moment so that the action taken is no longer one of habit but one which is informed through choice (Mason, 2002). Doerr (2006, p. 256) identified "ways of responding with pedagogic strategies" as one of three dimensions of teachers' knowledge. However, this might include a teacher responding relatively mechanistically with a *pedagogic strategy* read from a book or heard from a tutor. My interest is not so much responding *with* pedagogic strategies but having a response *informed by* a personal pedagogy developed through personal beliefs and theories. Smith (2003) discussed both public and personal theories, based on Mason's (1998) description of "outer research" and "inner research." The public theories coming from outside, such as the school, university, government, and the personal theories coming from their own sensitivities, experiences, and ideas. The same classroom can be seen in different ways according to the sensitivities and awareness of an observer. Ainley and Luntley (2007) argued that experienced teachers have what they call "attentional skills," which I interpret as resulting from the awareness they have of teaching and learning issues, and which inform the attention they give to classroom events and the readings they have of those events. Thus, teachers' actions are a result of judgement rather than rule following. They proposed that "teacher education has to be as much concerned with the development of attentional skills and the capacity for attention-dependent responses as it is with the ability to set appropriate learning objectives and plan lessons effectively" (ibid., p. 5). In trying to develop our student teachers' awareness and sensitivity to different ways in which they might respond to pupils' work, a number of experiences and activities were built into their course throughout the year. This was in the belief that it was through student teachers having experiences, and working on those experiences, that they will develop awareness and sensitivity. The experiences and activities were as follows:

1. Teaching placement 1 – Links lesson
2. University session – Ball and box activity and polyominoes activity
3. Teaching placement 2 – Silent lesson

The student teachers were asked on two occasions during their teaching placements to teach a certain form of lesson to a class (a "links" lesson during their first placement and a "silent" lesson during their second). Both forms of lessons were chosen so as to take the student teachers a little outside their normal comfort zone and work with pupils in a way which was different to the way in which they had normally worked with pupils up to that point in time. In this way we hoped they would come to know, through their own experience of working with pupils, different ways of working and extend the possibilities open to them in the future. The details of these lessons are given later. The other way of raising awareness was through a specific session run at the university between the two teaching placements. This involved two related activities which offered personal experiences related to feedback.

The "Links" Lesson

On their first teaching placement, student teachers were asked to teach a lesson where they did not attempt to teach anything directly themselves. We chose this due to some student teachers being initially concerned with their own "performance" in the classroom in terms of presenting and explaining mathematics. We felt that asking student teachers quite early on in their teaching experience to teach a lesson where they were not personally explaining or presenting some mathematics might contrast with the lessons they were teaching up to that point in time. We modeled the form of lesson with them in a session at the university where the word "Calculus" was written in the middle of the board and I asked student teachers what came to mind when they thought of this. Gradually I wrote up on the board, in the form of words and connecting lines originating from the word "Calculus," what was offered by each student teacher. After a period of time a web-like diagram emerged with various lines interconnecting the offerings made from the student teachers. Up to this point in the session, I had taken the role of a scribe, checking on occasions that what I wrote represented what they said and that interconnection lines were appropriate. I then asked whether there was anything written up on the board which someone would like to ask a question about and this led into questions being raised, student teachers clarifying terms or explaining some mathematics behind what was written on the board. My role was one of managing this process (one voice speaking at a time, checking whether what was said was helpful or whether more needed to be said, etc.) rather than actually offering any explanations myself. At times student teachers would come up to the board to draw or write something, or offer examples. The session ended with me coming out of role to reflect upon the "lesson" they had just experienced.

We discussed the form the lesson had taken and the possible purpose of such a lesson in relation to a series of lessons. Implicitly we were interested in our student teachers becoming aware that there was an alternative way to respond when a pupil does not know something, the teacher does not always have to start explaining. Many years ago I was observing a student teacher, whom I will call Richard (all names used from now are pseudonyms). He was teaching pupils who were used to exploring mathematical situations and were involved in discussing something they had seen on a computer. Richard was mainly observing and listening to the pupils and looked relaxed. Then a pupil asked why something had happened on the screen. Richard hesitated but then gave a short explanation. Another pupil said they did not understand what Richard said and this led to a series of increasingly long explanations from Richard with him becoming noticeably frustrated that some pupils were not understanding what he said. I noticed the posture of many pupils change from leaning forward in their chairs (as if being active in the lesson and interested in what was happening) to one where they were leaning backwards (as if relaxing and being passive within the lesson and less interested in what was happening). This led me to label such a scenario as *the explanation trap* (Hewitt, 1994). Once a teacher takes on the role of being the person who explains then this can lead to pupils shifting from active participants using their own awareness to try to account for something themselves to relative passive participants who await a "clear" explanation from the teacher. The more a teacher explains, the more a loop is established with pupils shifting from exploring, noticing, and accounting for what they notice and instead taking on the role of sitting back and expecting the teacher to explain, and so the more a teacher might feel the need to explain further, etc. In some cases it can lead to a teacher feeling that they are doing all the work and the students almost enjoying saying "I don't understand" after each explanation.

The current student teachers were told they had to teach a lesson of the form I had modeled, choosing the topic and class and 31 of them reported back at a review afternoon where they came back to the university in the middle of their first placement. In written reflections about the links lesson 14 student teachers wrote about how they used praise and positive comments and avoided negative comments, whilst 15 of them wrote of not being judgmental and avoiding saying yes or no to suggestions coming from the pupils and instead either deflecting questions or replying with questions themselves. For many student teachers this lesson raised the awareness of considering the use of questioning as a form of feedback rather than providing judgmental comments or explaining. For example, Aliya commented "I tried not to say 'no' or give the answer if the pupils were nearly right. I tried to bounce contributions off pupils, so was asking things like 'how do you mean?', 'what else?', 'so what about if I. . .?'. . . Instead of me giving them an answer they can bounce ideas and answers from each other and work it out themselves." Questioning was used by many student teachers to ask pupils to give examples or give further reasoning behind their answers. One student teacher commented "I would ask them a question about what they had just offered. I wasn't asking them to find a correct answer. I was still trying to respond in a way that made them think about their maths." A key aspect of the learning many student teachers expressed was an awareness that there is not a dichotomy of telling or questioning, but that the form of questioning is something to be explored – there are different types of questions. The nature of the questioning had shifted from just asking closed mathematical questions into using probing questions which invited greater depth of mathematical thought.

The nature of the activity the student teachers were asked to carry out resulted in less explaining from the student teachers and a greater depth of mathematical thought from the pupils. Esme commented she wanted to repeat this form of lesson again as she wanted to "get the pupils thinking and expressing themselves instead of me giving them all the explanations/answers." This was a change in the role most of the student teachers took in their lessons and also resulted in a change in the nature, quality and quantity of pupil contributions. The shift away from explaining to questioning brought an awareness of a shift in the balance of work being carried out by the student teachers and pupils. Some comments from student teachers indicated this

"I tried to ask more questions and do less of the work. . . it felt like they were doing the work more, instead of me." (Penny)

"Although you are in control the pupils are doing all the work." (Kevin)

"This actually got the pupils to do the hard work rather than me giving them definitions of things." (Aliya)

Many student teachers felt pupils were thinking harder about the mathematics and the balance of who is doing the work within a lesson changed from student teacher to pupils. This was, of course, one lesson and this does not imply that the awarenesses shown reflecting on this one lesson resulted in a change of usual practice in other lessons. Indeed in Esme's comment above, she wanted to repeat this particular lesson in order to explore pupils, thinking and expressing themselves and her doing less explaining, rather than seeing this as a potentially general way of responding to pupils throughout all her lessons. However, I claim that one aspect of learning is awareness raising and this is a necessary part of a journey to developing practice. So I do not expect changes in practice to happen immediately after awareness is raised since, as Griffin (1989) right points out, teaching takes place in time and learning takes place over time.

Ball and Box Activity

After the student teachers had completed their first teaching placement they returned to the university for a few weeks ahead of starting their second school placement. Whilst at university, one session involved two activities, the first of which was the ball and box activity. This involved one student teacher whose task was to throw five balls, one at a time, into a box. This student teacher, whom I will call the "thrower," sat with fellow student teachers in a circle and the task was repeated five times as follows:

1. The thrower was blind folded and did not know the position of the box;
2. The thrower was still blind folded but had feedback from a teacher – "Teacher 1." This teacher said after each ball thrown only whether it had landed in the box or not;
3. The thrower was still blind folded but had feedback from a different teacher – "Teacher 2." This teacher said after each ball was thrown, the position the ball landed with respect to the box (e.g. one metre to the right and three metres behind – relative to the thrower);
4. The thrower took off the blind fold and could see the box;
5. A third teacher – "Teacher 3" – offered some input ahead of the thrower throwing the first ball. This teacher asked the thrower to imagine a video having been taken of them throwing the ball successfully into the box and invited the thrower to imagine the video being played backwards with the ball coming out of the box and back into their hand. They were then asked to play in their mind the video going forwards again whilst throwing the ball.

After each of the five stages there was a short period of time where the thrower commented on the quality of feedback with occasional comments from other student teachers. At the end of the activity the group as a whole reflected upon the quality of feedback and also considered mathematical equivalents of such feedback. The whole group of student teachers were split into two groups with one group involved in this activity whilst the other was involved in a separate session run in a different room by my colleague, Pat Perks. The groups then swapped round. I will report on comments labeling them as coming from group 1 (Gp 1) or group 2 (Gp 2).

The initial activity with the thrower blindfolded and not knowing the position of the box was generally met with humour. Neither thrower from the two groups was successful with any of his/her throws. The balls were thrown in quite different places. As Latif commented "Brownian motion comes to mind."

The second set of throws had the thrower still blindfolded but now there was feedback from Teacher 1 as indicated above. The thrower (Gp 2) commented that the feedback was "pretty useless really. It didn't offer me anything where I might get it closer to the box next time." Latif (Gp 1) related this back to a previous assignment where they were asked to mark pupils' homework by just using ticks and crosses one week followed by marking using only written comments the next week. Information was collected from pupils about what they felt about each type of marking. Barry (Gp 2) also related this to marking pupils' work saying "it is the equivalent of ticks and crosses on your work. In some ways they know they have or haven't got it right but there is no idea of how to improve." The thrower in group 1 felt frustrated because she was not allowed to ask questions (a rule I had made in order that only a certain type of feedback was experienced at each stage) as she felt the feedback was not useful to her. This raised an issue of whether there is an environment in a classroom whereby pupils feel able to ask questions about their work and indeed are encouraged to consider what feedback they might find useful in order to help them improve. This does not mean that they will necessarily receive that feedback because it may be that a teacher decides for pedagogic reasons not to provide it. For example, a pupil might ask the teacher to tell them how to do a question and their teacher might have a belief that just telling pupils how to do a question will not help them develop their awareness of mathematics and might instead lead to that pupil seeing mathematics as a set of procedures to memorize. A different teacher might feel that offering an example of what needs to be done would be helpful for that pupil doing other questions of a similar type and might not want to give other types of feedback due to a sense of limited available time and a need to move on to new topics. Doerr (2006, p. 257) said that "It is precisely a teacher's perceptions and interpretations of classroom situations that influence when and why as well as what the teacher does." However, it is not just their perceptions but also their existing repertoire of ways to respond and a set of beliefs which guide selection from that repertoire which influence what a teacher does. Thus, pupils may not get what they want in terms of feedback as it might conflict with the pedagogic beliefs that teacher holds about how to help pupils learn mathematics.

The third set of throws had feedback from Teacher 2. The throwers from both groups got their balls much closer to the box but still none went in the box. Thrower 2 commented "After the first one I knew how far away I was so I could adjust the next one to try and get a bit closer. Rather than just throwing them aimlessly around the room. I felt like I was improving that time." Both throwers seemed very positive after their throws despite the fact that they got none in the box. Thrower 1 commented that "I like the way he (Teacher 2) said *she got zero but she was very close*. That's encouraged me a little bit." Feedback does not always have to concern the mathematical content but can also concern emotional support. Empathizing with the feeling of being stuck, for example, and sharing the fact that being stuck is a common experience when working on mathematics, can help that person deal emotionally with the sense of being stuck. This can help them shift from giving up on the mathematics due to the uncomfortable emotion of being stuck and seeing this as meaning they cannot do what is asked of them, into accepting such feelings as common experiences when working on mathematics and that there are strategies which can be taken to help shift from that state (see Mason, Burton, & Stacey, 1985, for some strategies).

The fourth set of throws involved the throwers taking off their blindfolds. They managed to get at least one of their balls into the box with the rest being close. Thrower 1 said "I could see what I was doing. I could see where the box was, what to aim for." There was a sense of someone feeling more in control, they could now work on the task more independently as they had a clarity of not only what they were aiming for but also an awareness of their own processes on the way to achieving that task. I asked Thrower 1 what sort of feedback she had got. The response of "I haven't got any. It is just what I see" indicated to me a sense of thinking about feedback only in terms of what was received from another person. She did not seem to consider that feedback might come from her own senses through her paying attention to what happened and adjusting, as a consequence, what she could do in future. The continuation of my questioning with Thrower 1 was as follows:

Dave:	Tell me about what you see. . .. Is that feedback?
Thrower 1:	*No* [said in a very definite tone]. *Well I guess it is in that the result is feedback. I only got one ball in the box. That is feedback that I could work out.*
Dave:	After you threw the first ball did you do anything different when you threw the second ball?
Thrower 1:	*Yes, I threw harder.*
Dave:	And why did you throw harder?
Thrower 1:	*Because the first ball did not quite make the box. So I threw the second ball harder but then that went too far.*
Dave:	OK. So that affected your third ball?
Thrower 1:	*Yep.*
Dave:	It sounds like you got feedback there.
Thrower 1:	*OK.*

The "OK" at the end did not sound particularly convincing! Initially she was very clear that she did not get feedback but then considered the result of getting one ball in the box as feedback. A classroom activity is itself only a pedagogic tool to help pupils in their learning – the real aim is learning, improving, not so much succeeding in the stated aim of the activity. Here the thrower saw feedback in terms of the number of balls in the box, the explicit stated *activity aim* of the activity, and was disappointed. However, she was successful in terms of an implicit *learning aim* of gaining more control over her ability to throw a ball. Tahta (1981) talked about activities having outer and inner meanings. The outer meaning here concerns the explicit task of throwing the ball into a box. However, a teacher might have an inner meaning for the task, in this case using the task as a vehicle to improve someone's control over their throwing. The inner meaning concerns the significant awarenesses and skills developed through focused engagement with the outer activity. The outer meaning may not be seen as of particular importance for a teacher choosing an activity. For example, I might offer an investigation such as asking pupils to find how many squares (of varying sizes) there are on a chessboard. It is of little importance to me whether pupils retain the answer to this question – it is not something on the mathematics curriculum and it is not a fact which is significant for further work in mathematics. The outer meaning is not of importance. However, the choice of such an activity is based on the inner meanings which are involved – being able to *see* different sized squares, developing strategies to tackle what might appear initially as too difficult a problem, having a way of counting, knowing that all squares have been counted and none have been counted twice, imposing a structure onto something which felt chaotic, seeing generality which can lead to extending the original scenario into larger sized squares, rectangles, going into three or more dimensions, and so on. As a teacher I do not really care about the outer meaning – actually knowing the number of squares. However, I do want my pupils to care, otherwise they will not engage in the activity and, therefore, will not gain inner meanings. The inner meanings are developed as a consequence of the focus being on the outer meaning. My questioning with Thrower 1 tried to focus attention onto the inner meaning of the activity and although she could identify ways in which she changed her throwing as a result of what she noticed she did not seem convinced that this was something to label as feedback.

With the fifth and last set of throws, Teacher 3 offered an image for the thrower prior to them actually throwing the balls. Thrower 1 did not get any of her balls into the box but still felt Teacher 3 was helpful. She said "once I was imagining the ball going in, I really did think it would." So gaining a belief in herself was considered helpful even though she did not actually get any into the box. Here the attention seemed to be on the inner meaning. This was also the case with Thrower 2 who despite having got three balls into the box felt that Teacher 3 was only a little helpful as he put his success down to getting used to "taking the shot... rather than [Teacher 3]'s mental picture." His attention seemed to be with his own sense of throwing rather than anything to do with the relative success he had in getting balls into the box.

Remarks were made in both groups about how Teacher 3's imagery helped the throwers make better use of their own feedback. Carol (Gp 1) said "when you sort of delve into an unknown area of mathematics you don't always know what to notice and what to adjust. The imagery isn't re-winding, undoing the mathematics you have just done and trying to replay it forward. The imagery, the equivalent imagery for a mathematics lesson wouldn't be undo this sum, it would be nurturing to help pupils know what to pay attention to."

This showed an awareness of what it can feel like as a learner. There may be lots of feedback from engagement in an activity itself which a learner has the potential to notice and which someone more skilled might certainly notice. However, without a sense of what to pay attention to can result in little from the potential actually being seen as relevant. With the ball activity, a thrower's own feedback may concern where they noticed the ball landing, yet their attention might need to be with their arm as much as the ball or box. Mason (1989) talked of the significance of where attention is placed in respect to the mathematical act of abstraction and generally where attention is placed is crucial in what pupils notice and work on. This is also the case with teachers and their craft of teaching. With our student teachers we deliberately set up a week's observation after they have taught for three weeks. This is because we feel that the experiences involved in teaching can help student teachers become aware of what it is they need to attend to when observing. The act of noticing requires a person to direct their attention onto certain aspects. A teacher can offer valuable feedback to a learner through directing where to place their attention, a more subject specific version of what Winne and Marx (1982) described as "orientating" – controlling the focus of pupils' attention. This will help them notice particular aspects which may be relevant to the success of the task at hand and which the learner might not have known to be relevant. Chazan offers an example where pupils he was working with were disagreeing over whether a zero should be included in a set of numbers when finding the average. He made a conscious decision to ask a question which shifted attention away from the zero, by asking what the average they have calculated (without the zero) would mean (Chazan and Ball, 1999). Such an act of directing attention was an attempt to offer something which might help and was different to playing the role of judge and telling pupils who was "right."

Student teachers were able to relate Teacher 1 to different mathematics contexts such as Tracy (Gp 1) who said "I thought of enlargement. You could just say *do it, get on with question*s and then just go tick, cross, tick, cross, so if one person got a tick you would go *Oh OK then, so I can do that one and you can try and work it out for yourself* but if you got them all wrong you still wouldn't have any inclination whatsoever if you had nothing but ticks and crosses."

Teacher 3 continued the issue of what my student teachers considered to be feedback. Some student teachers felt that what Teacher 3 offered was not feedback as it was said in advance of balls being thrown. The significance of what was offered was that it applied to the coming attempt rather than being a comment about the last attempt. Yet the reason for why it might be offered could be due to a previous attempt. For example, if all the balls had gone into the box in the previous attempt then it might not have been said. However, if the previous attempt was such that the teacher felt such an image might be useful then it was being said because of the previous attempt even though it applied to the next attempt. The idea of offering general strategies or directing pupils' attention are examples of ways of offering feedback which concern their future attempts.

I asked both groups for a quick response, either yes or no, as to whether they felt Teacher 1 (who had just given a kind of yes/no response saying whether the ball had landed in the box or not) was helpful in the feedback they gave. There was a resounding "no." This was significant for the next activity.

Polyominoes

Immediately following the ball and box activity I moved on to a different activity with the student teachers where I had the set of all 17 pentominoes and tetrominoes (see Fig. 1) on the right hand side of an interactive whiteboard.

I said there were some of these that I liked and some of these that I did not like and the student teachers' task was to find what I liked about the ones that I did. The activity involved them collectively deciding to drag one of the shapes to the left side of the board. If I liked it, the shape stayed there. If I did not like it, I moved it down to the bottom of the board. There was also a point system where points were taken off for every one which I rejected. Throughout the activity there was a lot of discussion about properties of these shapes, including possible paths from square to square within a shape, and symmetry. There was conversation about how rotating these shapes affected the discussion they had regarding paths. There was also discussion of logic issues such as which shape would give the most information to them given the possible rules they were considering. There was a sense of a lot of mathematics taking place. The student teachers were motivated and involved, with one of the two groups complaining when I brought the activity to a premature close before they had decided upon a rule. They reflected on the fact that my role, as teacher, was one where I only said "yes" or "no." This feedback was similar to Teacher 1 in the previous activity which they had almost unanimously decided was not helpful. Yet the same kind of feedback in this activity helped generate mathematical activity. There was a sense of the same kind of feedback might not be helpful in some contexts whilst being very helpful in others. So the form of feedback does not have a universal value but is context dependent. These activities had an inner aim of raising the students' awareness that there is variety in the form feedback can take, that the nature of feedback can be significant for a pupil's learning and that its usefulness is context dependent.

Silent Lesson

During the student teachers' second teaching placement we wanted to get them to teach a lesson which they would not normally have done. The aim was similar to

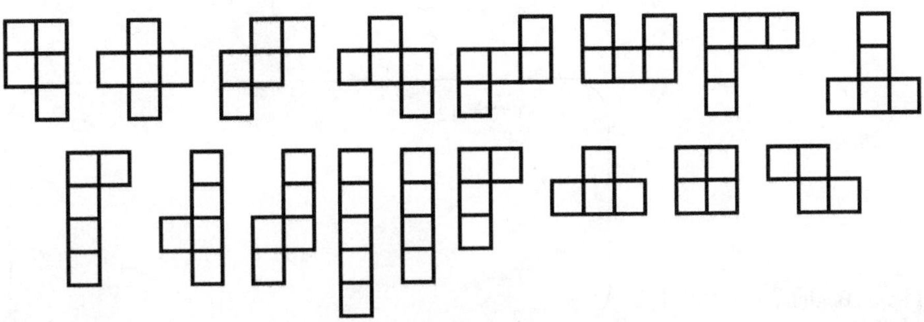

Fig. 1 a collection of pentominoes and tetrominoes

the "links" lesson in their first placement and once again I modeled the lesson with them at university before they adapted it for a particular class when on teaching placement. The lesson was one where neither the teacher nor the pupils spoke whilst a particular activity was carried out collectively as a whole class (for other accounts of such a lesson see Brown & Coles, 2008; Brown & Waddingham, 1982). I will give brief details of the lesson here.

After working on getting student teachers attentive and silent there was already a curiosity as I did so without speaking. I wrote up on the board, quite slowly, some numbers as in Fig. 2.

I passed the pen onto a student teacher and pointed to below the number 3 in Fig. 2. Initially the student teacher was unclear what was expected but still came up and wrote a number down; in this case it was number 8. I took the pen and drew an arrow from the 8 and then thought for a while after which I wrote 72. I passed the pen on to a different student teacher and pointed to below the 8 and that person came up and wrote the number 5. I took the pen, draw another arrow and paused looking at the group of student teachers offering the pen generally, rather than to a specific student teacher. After a while someone came up and wrote down what they thought the number would be. I took the pen and drew a ☺ if I was happy with that number or a ☹ if I was not. The lesson continued with me taking a more withdrawn role although I was always quick to respond (silently) if someone was about to speak. After a while, when I felt the majority of student teachers knew my rule, I wrote fractions or decimals as start numbers and later on wrote an "*n*" on the board and ended up with two rules offered. I then continued but this time wrote a number for the result and drew a backwards arrow for a possible start number. Figure 3 gives a simplified sense of what the board looked like at the end of the activity.

We discussed certain aspects of the lesson (now with everyone allowed to talk) and the student teachers discussed ways in which they might adapt this activity for particular classes. One feature which followed on from the polyominoes activity was that feedback was once again in the form yes/no and this feedback seemed to help motivate mathematical activity.

On a review day when student teachers returned to the university from their second teaching placement they shared how their silent lessons had gone. We had found this quite a powerful lesson for the student teachers as many of them were very apprehensive about running such a lesson mainly due to a sense that none of their

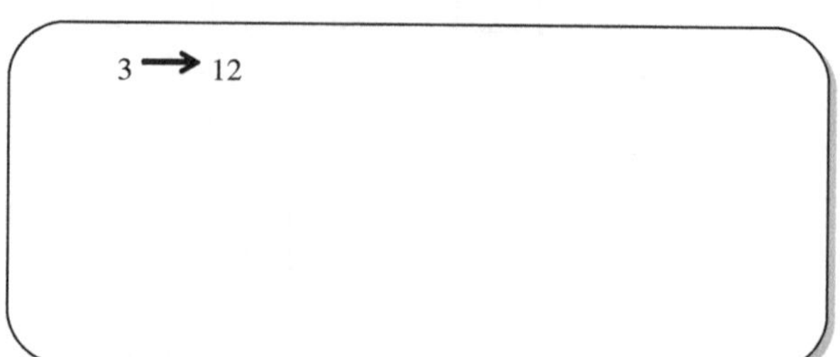

Fig. 2 Beginning of the silent lesson

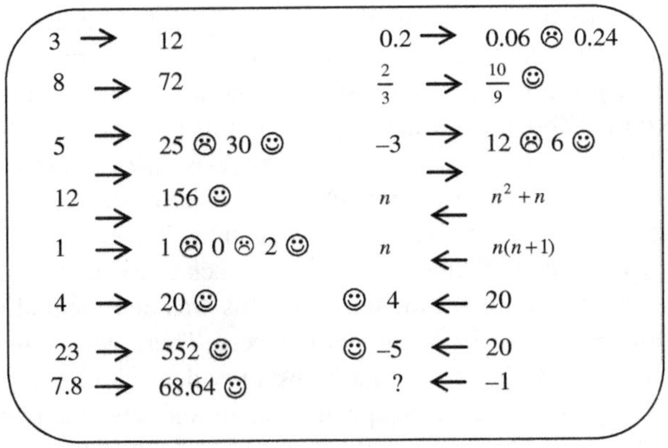

Fig. 3 A simplified idea of the board at end of the silent lesson

classes would be quiet. Nearly all student teachers reported a very positive experience and being amazed that the class was indeed quiet and that they had managed to run this activity in silence. Several weeks following this, I asked them to write some reflections down about this silent lesson in relation to the issue of feedback.

Several student teachers, from the 29 who gave written responses, commented how being silent was different to how they normally worked in the classroom. So there was a sense of change being forced – things had to be different to usual in this lesson. Different behaviors were developed as a consequence such as more exaggerated expressions, gestures and actions, scanning pupils' faces more regularly, greater eye contact, clapping and use of pointing were all mentioned as well as the use of smiley faces on the board. A sense of an expanded repertoire came through from their writing such as Carol who wrote "Previously feedback was mainly verbal, but this showed that the pupils can get feedback from me just from my facial expressions." Indeed, when considering the use of voice in later lessons Wendy commented that "More recently changes of voice tone have also played a big part in the feedback I give in class." So the absence of voice has brought with it an increased awareness of possible variation within the use of voice. Four student teachers commented that the lesson brought greater involvement from the pupils generally and also brought in pupils who in other lessons had not contributed so much.

There were three themes I noticed from within their writing. The first was noticing some benefits to offering visual feedback. Tracy commented that "Pupils can understand visual rather than auditory instructions" and Steve reflected on a particular class he taught through the placement saying that "All of them were SEN [pupils with identified Special Educational Needs] and in the past confusion has often come from my language – which many struggle with as they have communication difficulties. Visual prompts, however, left them with less confusion – so I have learnt that often words are not enough – or even not necessary at *all*" (his emphasis). Andy felt that the lack of verbal comments forced the pupils to pay greater attention to the visual feedback. Others talked about using visual along with verbal feedback at the same time. The use of pointing was mentioned by 11 of the student teachers with some writing explicitly about referring back to previous answers. George said that since the silent lesson he has based feedback more on previous contributions pupils have given in the lesson. Steve talked about using pointing with a particular group saying "Too often with this group pupils have been confused about not *what* to write but *how* or *where* to write it" (his emphasis). This reveals for me an increased awareness of the focus of feedback he might offer this group of pupils and that pointing, in this case, has a role to play in such feedback.

The second theme concerned the lack of explaining and the ability of students to work things out for themselves. This was commented on by well over half the 29 responses I got from student teachers. There was some surprise at what the pupils were able to learn without being provided with an explanation. Andy commented "The pupils didn't need specific details and lots of explaining in order to focus on an exercise and to successfully perform it." Indeed even if mistakes were made several student teachers wrote about how they now wait and see whether other pupils notice those mistakes. For example, Kay commented that "given time and freedom pupils can work out answers for themselves. It's not always necessary to jump in and help them." Aliya said that "If the pupil was half correct I'd wait until someone else came up and noticed what was wrong... I do not just give the answers anymore." Some student teachers commented on how they now use peer and self assessment in class, one student teacher commenting on how she then uses questioning to ask pupils why they think something is right or wrong. A sense of thinking about pupils as more independent learners came through in many of the writings with some explicitly mentioning this.

The third theme was the use of a delayed response. Seven student teachers wrote about how they found that after a pupil had written on the board a delay in putting a sad or happy face had a powerful effect and created productive tension. Steve commented that if "one pupil offers an answer and teacher gives a response – only the one pupil and the teacher are involved. If one pupil gives an answer and it is left there without response but with a visual prompt [head and eyes only] which suggest a response *will* happen eventually, then all the pupils in the room are potentially involved in the process just by having their own opinion" (his emphasis). Kevin wrote about what he had taken from the silent lesson and used in other lessons: "One thing it did give to me was the use of giving time to answer questions and waiting for response rather than just giving the feedback straight away."

These themes showed significant learning of the student teachers through the depth with which they now spoke about giving feedback. With the first lesson we asked them to teach ("links"), the pupils were recalling things they had learnt before and perhaps doing a little reminding themselves of some of those things. This time pupils were engaged in a new challenge and many student teachers learnt that pupils were capable of working things out for themselves. This realization is all the more powerful because it came from them observing what happened within their own classroom whilst they were teaching. This was not a story told to them by someone else, nor was it an observation of an experienced teacher's classroom. The ownership of the experience gives a sense of truth and also a sense of future possibilities.

Two Student Teachers' Lessons

I visited two student teachers, Steve and Dianne, towards the end of their two year PGCE course when they were finishing their second placement and video recorded one of each of their lessons. Following the recording I interviewed each of them separately, playing back selected moments from the lesson and getting them to talk about what thinking was behind what they said or did. These student teachers were chosen due to a mixture of factors: they were from a group I had visited earlier in their placements; they were already strong enough in the classroom to be sure of achieving the standards required to pass the course; I wanted one male and one female; and their timetables fitted in with my own existing commitments.

Steve's Lesson

Steve's lesson was with a Year 8 class (12–13 year olds). In this school the pupils were organized in ability classes or "sets," with Steve's class being set seven out of eight. He was working on properties of two-dimensional shapes in a number of different ways: a discussion with the whole class; an activity involving them getting into particular groups at the front of the class; and working on a task in pairs. I noted that Steve did not tell them much during the lesson but used questioning extensively. However, there were times when he did state something and at one point I noted he gave some definitions when he felt many of the pupils had already gained an idea of those definitions. Part of the interview was as follows:

Dave: Some people would say that is a crazy way round that when you think your class do know about something you do give them the definition and when they don't know about something you won't give them the definition.

Steve: *Yes I suppose it is a crazy way round but I think once they have got the definition then you are just emphasising a point whereas if they don't then there's some exploring to be done.*

An articulation of something which Steve felt the students were already aware is both a form of stressing but is also a way of re-forming what is already noticed into formal mathematical language. Steve commented that he wants to "attach the words to that meaning rather than the other way round" and so would not introduce mathematical words unless meaning which related to those words was already around – meaning first, words second. Another time in which Steve offered something was when two pupils were working together trying to decide where different cards, each with a polygon drawn on, should be placed within a grid on a large sheet of paper (see Fig. 4).

Two pupils were looking at two of the cards – a square and an isosceles trapezium (see Fig. 5) – which they had placed in the bottom right cell in the grid indicating both shapes satisfied the properties "some sides are equal length" and "more than 1 right angle." Steve begun the following dialogue with the pupils:

Fig. 4 Grid for placing shapes

	No Right Angles	1 Right Angle	More Than 1 Right Angle
No sides are equal lengths			
Some sides are equal length			

Fig. 5 Two of the shapes

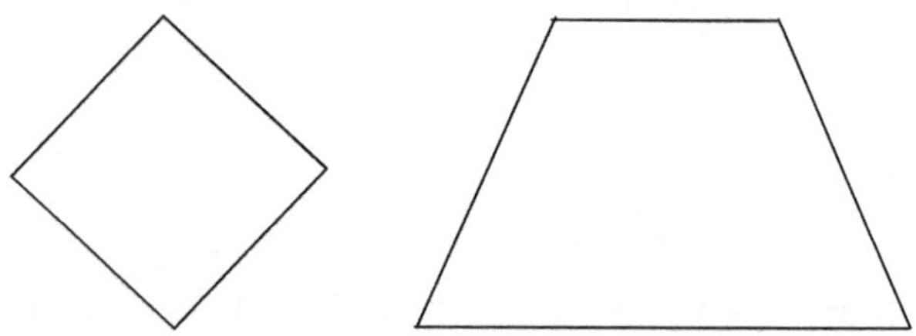

Steve: *That one* [the isosceles trapezium]*got more than one right angle?*

Pupil 1: Yes, it's got two.

Steve: *Show me where the right angle is.*

Pupil 2: There and there [pointing to the two obtuse angles].

Steve: *OK, so which were the right angles in this bit* [pointing at square]*?*

Pupil 2: There, there, there and there [pointing at the four corners of the square].

Steve: *What's special about a right angle? Is it possible that that can be a right angle and that can be a right angle* [pointing to one angle from the square and then one obtuse angle from the isosceles trapezium]*?*

Pupil 1: Yes.

Steve: *Can we have right angles which are different? So that one right angle which was different from another right angle?*

Pupil 1: That side is not straight [pointing to one of the "sloping" sides of the isosceles trapezium]. [Pupil 1 moves it to cell indicating it has "1 right angle" and "no sides are equal lengths."]

Steve: *Has it got one right angle as well? So this here* [pointing to a right angle in the square]*you say is a right angle, OK. There is only one particular type...* [Pupil 1 moves the isosceles trapezium card to the cell indicating it has "more than 1 right angle" and still "no sides of equal lengths"]. *Ah, so that one you are saying has no sides of equal lengths* [Pupil 2 moves it to "1 right and angle' and "some sides are equal length']. *I think you are just randomly placing it now. Take it off, take it off* [Pupil 1 points to the cell which would indicate "more than 1 right angle" and "some sides of equal length" and Pupil 2 moves it there]*and have another look at it* [Steve places it off the sheet] *and tell me, see if you can find some right angles there. Definitely right angles* [Pupil 2 picks it up and places it back in the cell indicating "more than 1 right angle" and "some sides of equal length"]. *OK show it to me again.*

Pupil 2: [Pupil 2 points to the obtuse angles of the isosceles trapezium] There and there.

Steve: *So let me show you something special about a right angle. OK? 90 degrees* [picks up an envelope and points to a corner].*Is that a right angle?*

Pupil 1: Yes.

Pupil 2: Yes.

Steve: *OK* [placing a corner of the envelope on a right angle of the square]. *So that, you are happy that that is a right angle?*

Pupil 1: Yes.

Pupil 2: Yes.

Steve: *What about this one* [pointing to the obtuse angle of the isosceles trapezium]*?*

Pupil 2: [Puts the envelope corner onto the obtuse angle to reveal that it does not fit the angle exactly] No.

Steve: *So that's not.*

Pupil 1: No right angles. Here [pointing to the cell indicating "no right angles" and "'no sides are equal lengths." Pupil 2 moves it there].

Steve: *OK.*

Pupil 2: [Moves it to the cell indicating "no right angles" and "some sides are equal length"] go on here [moves it back again].

[Steve now deals with another pupil and moves away from Pupil 1 and Pupil 2].

This interchange started with Steve questioning them but after a while he decided that something else was needed. Steve said

> "I wasn't actually, up until I picked up the envelope, offering him any sort of insight that would help him. My questions weren't helping him find whether this shape had right angles or not... He could see the right angles on the shape which didn't have any right angles, so he needed something more... The feedback there was actually to offer him a means or strategy to help him to establish whether this works. Again, I wasn't saying yes or no. I was just saying well here is a way of finding out."

The fact that the pupils appeared to be having difficulty with deciding whether an angle was a right angle or not might be as much to do with knowing the social convention of exactly what gets labeled "right angle" and what does not, as with their awareness of angle. Since "right angle" is a socially agreed term there is a sense of either someone having a consistent meaning for that term with the rest of the mathematics community or not. If not, then it is not necessarily a mathematical issue but an issue about conventions. Elsewhere (Hewitt, 1999) I have argued that pupils *need* to be informed of names and social conventions whether from a book, a teacher, the internet, or wherever. Pupils can invent names and invent conventions but the only way to know whether they have the same name or convention as the mathematics community is by being informed in some way. Rather than telling the pupils Steve found something which both he and the pupils agreed was a right angle. This then becomes the mediator in determining whether other angles were right angles as well. The way in which the corner was manipulated as it was carried from one place to another and held against the angles of the shape indicated a mathematical property of a right angle – it remains a right angle through rotation and translation. For example, right angles do not have to involve a horizontal line. So the envelope corner *and the way in which it was moved* forms the mediator for what is conventionally known as a right angle. This is another way in which a pupil can be informed – an example is found where all parties agree that indeed it is an example and this is then manipulated, preserving the relevant mathematical properties, to act as a tool to check whether other things have this property. Here the teacher would need to be the person who either controls the manipulation or watches over manipulations of that common object so as to ensure the relevant mathematical properties remain. In this case, Steve might allow rotation of the envelope but not allow it to be creased in a way which affects the angle at the corner.

Toward the end of the interview I asked Steve how he had and will continue to develop a range of possible strategies in responding to pupils. His reply was one of learning from the pupils he teaches:

> You say something and look for a reaction in a way. Certainly when a particular type of feedback has had a particularly good reaction in terms it has helped or created a discussion or something like that, you try it again... I think also the feedback you give, you know you can have a completely different perception of something than the kid and so the feedback can be completely meaningless to them... So actually that comes out by finding out what kind of misconceptions are, not necessary about the type of feedback you give and testing the feedback but about learning about certain misconceptions. So there has to be some sort of activity in which actually very little feedback is given and therefore you can judge where they are at and find out what the misconceptions were.

The misconceptions issue had arisen earlier in the interview when I suggested that there were different ways in which the statement "more than 1 right angle" can be interpreted (having more than one angle which is a right angle, and having one angle whose size is more than one right angle) and this led him to re-evaluate a response he had given as he could see that his feedback would not have made much sense if the pupil concerned had this different interpretation. His thoughts about feedback were consistent with much of what I observed during the lesson where he often just listened or observed what the pupils were doing or saying, and used questioning extensively to find out what they were thinking about. I gained a sense of someone who now was learning through listening to his pupils and that he will continue developing his skills within the classroom as he has a strong sense of knowing where to place his attention in order to continue learning within this area.

Dianne's Lesson

Dianne taught a Year 9 class (13–14 year olds) in a single sex girls' school. Again classes were organized in ability classes with her class being set two out of six. The lesson addressed angle properties involving lines and led into an activity on bearings. During the first part of the lesson when she was getting answers back from the pupils, she had a very fluent way of responding to pupils. She used questions but what I noticed immediately was the fluency and speed with which each question was asked following a response from the pupil. For example:

Dianne: *Angle A is what, [Pupil A]?*
Pupil A: 79 degrees.
Dianne: *And how did you get that?*
Pupil A: I took 79, mean I took 281 away from 360.
Dianne: *And why did you do that?*
Pupil A: Because 360 degrees is the, 360 degrees is how many you get all the way around.
Dianne: *Excellent, that is called angles round a point, [Pupil A]. Really, really good. Fantastic.*

This came out in the interview when Dianne said that "the three things I was interested in was the answer, the method and the key language." Her series of questions indeed brought out these three aspects and seemed to take the form of a pattern of questioning. This is a pattern of interaction (Voigt, 1995) which brings out all three aspects which Dianne wanted to focus on. Unlike funneling (Bauersfeld, 1988; Wood, 1998) this pattern did not begin to reduce the scope of pupil response but created a structure for Dianne to hear what a pupil thought about this question. It provided a series of questions which dug deeper into a pupil's thinking and stressed what Dianne felt was important in pupils' responses. Interestingly, later on a different pupil gave all three aspects within her initial response, perhaps having registered the form of response her teacher wanted.

Dianne was clear that it was not the answer which she was primarily interested in, instead she used questioning to get at the mathematics involved in getting the answer:

> That's the bit I am interested in, the explanation not the answer. So the answer is just the consequence of them actually understanding what they are doing. I think that they can prove to themselves they are right because of their explanation, they can prove to everyone else that they are right because of their explanation, they don't have to listen to me saying 'yes that is right', 'no that's wrong' because they worked it out for themselves.

Even if a teacher does not act as the judge of whether something is right or wrong they can play a significant role in pupils deciding the correctness of their work by using questioning to help bring out the explanations and the thinking behind an answer. Such explanations and thinking becomes the material with which pupils can then work in order to decide the correctness of their thinking. I noticed that once all three aspects Dianne was interested in were said – answer, method, and key language – she did confirm correctness by saying "Excellent... really, really good. Fantastic." Indeed she also did tell the pupil something by naming the property Pupil A used as "angles round a point." In respect to when she felt she would tell things, Dianne said:

> I think if they were getting confused with the language, with the angles, complementary, corresponding, alternative then yes I would because it is just an arbitrary name that has

> been given to the problem. But if they are getting confused with what the angles actually were and what the relationship between those angles were and which angles we are talking about then no I wouldn't, or I wouldn't be happy if I had made a statement about that because I don't feel that I need to make a statement about that because they can just work it out.

There was a sense here of her having taken on board the arbitrary/necessary divide (Hewitt, 1999) which was part of a session at University. In this divide, the arbitrary are names and conventions which pupils need to be informed about and need to be memorized, whereas the necessary are properties and relationships which is where mathematics lies and which can become known through awareness without having to be informed by someone else. This had now become a strong part of her practice and a way in which she decided what to tell and what not to tell.

Another aspect of her teaching I noticed was the fact that if a pupil did not give a correct response Dianne would stick with that pupil and ask further questions rather than moving onto a different pupil. In an extreme example, Dianne asked a series of 19 questions to one pupil until a correct answer and appropriate reasoning was given by that pupil. This was a significant change in her practice since her first school placement. She commented on what was behind her making this shift:

> Firstly, well there are lots of reasons I guess. Because I've had good feedback about doing that when I do it in lessons. The teacher picks up on it and says 'oh I like the way you stuck with that girl and kept on until she got the answer' and so there I think 'so therefore it must be a good thing. So I will do it again.' But also I mean I think it is educationally beneficial for them that they can see that the teacher isn't going to give up on them and that actually they do know it and they can get to the right answer and they can give an explanation that's valid.

What struck me from Dianne's comments was the combination of learning from pupils during her own practice in the classroom, listening to feedback from experienced teachers and, significantly, these are supported by her own set of beliefs. I know Dianne well enough to feel that she is not someone just to do what an experienced teacher tells her is a good thing to do. Her own developing pedagogy and her attention to what happens within her classroom are key to the development she makes within her practice as a teacher.

Final Remarks

On our PGCE course we offer our students activities and tasks which are designed to offer personal experiences relevant to the issues we are trying to raise with our students. This comes from a belief that having personal experience of something is more powerful than listening to the experience of others. However, as Mason (2002, p. 8) said "one thing we do not seem to learn from experience, is that we do not often learn from experience alone." Having an experience is one thing but working on that experience is required to develop awareness and sensitivities. We can only work with what we know about and as such we "force" our student teachers to undertake lessons outside their comfort zone which extend ways of working in the classroom. In the comments made following these sessions many of our students were surprised at how effective a shift in their practice was, from commonly offering explanations to giving none at all. Pupils engaged positively with the mathematical challenge and tried to work out things themselves. We consider that this personal experience of learning through changing the students' own actions in the classroom and them seeing the consequences is more powerful than being told such things are possible within a university session. For some of our students there was a considerable shift in practice from giving explanations to considering other forms of feedback, such as waiting to allow other pupils to give their thoughts, using body language, asking a series of questions, listening and allowing time for pupils to work things out for themselves. This did not mean that they never told pupils things but some of the student teachers developed a clear set of beliefs which informed the way in which they responded to pupils, including when to tell and when not to tell. From comments made it seems that there was a mix of experiences which have been influential, not only those experiences inside classrooms but also the university sessions on feedback and the arbitrary/necessary divide, along with an assignment which involved marking work in different ways.

Using Tahta's (1981) idea of inner and outer meanings again, the lessons we asked our student teachers to undertake had a certain outer meaning which, for example, might be "keeping silent" for the silent lesson. However, my own inner meaning for asking them to carry out such a lesson concerned students experiencing pupils working mathematically: having conjectures; testing out those conjectures; expressing rules; etc. All of this happening without a teacher explicitly telling pupils what to do. The lesson is a vehicle for our student teachers to experience pupils working in this way during one of their own lessons. Our hope is that this experience might lead to our student teachers working with their pupils in ways which uses these mathematical abilities of pupils more in future lessons. Likewise, the activities I used for the university sessions, such as the polyominoes activity, have an outer meaning of trying to find my rule but the inner meaning involves students experiencing involvement in mathematical activity when the teacher was giving no other feedback that yes/no, after already deciding that yes/no feedback was not helpful in the previous activity. Thus raising their awareness that judgmental decisions about the quality of feedback are context dependant. The activities at university and lessons the student teachers were asked to do in schools had as their focus outer meanings which provided a vehicle for the students to experience personally aspects of the desired inner meanings.

Finally, I return to Wheeler's (1998) metaphor of a kitbag – a set of knowhows for dealing with practical situations in a classroom. I wonder about the nature of what might be in this kitbag with regard to giving feedback to pupils. Rather than a collection of things to do, which might be carried out mechanically, I see the kitbag as an awareness of different possible ways of acting, along with a set of beliefs forming a strong personal pedagogy which will inform choices taken. Although the lessons we asked our student teachers to carry out and the activities they were given at university were designed to help them become aware of other possible ways of responding to pupils, in the end the extending of possibilities is only one aspect of what is important because this only gives an extended range. In the end a choice needs to be made so the "why choose this rather than that?" is as important as the awareness that there exists a "this" and "that."

References

Ainley, J., & Luntley, M. (2007). The role of attention in expert classroom practice. *Journal of Mathematics Teacher Education.* 10, 3–22.

Bauersfeld, H. (1988). 'Interaction, construction, and knowledge: Alternative perspectives for mathematics education'. In D. A. Grouws & T. J. Cooney (Eds.), *Perspectives on research on effective mathematics teaching* (pp. 27–46). Reston, VA: NCTM.

Brown, L., & Coles, A. (2008). *Hearing silence: Steps to teaching mathematics.* Cambridge: Black Apollo Press.

Brown, L., & Waddingham, J. (1982). *An addendum to Cockroft*, Bristol: Resources for Learning Development Unit.

Calderhead, J. (1984). *Teachers' classroom decision-making.* London: Holt, Rinehart and Winston.

Chazan, D., & Ball, D. (1999). Beyond being told not to tell. *For the Learning of Mathematics*, 19, 2–10.

Doerr, H. M. (2006). Teachers' ways of listening and responding to students' emerging mathematical models, *ZDM*, *38*, 255–268.

Griffin, P. (1989). Teaching takes place in time, learning takes place over time. *Mathematics Teaching*, 126, 12–13.

Hewitt, D. (1994). *The principle of economy in the teaching and learning of mathematics*, Unpublished PhD dissertation, The Open University, Milton Keynes.

Hewitt, D. (1999). Arbitrary and necessary: Part 1 a way of viewing the mathematics curriculum. *For the Learning of Mathematics*, 19, 2–9.

Iannone, P., & Nardi, E. (2005). On the pedagogical insight of mathematicians: 'Interaction' and 'transition from the concrete to the abstract'. *Journal of Mathematical Behavior*, 24, 191–215.

Kluger, N., & DeNisi, A. (1996). The effects of interventions on performance: a historical review, a meta-analysis, and a preliminary feedback intervention theory. *Psychological Bulletin*, 19, 254–284.

Mason, J. (1989). Mathematical abstraction as the result of a delicate shift of attention. *For the Learning of Mathematics*, 9, 2–8.

Mason, J. (1998). 'Researching from the inside in mathematics education'. In A. Sierpinska & J. Kilpatrick (Eds.), *Mathematics education as a research domain: a search for identity* (Book 2, pp. 357–378.) Dordrecht, Netherlands: Kluwer.

Mason, J. (2002). *Researching your own practice: The Discipline of Noticing.* London: Routledge Falmer.

Mason, J., Burton, L., & Stacey, K. (1985). *Thinking mathematically.* Wokingham: Addison-Wesley.

Scherer, P., & Steinbring, H. (2006). Noticing children's learning processes – teachers jointly reflect on their own classroom interaction for improving mathematics teaching. *Journal of Mathematics Teacher Education*, *9*, 157–185.

Shulman, L. (1986). Those who understand: Knowledge growth in teaching. *Educational Researcher*, *15*, 4–14.

Smith, T. J. (2003). Connecting theory and reflective practice through the use of personal theories. In N. A. Pateman, B. J. Dougherty, & J. Zilliox (Eds.), *Proceedings of the 2003 Joint meeting of the international group for the psychology of mathematics education (PME) and PME North America chapter (PME-NA)* (Vol. 4, pp. 215–222). Honolulu, USA: College of Education, University of Hawai'i.

Tahta, D. (1981). Some thoughts arising from the new Nicolet films. *Mathematics Teaching, 94*, 25–29.

Voigt, J. (1995). Thematic patterns of interactions and sociomathematical norms. In P. Cobb & H. Bauersfeld (Eds.), *Emergence of mathematical meaning: Interactions in classroom culture* (pp. 163–201). Hillsdale, NJ: Erlbaum.

Wheeler, D. (1998). The commonsense of teaching. In Y. Pothier (Ed.), *Proceedings of the 1998 annual meeting of the Canadian Mathematics Education Study Group* (pp. 93–99). Halifax, Nova Scotia, Canada: Mount Saint Vincent University Press.

Winne, P. H., & Marx, R. W. (1982). Students' and teachers' views of thinking processes for classroom learning. *The Elementary School Journal, 92*, 493–519.

Wood, T. (1998). Alternative patterns of communication in mathematics classes: Funneling or focusing. In H. Steinbring, A. Sierpinska, & M. G. Bartolini-Bussi (Eds.), *Language and communication in the mathematics classroom* (pp. 167–178). Reston, VA: NCTM.

Conclusion

In introducing the chapters in this volume we chose to employ the notions of mathematical pedagogy and pedagogical mathematics because, in several cases, we had difficulty separating mathematics learned through and for teaching from pedagogy. However, the chosen categories of classification only reinforced the realization that mathematics and pedagogy are closely intertwined, creating a kind of a circular approach, interchanging the focus of attention and the supporting background.

In our opening chapter (Leikin & Zazkis), we described a number of factors that support or impede learning through teaching. Here, in our conclusion, we highlight two important themes that traverse the chapters. These are the notions of communication and of "critical event."

Communication and Interaction

Undoubtedly, teachers' interactions with their students are the main source of learning through teaching in the majority of chapters in this volume. Some chapters place special emphasis on the analysis of communicative practices. For example, communication is a central issue in the study of Doerr and Lerman. They introduce a teacher named Cassie and discuss the change in her views and pedagogical practices related to reading, writing, and the use of oral language in a mathematics classroom.

Focus on the communication of mathematical ideas by and among students has been mentioned by Liljedahl, as a development of Mary's pedagogical strategies. Leikin and Zazkis used a six-step interaction model, developed by Leikin (2005), to analyze the case of Einat – a teacher who diverted from a preplanned lesson and, challenged by the need to refute her students' incorrect conjecture, developed a new pedagogical approach. In Yerushalmy and Elikan's chapter, the importance that teachers assign to a classroom algebra-focused discussion, to communication among students and to listening to students, appears to be one of the main differences between the "traditional" and "reform" teachers. Yerushalmy and Elikan employed an IRE (Initiation-Response-Evaluation) model (Cazden, 1988) of instructional interactions to analyze LTT.

A related issue is that of miscommunication in language usage – or assigning a different meaning to the mathematical concept of divisor by a teacher and a learner,

R. Leikin, R. Zazkis (eds.), *Learning Through Teaching Mathematics*, Mathematics Teacher Education 5, DOI 10.1007/978-90-481-3990-3,

and what can be learned from it – is discussed in the case of Lora in Leikin and Zazkis' chapter.

It is not surprising that interaction with students is one of the most important factors in teachers' learning. After all, even when interpreting teaching broadly, to include preparation and marking, the heart of teaching is interaction with students. However, while interaction with students is mentioned by most authors as a trigger for learning, there are other kinds of interaction that support learning and intensify reflection. This includes interaction among teachers when they are "collaborating in their inquiries" (Mason, Doerr & Lerman) and interaction with a researcher or a research team (Doerr & Lerman, Kieran & Guzman, Leikin, Liljedahl, Marcus, & Chazan, and Hewitt). In particular, Tzur highlights the role of teacher educator in prompting "teachers' noticing in situations that would other wise go unnoticed," acknowledging the sense of threat that such an interaction may elicit.

Though interaction was first considered among humans, interaction between teaching and research is the theme explicitly developed in Alcock's chapter, where the author skillfully situates her development of teaching in the contemporary research literature. Furthermore, technology puts a different spin on the issues of interaction. It serves as a vehicle to promote interaction among teachers, among students, and between teachers and researchers. This is the case for teachers who are analyzing videos of the lessons located in the same classroom (Yerushalmy & Elikan) as well as for teachers located thousands miles apart and using online-communication for their discussion (Borba & Zulatto). The study of Kieran and Guzman demonstrates that interaction with a computer output promotes students' conjecturing and prepares a background for proving activities. These student activities are the foundation for their teacher's learning (Michael). These researchers acknowledge the centrality of CAS (Computer Algebra System) in Michael's learning, as the task could not have been managed without it. The centrality of technology in the Borba and Zullatto study is twofold; online communication allows interaction at a distance to present a mathematical problem and Dynamic Geometry software is assisting the instructor's learning to eventually resolve the problem. Similarly, in Yerushalmy and Elikan's study, computer software promotes one teacher's learning within a reform curriculum and a video-recording of those lessons aids in the learning of other teachers through a discussion and analysis of teaching.

Furthermore, Jackiw and Sinclair present a rather uncommon view on communication; they discuss the discursive interaction of a user of a Geometer's Sketchpad Dynamic Geometry environment. They demonstrate that the characteristics of this interaction position a user as a teacher, whose learning takes place while teaching or interacting with the computer software.

Critical Event and Search for Equilibrium

We agree with Ball and Bass (2000) that

> No repertoire of pedagogical content knowledge, no matter how extensive, can adequately anticipate what it is that students may think, how some topic may evolve in a class, the need for a new representation or explanation for a familiar topic (ibid., p. 88).

Further, we believe that the same claim should not be restricted to pedagogical content knowledge, as it applies similarly to knowledge of subject matter. This need for something "new" is an opportunity for learning. The chapters in this volume reinforce the idea explicitly mentioned by Leikin in Part I: Teachers' pedagogical knowledge and their mathematical knowledge are mutually related, where strength in one contributes towards developing strength in the other.

It was mentioned repeatedly in the chapters that teachers' mathematical knowledge, their attentiveness to students, and instructional interaction are some of the supporting factors of teachers' learning through their practice. However, all these cannot initiate change until there is a *triggering event* followed by a reflection on this event. Mason calls such an event an "experience of disturbance." More explicitly, Mason (2002) stated.

> Most frequently there is some form of disturbance which starts things off. It may be a surprise remark in a lesson, a particularly poor showing on a test, something said by a colleague, something asserted in a journal or book, or a moment of insight (ibid., p. 10).

Tzur describes the phenomenon as a "misfit" between an effect on students and teachers' anticipation and suggests that learning occurs as a result of "shaking" the teachers' "deeply entrenched anticipations." Echoing these ideas, Zazkis refers to memorable and unexpected experiences that "shake the routine."

The triggering event can be short, such as a student's question (Borba & Zulatto, Zazkis), a student's suggestion (Leikin & Zazkis, Leikin), a student's mistake, misunderstanding or identified difficulty (Doerr & Lerman and Marcus & Chazan), a student's success in providing an alternative unexpected solution when dealing with challenging material (Leikin & Zazkis, Leikin and Tzur), or unexpected feedback from a computer (Jackiw & Sinclair and Borba & Zulatto). It can also be longer, consisting of teachers' involvement in professional development initiatives (Liljedahl, Tzur, Hewitt), implementation of new curricula (Yerushalmy & Elikan and Marcus & Chazan), new teaching approaches or implementation of novel tasks (Leikin, Hewitt, Kieran & Guzman, Liljedahl), teachers' participation in research projects (Doerr & Lerman, Kieran & Guzman), or a self-motivated desire to improve one's teaching and students' understanding of mathematics (Alcock).

As an umbrella to these descriptions, a Piagetian construct of disequilibration comes to mind. In Piaget's theory of child development, disequilibration and a desire to achieve equilibrium are the deriving forces for the continuous reconstruction of knowledge. A similar claim can be made about teachers' learning in their practice: A triggering event creates a disequilibration and a state of equilibrium is achieved by developing new knowledge, whether within mathematical pedagogy or pedagogical mathematics.

However, in order to unsettle a teacher's equilibrium, the teachers need, according to Tzur, a "predisposition toward unexpected situations as an opportunity, not a threat." How to achieve such a predisposition? Further research will focus on this question explicitly. Though the chapters in this volume describe valuable ideas, elaborate on contributing and impeding factors, and provide illustrative evidence for teachers' learning, it remains an "unrealized potential." Realizing this potential is an ongoing effort for teachers, educators, and researchers alike. This book lays a foundation and opens the gate for a deliberate and more informed investigation in this direction.

References

Ball, D., & Bass, H. (2000). Interweaving content and pedagogy in teaching and learning to teach: Knowing and using Mathematics. In J. Boaler (Ed.), *Multiple perspectives on mathematics teaching and learning*, (pp. 83–104). Westport, CT: Ablex.

Cazden, C. B. (1988). *Classroom discourse. The language of teaching and learning*. Portsmouth, NH: Heinemann.

Leikin, R. (2005). Teachers' learning in teaching: Developing teachers' mathematical knowledge through instructional interactions. The paper presented at the 15th *ICMI Study: The Professional Education and Development of Teachers of Mathematics*. Retrieved in March 2009 from: http://stwww.weizmann.ac.il/G-math/ICMI/log_in.html

Mason. J. (2002). *Researching your own practice: The discipline of noticing*. London, UK: Routledge Falmer.

Name Index

Subject Index